The Definitive Guide to Spring Batch

Modern Finite Batch Processing in the Cloud

Second Edition

Michael T. Minella

Foreword by Dave Syer, Spring Batch Project Founder

Apress®

The Definitive Guide to Spring Batch: Modern Finite Batch Processing in the Cloud

Michael T. Minella
Chicago, IL, USA

ISBN-13 (pbk): 978-1-4842-3723-6 ISBN-13 (electronic): 978-1-4842-3724-3
https://doi.org/10.1007/978-1-4842-3724-3

Managing Director, Apress Media LLC: Welmoed Spahr
Acquisitions Editor: Steve Anglin
Development Editor: Matthew Moodie
Coordinating Editor: Mark Powers

Cover designed by eStudioCalamar

Cover image designed by Freepik (www.freepik.com)

Distributed to the book trade worldwide by Springer Science+Business Media New York, 233 Spring Street, 6th Floor, New York, NY 10013. Phone 1-800-SPRINGER, fax (201) 348-4505, e-mail orders-ny@springer-sbm.com, or visit www.springeronline.com. Apress Media, LLC is a California LLC and the sole member (owner) is Springer Science + Business Media Finance Inc (SSBM Finance Inc). SSBM Finance Inc is a **Delaware** corporation.

For information on translations, please e-mail editorial@apress.com; for reprint, paperback, or audio rights, please email bookpermissions@springernature.com.

Apress titles may be purchased in bulk for academic, corporate, or promotional use. eBook versions and licenses are also available for most titles. For more information, reference our Print and eBook Bulk Sales web page at http://www.apress.com/bulk-sales.

Any source code or other supplementary material referenced by the author in this book is available to readers on GitHub via the book's product page, located at www.apress.com/9781484237236. For more detailed information, please visit http://www.apress.com/source-code.

Printed on acid-free paper

To my daughter, Addison. If Daddy can do this, you can do anything.

Contents

About the Author xiii

About the Technical Reviewers xv

Acknowledgments xvii

Foreword xix

■Chapter 1: Batch and Spring 1

A History of Batch Processing 2

Batch Challenges 3

Why Do Batch Processing in Java? 4

Other Uses for Spring Batch 5

The Spring Batch Framework 8

    Defining Jobs with Spring 9

    Managing Jobs 10

    Local and Remote Parallelization 10

    Standardizing I/O 10

    The Rest of the Spring Batch Ecosystem 10

    And All the Features of Spring 11

How This Book Works 11

Summary 12

■Chapter 2: Spring Batch 101 13

The Architecture of Batch 13

    Examining Jobs and Steps 14

    Job Execution 15

Parallelization 17

Documentation 20

Project Setup.. 21

Obtaining Spring Batch.. 21

It's the Law: Hello, World! 24

Running Your Job 27

Summary.. 28

■Chapter 3: Sample Job 29

Understanding Agile Development 29

Capturing Requirements with User Stories 30

Capturing Design with Test-Driven Development 32

Using a Version-Control System 32

Working with a True Development Environment... 32

Understanding the Requirements of the Statement Job 32

Designing a Batch Job 38

Job Description... 39

Understanding the Data Model 40

Summary.. 41

■Chapter 4: Understanding Jobs and Steps 43

Introducing a Job 43

Tracing a Job's Lifecycle 44

Configuring a Job 46

Basic Job Configuration... 46

Job Parameters 47

Working with Job Listeners 61

ExecutionContext... 64

Manipulating the ExecutionContext 65

Working with Steps 69

Tasklet vs. Chunk Processing 69

Step Configuration... 70

Understanding the Other Types of Tasklets...71

Step Flow...88

Summary.. 105

■Chapter 5: JobRepository and Metadata 107

What Is the Job Repository?.. 107

Using a Relational Database.. 107

The In-Memory Job Repository 111

Configuring the Batch Infrastructure... 112

The BatchConfigurer Interface.. 112

Customizing the JobRepository... 112

Customizing the TransactionManager 114

Customizing the JobExplorer.. 115

Customizing the JobLauncher 116

Database Configuration 117

Using Job Metadata 118

The JobExplorer.. 118

Summary.. 122

■Chapter 6: Running a Job 123

Starting a Job with Spring Boot 123

Launching a Job via REST 125

Scheduling with Quartz 130

Stopping a Job 134

The Natural End 134

Programmatic Ending 135

Error Handling.. 148

Controlling Restart 150

Preventing a Job from Being Rerun... 150

Configuring the Number of Restarts 151

Rerunning a Complete Step 152

Summary.. 153

■**Chapter 7: ItemReaders**... **155**

The ItemReader Interface.. 155

File Input 156

    Flat Files 156

    XML.. 183

JSON.. 188

Database Input 191

    JDBC.. 191

    Hibernate 198

    JPA... 202

    Stored Procedures 204

    Spring Data.. 206

Existing Services... 211

Custom Input 213

Error Handling 218

    Skipping Records... 218

    Logging Invalid Records 220

    Dealing with No Input 222

Summary.. 223

■**Chapter 8: ItemProcessors** **225**

Introduction to ItemProcessors 225

Using Spring Batch's ItemProcessors 227

    ValidatingItemProcessor... 227

    ItemProcessorAdapter 234

    ScriptItemProcessor 236

    CompositeItemProcessor.. 237

Writing Your Own ItemProcessor 242

    Filtering Items.. 242

Summary.. 244

■Chapter 9: ItemWriters 245

Introduction to ItemWriters 246

File-Based ItemWriters.. 247

　FlatFileItemWriter 247

　StaxEventItemWriter... 259

Database-Based ItemWriters 263

　JdbcBatchItemWriter... 263

　HibernateItemWriter 270

　JpaItemWriter.. 275

Spring Data ItemWriters... 279

　MongoDB 279

　Neo4J 282

　Pivotal Gemfire and Apache Geode... 286

　Repository 291

Alternative Output Destination ItemWriters................................. 294

　ItemWriterAdapter 294

　PropertyExtractingDelegatingItemWriter 297

　JmsItemWriter.. 299

　SimpleMailMessageItemWriter.. 304

Multipart ItemWriters 310

　MultiResourceItemWriter.. 310

　CompositeItemWriter... 319

　ClassifierCompositeItemWriter 322

Summary... 326

■Chapter 10: Sample Application 327

Reviewing the Statement Job 327

Setting Up a New Project 328

Importing Customer Updates 330

　Validating Customer ID 338

　Writing Customer Updates.. 339

Importing Transactions 343

    Reading Transactions 345

    Writing Transactions 346

Applying Transactions to Current Balance 347

    Reading the Transaction Data.. 348

    Updating the Account Balance 349

Generating Monthly Statement 349

    Reading the Statement Data... 350

    Enrich the Statement with Accounts 353

    Writing Statements... 355

Summary 359

■Chapter 11: Scaling and Tuning 361

Profiling Your Batch Process 361

    A Tour of VisualVM 362

    Profiling Spring Batch Applications 366

Scaling a Job... 374

    Multithreaded Steps 374

    Parallel Steps... 376

    AsyncItemProcessor and AsyncItemWriter.. 381

    Partitioning 384

    Remote Chunking 399

Summary 405

■Chapter 12: Cloud Native Batch 407

Twelve Factor Applications.. 408

    Codebase.. 408

    Dependencies... 408

    Config 409

    Backing Services 409

    Build, Release, Run 409

    Processes 409

Port Binding 409

Concurrency 410

Disposability 410

Dev/Prod Parity 410

Logs 410

Admin Processes 411

A Simple Batch Job 411

Circuit Breaker 417

Externalizing Configuration 420

Spring Cloud Config 421

Service Binding via Eureka 423

Orchestrating Batch Processes 427

Spring Cloud Data Flow 427

Spring Cloud Task 429

Registering and Running a Task 430

Summary 434

■Chapter 13: Testing Batch Processes 435

Unit Tests with JUnit and Mockito 435

JUnit 436

Mock Objects 438

Mockito 439

Integration Tests with Spring Classes 443

General Integration Testing with Spring 443

Testing Spring Batch 445

Summary 454

Index 455

About the Author

Michael T. Minella is a software engineer, author, and speaker with over 18 years of professional experience. Michael is a Director of Software Engineering, leading the Spring Batch and Spring Cloud Task projects for Pivotal. He was also on the expert group for JSR-352 (Java Batch). He is a Java Champion and JavaOne Rockstar having spoken internationally at many different Java conferences.

Outside his normal day job, Michael serves as the "curmudgeon at large" on the regular podcast OffHeap (`https://www.javaoffheap.com`). He also has a personal passion for InfoSec topics. With hobbies including photography (`https://500px.com/michael160`) and woodworking, the most important jobs in Michael's life are as husband to Erica and father to Addison.

About the Technical Reviewers

Wayne Lund was one of the original creators of Spring Batch while working in the global architecture group for Accenture and delivered Spring Batch to the Java community at JavaOne 2007. While at Accenture, he served as a Master Technology Architect and worked in a global architecture group focused on OSS projects helping clients adopt Spring as a platform of choice for abstracting away JEE and preferring lightweight frameworks. After Spring was purchased by VMWare, he joined the vFabric group (now part of Pivotal services) that supported Spring Source, RabbitMQ, Gemfire, and other OSS lightweight frameworks in the sales channel. He is currently working as an Advisory Platform Architect for Pivotal Data Services where he helps provide solutions for Spring-enabled data products that include Spring Cloud Data Flow (Spring Cloud Stream, Spring Cloud Task, and Spring Batch), Spring Cloud Stream, and messaging with RabbitMQ and Kafka.

Felipe Gutierrez is a solutions software architect, with bachelor's and master's degrees in computer science from Instituto Tecnologico y de Estudios Superiores de Monterrey Campus Ciudad de Mexico. Gutierrez has over 20 years of IT experience and has developed programs for companies in multiple vertical industries, such as government, retail, healthcare, education, and banking. He is currently working as a Platform and Solutions Architect for Pivotal, specializing in Cloud Foundry PAS and PKS, Spring Framework, Spring Cloud Native Applications, Groovy, and RabbitMQ, among other technologies. He has worked as a solutions architect for big companies like Nokia, Apple, Redbox, Qualcomm, and others. He is also the author of *Spring Boot Messaging* and *Introducing Spring Framework*, both published by Apress Media, LLC.

Acknowledgments

When I take a moment and look back between the writing of my first book and now, a lot has changed in my life, much of which has had material impacts on my career, allowing me to be in the position to write this second book on Spring Batch. I'd like to take this space to do what little I can to show my appreciation for those who have made some of those impacts.

First, I'd like to thank Dave Syer, not only as the founder of this framework that I have been fortunate enough to be the steward of for the past 6+ years and write two books about, but as someone I can look up to as an open source practitioner. Looking back, it was right after I wrote my first book that I met him to give him a copy. He told me about the efforts by the Java Community Process (JCP) to create a Java Specification Request (JSR) around batch processing (JSR-352), which leads me to the next person I'd like to thank.

One of the original members of the team and one of the tech editors for this book, Wayne Lund is largely responsible for me being in the position I am today. It was on the JSR-352 expert group that we met and worked together to improve the design of the Java batch specification based on our experience with Spring Batch. During our time working on the JSR, Wayne let me know that the Spring team was looking for a new lead for the Spring Batch project and asked if I was interested. I still believe that most of my non-technical family and friends do not understand what being asked to join the Spring team meant. I've never been happier in a position than I am working on the Spring engineering team. Thank you, Wayne, for your initial belief in me and your continued support.

I'd also like to thank my manager at Pivotal, Brian Dussault. I have had the privilege of working for a number of amazing managers in my career, many of whom I'd work for or with again in a heartbeat. But none have been able to give me the support and trusted me at the level that Brian has. They say you don't leave companies, you leave managers. If that is the case, Brian may be stuck with me for a very long time.

There are two other groups of people I'd like to thank here. The first is the team at Apress. Steve Anglin and Mark Powers both have been amazingly understanding during this long process. I'm sure I haven't been the easiest author to work with; however, I am fortunate to have had them as my editors on this project. Without their continued support throughout this project, it wouldn't have been possible. I'd also like to thank my second technical editor, Felipe Guiterrez. His reviews and encouragement on the book have both made a world of difference on the final results.

Finally, and above all, my family. Anyone who has written a book knows that the process takes a toll on everyone involved. The author dedicates time, energy, and emotion to the project. Those same authors are human and have only finite amounts of those three things to give. It is our families that provide the much-needed support in order for us to make these books possible. To my daughter Addison who inspires me every day with her passion to help others, endless curiousity, and caring personality. And to my wife Erica who really pushed me through this book. I would have given up on it half way through if it wasn't for her steady encouragement and support. To both of them, you mean the world to me and thank you.

Foreword

Spring Batch was the first open source project that I was more than peripherally involved in. I guess your first child is always special in some ways, and it was a bit like that with Spring Batch. It was a tad longer in gestation than we had thought it might be—more of an elephant in the end than a mouse—but it always behaved well and was a credit to its parents. I like to think we were at least partly responsible for that; after all, nurture has to beat nature sometimes. As I remember it, there were two things that dragged out the delivery of 1.0: one was to make extra sure that the quality of service features actually worked in the field, and the other was the level of care lavished on the API design. Mistakes were inevitably made anyway, but I think we can at least say we gave it our best shot at a good start in life.

If we were to look at the genealogy of Spring Batch, we would find, of course, that it has its origins in the field, born of the long and repetitive invention and re-invention of all the features in many businesses around the world. The first time I saw any of the code in 2006, it was a tiny prototype that Rob Harrop had written while he was on a consulting gig at one of the banks in London. Those pieces eventually landed in Spring Retry, after we split out some of the useful features of Spring Batch to be shared across other projects. The bulk of the rest of Spring Batch, and its state-machine-oriented view of the world, came from the collaboration with Accenture. There are too many contributors from that time to list them all, but Lucas Ward deserves a special mention as the other parent and main carer in the early years. Also I remember Robert Kasanicky and Dan Garrette were heavily invested in the success in the years before leading up to the launch in 2008.

They were also instrumental in the release of Spring Batch 2.0 in 2010, where we introduced the concept of a "chunk," as well as features to support distributed processing, parallel processing, and new language features from Java 5. I can still hear Lucas telling me "we can't call it a 'chunk,'" citing Wayne's World as prior art, but he never came up with a better name so it stuck. A chunk is a group of items that can be processed together, allowing excellent opportunities for increased efficiency and scalability. Spring Batch 2.0 was the state of the art for quite a while and fed into the JSR-352 specification when that work kicked off. Wayne Lund from Accenture was involved in the Spring Batch project from the early days, also sat on the JSR-352 Expert Group, and now works as a Platform Architect at Pivotal.

Michael Minella was a Young Turk in those days; also on the Expert Group, he had used Spring Batch a lot in real life and actually written a book on it. When he joined the Spring team in 2012, it was just in time to start work on the 3.0 release, which was where we first saw the "@EnableBatchProcessing" annotation and a shift of emphasis from XML configuration to Java. He quickly took over as project lead and shepherded the project through the 3.x series and on through to 4.0 where Java 8 became the baseline, and some new fluent-style configuration builders were added. The connection with Spring Cloud Data Flow and the industrialization of distributed processing also took place in this period. In early 2018, Mahmoud Ben Hassine came on board as the new project co-lead, and he has been helping Michael to drive the project and listen carefully to feedback from the many users.

So at the time of writing, Spring Batch has just turned 10 years old and has some proud new parents. Or is it some other relation? I don't know. Anyway the old parents, and grandparents if that's what they are, are equally proud of the way she turned out and of the new guardians. She surely has a lot more to give in the years to come, because batch processing never seems to go away. Funny how that works.

Dave Syer, Spring Batch Project Founder
London 2019

CHAPTER 1

■ ■ ■

Batch and Spring

If you read the latest press, the topic of batch processing will hardly come up. A quick scan of the largest Java conferences will have virtually zero talks dedicated to the topic outright. Rooms are filled with attendees learning about stream processing. Data science talks gather large crowds. Blog posts on cloud native applications focused on web-based systems (REST, etc.) get the highest number of views. However, under all of it all, batch is still there.

Your bank and 401k statements are all generated via batch processes. The e-mails you receive from your favorite stores with coupons in them? Probably sent via batch processes. Even the order in which the repair guy comes to your house to fix your laundry machine is determined by batch processing. Those data science models that recommend what products to show in the associated products on sites like Amazon, generated via batch processing. Orchestrating big data tasks, that's batch too. In a time when we get our news from Twitter, Google thinks that waiting for a page refresh takes too long to provide search results, and YouTube can make someone a household name overnight, why do we need batch processing at all?

There are a number of good reasons:

- You don't always have all the required information immediately. Batch processing allows you to collect information required for a given process before starting the required processing. Take your monthly bank statement as an example. Does it make sense to generate the file format for your printed statement after every transaction? It makes more sense to wait until the end of the month and look back at a vetted list of transactions from which to build the statement.

- Sometimes it makes good business sense. Although most people would love to have what they buy online put on a delivery truck the second they click Buy, that may not be the best course of action for the retailer. If a customer changes their mind and wants to cancel an order, it's much cheaper to cancel if it hasn't shipped yet. Giving the customer a few extra hours and batching the shipping together can save the retailer large amounts of money

- It can be a better use of resources. Data science use cases are a good example here. Typically, data model processing is broken up into two phases. The first is the generation of the model. This requires intensive mathematical processing of large volumes of data, which can take time. The second phase is evaluating or scoring new data against that generated model. The second phase is extremely fast. The first phase makes sense to do outside of a streaming use case via batch with the results of the batch process (the data model) to be utilized by a streaming system real-time.

This book is about batch processing with the framework Spring Batch. This chapter looks at the history of batch processing, calls out the challenges in developing batch jobs, makes a case for developing batch using Java and Spring Batch, and finally provides a high-level overview of the framework and its features.

M. T. Minella, *The Definitive Guide to Spring Batch*, https://doi.org/10.1007/978-1-4842-3724-3_1

A History of Batch Processing

A look at the history of batch processing is really a look into the history of computing itself.

The time was 1951. The UNIVAC became the first commercially produced computer. Prior to this point, computers were each unique, custom-built machines designed for a specific function (e.g., in 1946 the military commissioned a computer to calculate the trajectories of artillery shells, the ENIAC, at a cost of about $5 million in 2017 dollars). The UNIVAC consisted of 5,200 vacuum tubes, weighed at over 14 tons, had a blazing speed of 2.25 MHz (compared to the iPhone 7, which has a 2.34 GHz processor) and ran programs that were loaded from tape drives. Pretty fast for its day, the UNIVAC was considered the first commercially available batch processor.

Before going any further into history, we should define what, exactly, batch processing is. Most of the applications you develop have an element of interaction, whether it's a user clicking a link in a web app, typing information into a form on a thick client, receiving a message via middleware of some kind, or tapping around on phone and tablet apps. Batch processing is the exact opposite of those types of applications. *Batch processing*, for this book's purposes, is defined as the processing of a finite amount of data without interaction or interruption. Once started, a batch process runs to some form of completion without any intervention.

Four years passed in the evolution of computers and data processing before the next big change: high-level languages. They were first introduced with Lisp and Fortran on the IBM 704, but it was the Common Business Oriented Language (COBOL) that has since become the 800-pound gorilla in the batch-processing world. Developed in 1959 and revised in 1968, 1974, 1985, 2002, and 2014, COBOL still runs batch processing in modern business. A ComputerWorld survey[1] in 2012 stated that over 53% of those enterprises surveyed used COBOL for new business development. That's interesting when the same survey also noted that the average age of their COBOL developers is between 45 and 55 years old.

COBOL hasn't seen a significant revision that has been widely adopted in a quarter of a century.[2] The number of schools that teach COBOL and its related technologies has declined significantly in favor of newer technologies like Java and .NET. The hardware is expensive, and resources are becoming scarce.

Mainframe computers aren't the only places that batch processing occurs. Those e-mails I mentioned previously are sent via batch processes that probably aren't run on mainframes. And the download of data from the point-of-sale terminal at your favorite fast food chain is batch, too. But there is a significant difference between the batch processes you find on a mainframe and those typically written for other environments (C++ and UNIX, for example). Each of those batch processes is custom developed, and they have very little in common. Since the takeover by COBOL, there has been very little in the way of new tools or techniques. Yes, cron jobs have kicked off custom-developed processes on UNIX servers and scheduled tasks on Microsoft Windows servers, but there have been no new industry-accepted tools for doing batch processes.

Until Spring. In 2007, driven by Accenture's rich mainframe and batch processing practices, Accenture partnered with Interface21 (the original authors of the Spring Framework, now part of Pivotal) to create an open source framework for enterprise batch processing. Inspired by concepts that had been considered a mainstay of Accenture architecture for years,[3] the collaboration yielded what would become the de facto standard for batch processing on the JVM.

As Accenture's first formal foray into the open source world,[4] it chose to combine its expertise in batch processing with Spring's popularity and feature set to create a robust, easy-to-use framework. At the end of March 2008, the Spring Batch 1.0.0 release was made available to the public; it represented

---

[1]http://www.computerworld.com/article/2502430/data-center/cobol-brain-drain--survey-results.html
[2]There have been revisions in COBOL 2002 (including object oriented COBOL) and 2014 COBOL, but their adoption has been significantly less than for previous versions.
[3]The reference architecture that was used was from the book *Netcentric and Client/Server Computing: A Practical Guide, 1999*. Key components within the book included scheduling, restart/recovery, batch balancing, reporting, driver program (job), batch logging systems, and more.
[4]https://www.cnet.com/news/accenture-jumps-into-open-source-in-a-big-way/

the first standards-based approach to batch processing in the Java world. Slightly more than a year later, in April 2009, Spring Batch went 2.0.0, adding features like replacing support for JDK 1.4 with JDK 1.5+, chunk-based processing, improved configuration options, and significant additions to the scalability options within the framework. 3.0.0 came along in the spring of 2014, bringing with it the implementation of the new Java batch standard, JSR-352. Finally 4.0.0 embracing Java-based configuration in a Spring Boot world.

Batch Challenges

You're undoubtedly familiar with the challenges of GUI-based programming (thick clients and web apps alike). Security issues. Data validation. User-friendly error handling. Unpredictable usage patterns causing spikes in resource utilization (have a link from a blog post you write go viral on Twitter to see what I mean here). All of these are by-products of the same thing: the ability of users to interact with your software.

However, batch is different. I said earlier that a batch process is a process that can run without additional interaction to some form of completion. Because of that, most of the issues with GUI applications are no longer valid. Yes, there are security concerns, and data validation is required, but spikes in usage and friendly error handling either are predictable or may not even apply to your batch processes. You can predict the load during a process and design accordingly. You can fail quickly and loudly with only solid logging and notifications as feedback, because technical resources address any issues.

So everything in the batch world is a piece of cake and there are no challenges, right? Sorry to burst your bubble, but batch processing presents its own unique twist on many common software development challenges. Software architecture commonly includes a number of *ilities*: maintainability, usability, scalability, etc. These and other ilities are all relevant to batch processes, just in different ways.

The first three ilities—usability, maintainability, and extensibility—are related. With batch, you don't have a user interface to worry about, so usability isn't about pretty GUIs and cool animations. No, in a batch process, usability is about the code: both its error handling and its maintainability. Can you extend common components easily to add new features? Is it covered well in unit tests so that when you change an existing component, you know the effects across the system? When the job fails, do you know when, where, and why without having to spend a long time debugging? These are all aspects of usability that have an impact on batch processes.

Next is scalability. Time for a reality check: When was the last time you worked on a web site that truly had a million visitors a day? How about 100,000? Let's be honest: most web sites developed in the enterprise aren't viewed nearly that many times. However, it's not a stretch to have a batch process that needs to process a million or more transactions in a night. Let's consider 8 seconds to load a web page to be a solid average.[5] If it takes that long to process a transaction via batch, then processing 100,000 transactions will take more than 9 days (and over 3 months for 1 million). That isn't practical for any system in the modern enterprise. The bottom line is that the scale that batch processes need to be able to handle is often one or more orders of magnitude larger than that of the web or thick-client applications you've developed in the past.

Third is availability. Again, this is different from the web or thick-client applications you may be used to. Batch processes typically aren't 24/7. In fact, they typically have an appointment. Most enterprises schedule a job to run at a given time when they know the required resources (hardware, data, and so on) are available. For example, take the need to build statements for retirement accounts. Although you can run the job at any point in the day, it's probably best to run it some time after the market has closed, so you can use the closing fund prices to calculate balances. Can you run when you need to? Can you get the job done in the time allotted so you don't impact other systems? These and other questions affect the availability of your batch system.

Finally you must consider security. Typically, in the batch world, security doesn't revolve around people hacking into the system and breaking things. The role a batch process plays in security is in keeping data secure. Are sensitive database fields encrypted? Are you logging personal information by accident? How

---

[5] `https://think.storage.googleapis.com/docs/mobile-page-speed-new-industry-benchmarks.pdf`

about access to external systems—do they need credentials, and are you securing those in the appropriate manner? Data validation is also part of security. Generally, the data being processed has already been vetted, but you still should be sure that rules are followed.

As you can see, plenty of technological challenges are involved in developing batch processes. From the large scale of most systems to security, batch has it all. That's part of the fun of developing batch processes: you get to focus more on solving technical issues than debugging the latest JavaScript front end framework. The question is, with the existing infrastructures on mainframes and all the risks of adopting a new platform, why do batch in Java?

Why Do Batch Processing in Java?

With all the challenges just listed, why choose Java and an open source tool like Spring Batch to develop batch processes? I can think of six reasons to use Java and open source for your batch processes: maintainability, flexibility, scalability, development resources, support, and cost.

Maintainability is first. When you think about batch processing, you have to consider maintenance. This code typically has a much longer life than your other applications. There's a reason for that: no one sees batch code. Unlike a web or client application that has to stay up with the current trends and styles, a batch process exists to crunch numbers and build static output. As long as it does its job, most people just get to enjoy the output of their work. Because of this, you need to build the code in such a way that it can be easily modified without incurring large risks.

Enter the Spring framework. Spring was designed for a couple of things you can take advantage of: testability and abstractions. The decoupling of objects that the Spring framework enables with dependency injection and the extra testing tools the Spring portfolio provides allow you to build a robust test suite to minimize the risk of maintenance down the line. And without yet digging into the way Spring and Spring Batch work, Spring provides facilities to do things like file and database I/O declaratively. You don't have to write JDBC code or manage the nightmare that is the file I/O API in Java. Spring Batch brings things like transactions and commit counts to your application, so you don't have to manage where you are in the process and what to do when something fails. These are just some of the maintainability advantages that Spring Batch and Java provide for you.

The flexibility of Java and Spring Batch is another reason to use them. In the mainframe world, you have one option: run COBOL or CICS a mainframe. That's it. Another common platform for batch processing is C++ on UNIX. This ends up being a very custom solution because there are no industry-accepted batch-processing frameworks. Neither the mainframe nor the C++/UNIX approach provides the flexibility of the JVM for deployments and the feature set of Spring Batch. Want to run your batch process on a server, desktop, or mainframe with *nix or Windows? It doesn't matter. Want to deploy it to an application server, Docker containers, the cloud? Choose the one that fits your needs. Thin WAR, fat JAR, or whatever the next new hotness is down the line? All are okay by Spring Batch.

However, the "write once, run anywhere" nature of Java isn't the only flexibility that comes with the Spring Batch approach. Another aspect of flexibility is the ability to share code from system to system. You can use the same services that already are tested and debugged in your web applications right in your batch processes. In fact, the ability to access business logic that was once locked up on some other platform is one of the greatest wins of moving to this platform. By using POJOs to implement your business logic, you can use them in your web applications, in your batch processes—literally anywhere you use Java for development.

Spring Batch's flexibility also goes toward the ability to scale a batch process written in Java. Let's look at the options for scaling batch processes:

- *Mainframe*: The mainframe has limited additional capacity for scalability. The only true way to accomplish things in parallel is to run full programs in parallel on the single piece of hardware. This approach is limited by the fact that you need to write and maintain code to manage the parallel processing and the difficulties associated with it, such as error handling and state management across programs. In addition, you're limited by the resources of a single machine.

- *Custom processing*: Starting from scratch, even in Java, is a daunting task. Getting scalability and reliability correct for large amounts of data is very difficult. Once again, you have the same issue of coding for load balancing. You also have large infrastructure complexities when you begin to distribute across physical devices or virtual machines. You must be concerned with how communication works between pieces. And you have issues of data reliability. What happens when one of your custom-written workers goes down? The list goes on. I'm not saying it can't be done; I'm saying that your time is probably better spent writing business logic instead of reinventing the wheel.

- *Java and Spring Batch*: Although Java by itself has the facilities to handle most of the elements in the previous item, putting the pieces together in a maintainable way is very difficult. Spring Batch has taken care of that for you. Want to run the batch process in a single JVM on a single server? No problem. Your business is growing and now needs to divide the work of bill calculation across five different nodes to get it all done overnight? You're covered. Have a spike once a month and want to be able to scale on that one day using cloud resources? Check. Data reliability? With little more than some configuration and keeping some key principles in mind, you can have transaction rollback and commit counts completely handled.

As you will see as you dig into the Spring Batch framework and its related ecosystem, the issues that plague the previous options for batch processing can be mitigated with well-designed and tested solutions. Up to now, this chapter has talked about technical reasons for choosing Java and open source for your batch processing. However, technical issues aren't the only reasons for a decision like this. The ability to find qualified development resources to code and maintain a system is important. As mentioned earlier, the code in batch processes tends to have a significantly longer lifespan than the web apps you may be developing right now. Because of this, finding people who understand the technologies involved is just as important as the abilities of the technologies themselves. Spring Batch is based on the extremely popular Spring framework. It follows Spring's conventions and uses Spring's tools as well as any other Spring-based application. It is a part of Spring Boot. So, any developer who has Spring experience will be able to pick up Spring Batch with a minimal learning curve. But will you be able to find Java and, specifically, Spring resources?

One of the arguments for doing many things in Java is the community support available. The Spring family of frameworks enjoy a large and very active community online through Github, StackOverflow, and related resources. The Spring Batch project in that family has a mature community around it. Couple that with the strong advantages associated with having access to the source code and the ability to purchase support if required, and all support bases are covered with this option.

Finally you come to the cost. Many costs are associated with any software project: hardware, software licenses, salaries, consulting fees, support contracts, and more. However, not only is a Spring Batch solution the most bang for your buck, but it's also the cheapest overall. Using cloud resources and open source frameworks, the only recurring costs are for development salaries, support contracts, and infrastructure—much less than the recurring licensing costs and hardware support contracts related to other options.

I think the evidence is clear. Not only is using Spring Batch the most sound route technically, but it's also the most cost-effective approach. Enough with the sales pitch: let's start to understand exactly what Spring Batch is.

Other Uses for Spring Batch

I bet by now you're wondering if replacing the mainframe is all Spring Batch is good for. When you think about the projects you face on an ongoing basis, it isn't every day that you're ripping out COBOL code. If that was all this framework was good for, it wouldn't be a very helpful framework. However, this framework can help you with many other use cases.

The most common use case for Spring Batch is probably ETL processing or extract, transform, load. Moving data around from one format to another is a large part of enterprise data processing. Spring Batch's chunk-based processing and extreme scaling capabilities make it a natural fit for ETL workloads.

Another use case is data migration. As you rewrite systems, you typically end up migrating data from one form to another. The risk is that you may write one-off solutions that are poorly tested and don't have the data-integrity controls that your regular development has. However, when you think about the features of Spring Batch, it seems like a natural fit. You don't have to do a lot of coding to get a simple batch job up and running, yet Spring Batch provides things like commit counts and rollback functionality that most data migrations should include but rarely do.

A third common use case for Spring Batch is any process that requires parallel processing. As chipmakers approach the limits of Moore's Law, developers realize that the only way to continue to increase the performance of apps is not to process single operations faster, but to process more operations in parallel. Many frameworks have recently been released that assist in parallel processing. Most of the big data platforms like Apache Spark, YARN, GridGain, Hazlecast, and others have come out in recent years to attempt to take advantage of both multicore processors and the numerous servers available via the cloud. However, frameworks like Apache Spark require you to alter your code and data to fit their algorithms or data structures. Spring Batch provides the ability to scale your process across multiple cores or servers (as shown in Figure 1-1 with master/worker step configurations) and still be able to use the same objects and datasources that your web applications use.

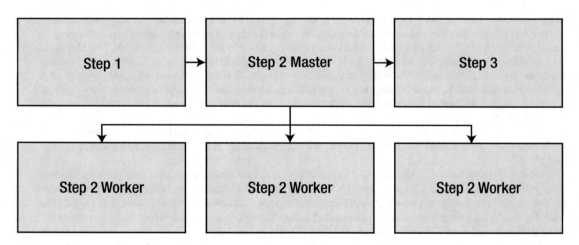

Figure 1-1. *Simplifying parallel processing*

Orchestration of workloads is another common use case for Spring Batch. Typically an enterprise batch process isn't just a single step. It requires the coordination of many, decoupled, steps to be orchestrated. Perhaps a file needs to be loaded, then two independent types of processing on that data occurs, followed up by a single export of the results. The orchestration of these tasks is a use case that Spring Batch addresses well. An example of that is Spring Cloud Data Flow and its use of Spring Batch to handle "composed tasks." Here, Spring Batch calls Spring Cloud Data Flow to launch other functionality and keeps track of what is done and what still needs to be done. Figure 1-2 illustrates the drag-and-drop user interface provided by Spring Cloud Data Flow for constructing "composed tasks."

Tasks

This section allows for creation of composed tasks.

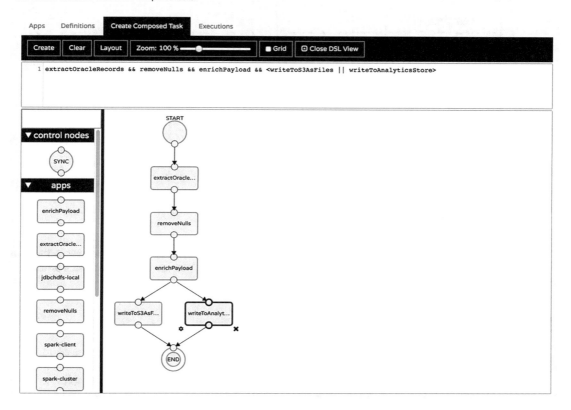

Figure 1-2. *Orchestrating tasks via Spring Cloud Data Flow*

Finally you come to constant or 24/7 processing. In many use cases, systems receive a constant or near-constant feed of data. Although accepting this data at the rate it comes in is necessary for preventing backlogs, when you look at the processing of that data, it may be more performant to batch the data into chunks to be processed at once (as shown in Figure 1-2). Spring Batch provides tools that let you do this type of processing in a reliable, scalable way. Using the framework's features, you can do things like read messages from a queue, batch them into chunks, and process them together in a never-ending loop. Thus you can increase throughput in high-volume situations without having to understand the complex nuances of developing such a solution from scratch.

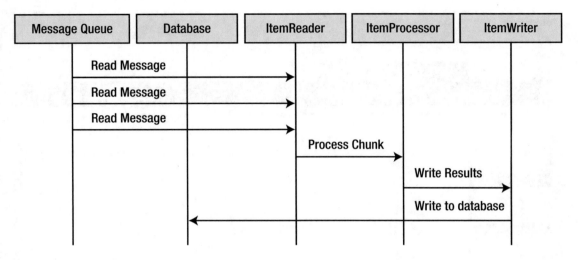

Figure 1-3. *Batching message processing to increase throughput*

As you can see, Spring Batch is a framework that, although designed for mainframe-like processing, can be used to simplify a variety of development problems. With everything in mind about what batch is and why you should use Spring Batch, let's finally begin looking at the framework itself.

The Spring Batch Framework

The Spring Batch framework (Spring Batch) was developed as a collaboration between Accenture and SpringSource as a standards-based way to implement common batch patterns and paradigms.

Features implemented by Spring Batch include data validation, formatting of output, the ability to implement complex business rules in a reusable way, and the ability to handle large data sets. You'll find as you dig through the examples in this book that if you're familiar at all with Spring, Spring Batch just makes sense.

Let's start at the 30,000-foot view of the framework, as shown in Figure 1-4.

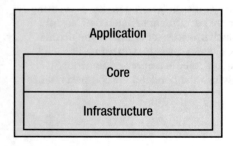

Figure 1-4. *The Spring Batch architecture*

Spring Batch consists of three tiers assembled in a layered configuration. At the top is the *application layer*, which consists of all the custom code and configuration used to build out your batch processes. Your business logic, services, and so on, as well as the configuration of how you structure your jobs, are all considered the application. Notice that the application layer doesn't sit on top of but instead wraps the other two layers, core and infrastructure. The reason is that although most of what you develop consists of the application layer working with the core layer, sometimes you write custom infrastructure pieces such as custom readers and writers.

The application layer spends most of its time interacting with the next layer, the core. The *core layer* contains all the pieces that define the batch domain. Elements of the core component include the Job and Step interfaces as well as the interfaces used to execute a Job: `JobLauncher` and `JobParameters`.

Below all this is the *infrastructure layer*. In order to do any processing, you need to read and write from files, databases, and so on. You must be able to handle what to do when a job is retried after a failure. These pieces are considered common infrastructure and live in the infrastructure component of the framework.

■ **Note** A common misconception is that Spring Batch is or has a scheduler. It doesn't. There is no way within the framework to schedule a job to run at a given time or based on a given event. There are a number of ways to launch a job, from a simple cron script to Quartz or even an enterprise scheduler like Control-M, but none within the framework itself. Chapter 4 covers launching a job.

Let's walk through some features of Spring Batch.

Defining Jobs with Spring

Batch processes have a number of different domain-specific concepts. A *job* is a process that executes from start to finish without interruption or interaction. A job can consist of a number of steps. There may be input and output related to each step. When a step fails, it may or may not be repeatable. The flow of a job may be conditional (e.g., execute the bonus calculation step only if the revenue calculation step returns revenue over $1,000,000). Spring Batch provides classes, interfaces, XML schemas, and Java configuration utilities that define these concepts using Java to divide concerns appropriately and wire them together in a way familiar to those who have used Spring. Listing 1-1, for example, shows a basic Spring Batch job configured in Java configuration. The result is a framework for batch processing that you can pick up very quickly with only a basic understanding of Spring as a prerequisite.

Listing 1-1. Sample Spring Batch Job Definition

```
@Bean
public AccountTasklet accountTasklet() {
    return new AccountTasklet();
}

@Bean
public Job accountJob() {
    Step accountStep =
        this.stepBuilderFactory
            .get("accountStep")
            .tasklet(accountTasklet())
            .build();

    return this.jobBuilderFactory
                .get("accountJob")
                .start(accountStep)
                .build();
}
```

In the configuration listed in Listing 1-1, two beans are created. The first is an `AccountTasklet`. The `AccountTasklet` is a custom component where the business logic for the step will live. Spring Batch will call its single method (execute) over and over, each call in a new transaction, until the `AccountTasklet` indicates that it is done.

The second bean is the actual Spring Batch `Job`. In this bean definition, we create a single `Step` out of the `AccountTasklet` we just defined using the builders provided by the factory. We then use the builders provided to create a `Job` out of the `Step`. Spring Boot will find this `Job` and execute it automatically on the startup of our application.

Managing Jobs

It's one thing to be able to write a Java program that processes some data once and never runs again. But mission-critical processes require a more robust approach. The ability to keep the state of a job for re-execution, maintaining data integrity when a job fails through transaction management and saving performance metrics of past job executions for trending, are features that you expect in an enterprise batch system. These features are included in Spring Batch, and most of them are turned on by default; they require only minimal tweaking for performance and requirements as you develop your process.

Local and Remote Parallelization

As discussed earlier, the scale of batch jobs and the need to be able to scale them is vital to any enterprise batch solution. Spring Batch provides the ability to approach this in a number of different ways. From a simple thread-based implementation, where each commit interval is processed in its own thread of a thread pool, to running full steps in parallel, to configuring a grid of workers that are fed units of work from a remote master via partitioning, Spring Batch and its related ecosystem provide a collection of different options, including parallel chunk/step processing, remote chunk processing, and partitioning.

Standardizing I/O

Reading in from flat files with complex formats, XML files (XML is streamed, never loaded as a whole), databases or NoSQL stores, or writing to files or XML can be done with only simple configuration. The ability to abstract things like file and database input and output from your code is an attribute of the maintainability of jobs written in Spring Batch.

The Rest of the Spring Batch Ecosystem

Like most projects within the Spring portfolio, Spring Batch does not sit in isolation. It is part of an ecosystem where other projects extend and complement it to provide a more robust solution. Some of the other projects in the portfolio that work with Spring Batch are as follows.

Spring Boot

Introduced in 2014, Spring Boot takes an opinionated approach to developing Spring applications. Now virtually *the* standard way of developing Spring applications, Spring Boot provides facilities for easily packaging, deploying, and launching all Spring workloads including batch. It also serves as a pillar in the cloud native story provided by Spring Cloud. As such, Spring Boot will be the primary method for developing batch applications for this book.

Spring Cloud Task

Spring Cloud Task is a project under the Spring Cloud umbrella that provides facilities for the execution of finite tasks in a cloud environment. As a framework that targets finite workloads, batch processing is a processing style that integrates well with Spring Cloud Task. Spring Cloud Task provides a number of extensions to Spring Batch including the publishing of informational messages (a job starts/finishes, a step starts/finishes, etc.), as well as the ability to scale batch jobs dynamically (instead of the various static ways provided by Spring Batch directly).

The Spring Cloud Data Flow

Writing your own batch-processing framework doesn't just mean having to redevelop the performance, scalability, and reliability features you get out of the box with Spring Batch. You also need some form of administration and orchestration toolset to do things like start and stop jobs and view the statistics of previous job runs. However, if you use Spring Batch, it includes all that functionality as well as a newer addition: the Spring Cloud Data Flow project. The Spring Cloud Data Flow project is a tool for orchestrating microservices on a cloud platform (CloudFoundry, Kubernetes, or Local). Developing your batch applications as microservices will allow you to deploy them in a dynamic way using Spring Cloud Data Flow.

And All the Features of Spring

Even with the impressive list of features that Spring Batch includes, the greatest thing is that it's built on Spring. With the exhaustive list of features that Spring provides for any Java application, including dependency injection, aspect-oriented programming (AOP), transaction management, and templates/helpers for most common tasks (JDBC, JMS, e-mail, and so on), building an enterprise batch process on a Spring framework offers virtually everything a developer needs.

As you can see, Spring Batch brings a lot to the table for developers. The proven development model of the Spring framework, scalability, and reliability features as well as an administration application are all available for you to get a batch process running quickly with Spring Batch.

How This Book Works

After going over the what and why of batch processing and Spring Batch, I'm sure you're chomping at the bit to dig into some code and learn what building batch processes with this framework is all about. Chapter 2 goes over the domain of a batch job, defines some of the terms I've already begun to use (*job*, *step*, and so on), and walks you through setting up your first Spring Batch project. You honor the computer science gods by writing a "Hello, World!" batch job and see what happens when you run it.

One of my main goals for this book is to not only provide an in-depth look at how the Spring Batch framework works, but also show you how to apply those tools in a realistic example. Chapter 3 provides the requirements and technical architecture for a project that you implement in Chapter 10.

The code examples for this book can be found on Github. I encourage you to download that repository and refernce it as you work your way through this book. It can be found at `https://github.com/Apress/def-guide-spring-batch`.

Summary

This chapter walked through a history of batch processing. It covered some of the challenges a developer of a batch process faces as well as justified the use of Java and open source technologies to conquer those challenges. Finally, you began an overview of the Spring Batch framework by examining its high-level components and features. By now, you should have a good view of what you're up against and understand that the tools to meet the challenges exist in Spring Batch. Now, all you need to do is learn how. Let's get started.

CHAPTER 2

∎ ∎ ∎

Spring Batch 101

Assembling a computer is an easy task. Many developers do it at some point in their careers. But it's really only easy once you understand what each part does and how it fits into the larger system. If I gave a bag of computer parts to someone that didn't know what a computer did and told them to put it together, things may not go so well.

In the enterprise Java world, there are many domains that transfer well. The MVC pattern common in most web frameworks is an example. Once you know one MVC framework, picking up another is just a matter of understanding the syntax for the various pieces. However, there are not many batch frameworks out there. Because of that, this domain may be a bit new to you. You may not know what a job or a step is. Or how an `ItemReader` relates to an `ItemWriter`. And what the heck is a `Tasklet` anyways?

This chapter should answer those questions. In it, we'll walk through the following topics:

- *The architecture of batch*: This section begins to dig a bit deeper into what makes up a batch process and defines terms that you'll see throughout the rest of the book.

- *Project setup*: I learn by doing. This book is assembled in a way that shows you examples of how the Spring Batch framework functions, explains why it works the way it does, and gives you the opportunity to code along. This section covers the basic setup for a Maven-based Spring Batch project.

- *Hello, World!* The first law of thermodynamics talks about conserving energy. The first law of motion deals with how objects at rest tend to stay at rest unless acted upon by an outside force. The first law of computer science seems to be that whatever new technology you learn, you must write a "Hello, World!" program using said technology. Here we will obey that law.

- *Running a job*: How to execute your first job may not be immediately apparent, so I'll walk you through how jobs are executed as well as how to pass in basic parameters.

With all of that in mind, what is a job, anyway?

The Architecture of Batch

The last chapter spent some time talking about the three layers of the Spring Batch framework: the application layer, the core layer, and the infrastructure layer. The application layer represents the code you develop, which for the most part interfaces with the core layer. The core layer consists of the actual components that make up the batch domain. Finally, the infrastructure layer includes item readers and writers as well as the required classes and interfaces to address things like restartability.

This section goes deeper into the architecture of Spring Batch and defines some of the concepts referred to in the last chapter. You then learn about some of the scalability options that are key to batch processing and what makes Spring Batch so powerful. Finally, the chapter discusses outline administration options as well as where to find answers to your questions about Spring Batch in the documentation. You start with the architecture of batch processes, looking at the components of the core layer.

Examining Jobs and Steps

Figure 2-1 shows the essence of a job. Configured via Java or XML, a batch *job* is a collection of states and transitions from one to the next. In essence, a Spring Batch job is nothing more than a state machine. Since steps are the most common form of state used in Spring Batch, we'll focus on them for now.

Using the use case of the nightly processing of a user's bank account as an example, step 1 could be to load in a file of transactions received from another system. Step 2 would apply all credits to the account. Finally, step 3 would apply all debits to the account. The job represents the overall process of applying transactions to the user's account.

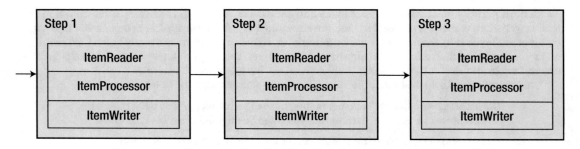

Figure 2-1. *A batch job*

When you look deeper at an individual step, you see a self-contained unit of work that is the main building block of a job. There are two main types of steps: a tasklet step and a chunk based step. A tasklet-based step is the more simple of the two. It takes a `Tasklet` implementation and runs its `execute` (`StepContribution contribution, ChunkContext chunkContext`) method within the scope of a transaction over and over until the execute method tells the step to stop (each call to the execute method gets its own transaction). It's commonly used for things like initialization, running a stored procedure, sending notifications, and so on.

A chunk-based step is a bit more rigid in its structure, but is intended for item-based processing. Each chunk-based step has up to three main parts: an `ItemReader`, an `ItemProcessor`, and an `ItemWriter`. Note that I stated a step has up to three parts. A step isn't required to have an `ItemProcessor`. It is okay to have a step that consists of just an `ItemReader` and an `ItemWriter` (common in data-migration jobs, for example). Table 2-1 walks through the interfaces that Spring Batch provides to represent these concepts.

Table 2-1. *The Interfaces That Make Up a Batch Job*

Interface	Description
`org.springframework.batch.core.Job`	The object representing the job, as configured within the `ApplicationContext`.
`org.springframework.batch.core.Step`	Like the job, represents the step as configured.
`org.springframework.batch.core.step.tasklet.Tasklet`	A strategy interface that provides the ability to execute logic within the scope of a transaction.
`org.springframework.batch.item.ItemReader<T>`	A strategy interface for providing the input of a step.
`org.springframework.batch.item.ItemProcessor<I,O>`	A facility to apply business logic, validation, etc. to an individual item as provided.
`org.springframework.batch.item.ItemWriter<T>`	A strategy interface for the persistence of items within a step.

One of the advantages of the way Spring has structured a job is that it decouples each step into its own independent processor. Each step is responsible for obtaining its own data, applying the required business logic to it, and then writing the data to the appropriate location. This decoupling provides a number of features:

- *Flexibility*: The ability to configure complex flows of work based on complex logic is something that is difficult to implement on your own in a reusable way. Yet Spring Batch provides a nice set of builders to do just that. The ability to use its fluent Java API as well as traditional XML to configure your batch applications is a powerful tool.

- *Maintainability*: With the code for each step decoupled from the steps before and after it, steps are easy to unit-test, debug, and update with virtually no impact on other steps. Decoupled steps also make it possible to reuse steps in multiple jobs. As you'll see in upcoming chapters, steps are nothing more than Spring beans and can be reused just like any other bean in Spring.

- *Scalability*: Independent steps in a job provide a number of options to scale your jobs. You can execute steps in parallel. You can divide the work within a step across threads and execute the code of a single step in parallel. Any of these abilities lets you meet the scalability needs of your business with minimum direct impact on your code.

- *Reliability*: The different phases of a step (reading via the `ItemReader`, processing via the `ItemProcessor`, etc.) all allow for facilities to do things like retrying an operation or skipping an item if an exception is thrown providing robust error handling options.

Job Execution

When a job is executed, a number of components interact to provide the resiliency that Spring Batch provides. Let's take a look at those components at a high level and how they interact.

We can begin with the main shared piece of the architecture, the `JobRepository`. Shown in Figure 2-2, this component is responsible for maintaining the state of a job as well as various processing metrics (start time, end time, status, number of reads/writes, etc.). Typically backed by a relational database, this component is shared by virtually all of the main components within Spring Batch.

15

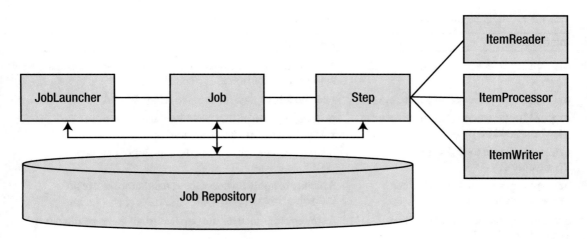

Figure 2-2. *The job components and their relationships*

Next is the JobLauncher. This is responsible for the execution of a Job. Tasks the JobLauncher does beyond just calling Job.execute can include things like validating that restarting a job is valid (not all jobs are restartable), how to execute the job (on the current thread, via a thread pool, etc.), validating parameters, and so on. All of these tasks, however, are dependent on the implementation. In the Spring Boot world, this is a component that you typically do not need to directly work with because Spring Boot provides facilities of launching a job out of the box. We'll see this in action a bit later.

Once a Job is launched, the Job executes each Step. As each Step is executed, the JobRepository is updated with the current state. What Step is executed, its current status, how many items were read, processed, written, and so on, are all stored in the JobRepository.

The processing of a Job and a Step are very similar. A Job goes through the list of Steps it has been configured to run, executing each one. As a chunk of items within a Step completes, Spring Batch updates the JobExecution or StepExecution in the repository with the current state. A Step goes through a list of items as read in by the ItemReader. As the Step processes each chunk of items, the StepExecution in the repository is updated with where it is in the Step. Things like current commit count, start and end times, and other information are stored in the repository. When a Job or Step is complete, the related execution is updated in the repository with the final status.

We've mentioned the JobExecution and StepExecution a few times now, so let's take a minute to understand what those are and how they relate. Figure 2-3 illustrates the components in this relationship. A JobInstance is a logical execution of a Spring Batch job. It's identified by the job name and the unique set of identifying parameters that are provided for that job's logical execution. If we use a statement-generating job named statementGenerator as an example, a new JobInstance would be created every time the statementGenerator job is launched with the same parameters. So in this case, running statementGenerator with a date of May 7, 2017, as a parameter would create a new JobInstance.

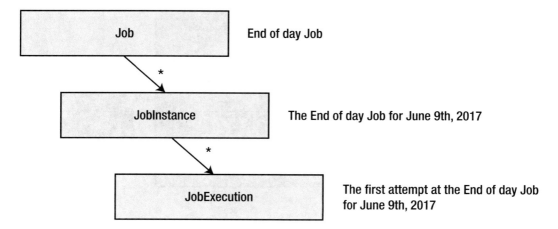

Figure 2-3. *Relationship of JobInstance, JobExecution, and StepExecution*

A JobExecution is a physical execution of a Spring Batch job. Every time you launch a Job, you'll get a new JobExecution. You may not get a new JobInstance. The obvious example for this is restarting a failed job. When a job is run the first time, you get a new JobInstance and a JobExecution. If that execution fails, on the restart, you won't get a new JobInstance since it's still the same logical run (executed with the same identifying parameters). However, you will get a new JobExecution to track the second physical run. So a JobInstance can have multiple JobExecutions within it.

Finally a StepExecution is the physical execution of a Step. There is no concept of a StepInstance in Spring Batch. A JobExecution can (and usually does) have multiple StepExecution instances associated with it.

Parallelization

A simple batch process's architecture consists of a single-threaded process that executes a job's steps in order from start to finish. However, Spring Batch provides a number of parallelization options that you should be aware of as you move forward (Chapter 11 covers these options in detail). There are five different ways to parallelize your work: dividing work via multithreaded steps, parallel execution of full steps, asynchronous ItemProcessor/ItemWriter configurations, remote chunking, and partitioning.

Multithreaded Steps

The first approach to achieving parallelization is the division of work via multithreaded steps. In Spring Batch, a job is configured to process work in blocks called *chunks*, each chunk being wrapped within its own transaction. Normally, each chunk is processed in series. If you have 10,000 records, and the commit count is set at 50 records, your job will process records 1 to 50 and then commit, process 51 to 100 and commit, and so on, until all 10,000 records have been processed. Spring Batch allows you to execute chunks of work in parallel to improve performance. With three threads, you can increase your theoretical throughput threefold, as shown in Figure 2-4.[1]

---

[1]This is a theoretical throughput increase. Many factors can prevent the ability of a process to achieve linear parallelization like this.

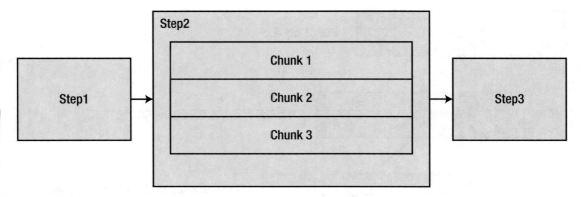

Figure 2-4. *Multithreaded steps*

Parallel Steps

The next approach you have available for parallelization is the ability to execute steps in parallel, as shown in Figure 2-5. Let's say you have two steps, each of which loads an input file into your database, but there is no relationship between the steps. Does it make sense to have to wait until one file has been loaded before the next one is loaded? Of course not, which is why this is a classic example of when to use the ability to process steps in parallel.

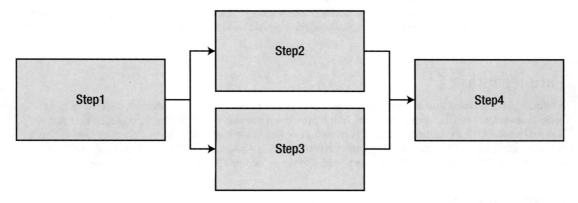

Figure 2-5. *Parallel step processing*

Asynchronous ItemProcessor/ItemWriter

In some use cases, the ItemProcessor within a step can be the bottleneck. For example, you may have a complex math calculation that needs to be done or remote services to call to enrich the data the ItemReader provided. The ability to parallelize that piece of the step can be useful. The AsynchonousItemProcessor is a decorator for an ItemProcessor implementation that executes each call to the ItemProcessor in its own thread. Instead of returning the result of the ItemProcessor call, the AsynchronousItemProcessor returns a java.util.concurrent.Future for each call. The list of Futures returned within the current chunk are then handed off to the AsynchronousItemWriter. This ItemWriter, also a decorator for the ItemWriter your Step actually needs to use, will unwrap the Future and pass the real results to the delegating ItemWriter.

Remote Chunking

The last two approaches to parallelization allow you to spread processing across multiple JVMs. In all cases previously, the processing was performed in a single JVM, which can seriously hinder the scalability options. When you can scale any part of your process horizontally across multiple JVMs, the ability to keep up with large demands increases.

The first remote-processing option is *remote chunking*. In this approach, input is performed using a standard ItemReader in a master node; the input is then sent via a form of durable communication (a message broker like RabbitMQ or ActiveMQ, for example) to a remote worker ItemProcessor that is configured as a message-driven POJO. When the processing is complete, the worker either sends the updated item back to the master for writing or writes it itself. Because this approach reads the data at the master, processes it at the worker, and then sends it back, it's important to note that it can be very network intensive. This approach is good for scenarios where the cost of I/O is small compared to the actual processing.

Partitioning

The final method for parallelization within Spring Batch is partitioning, shown in Figure 2-6. Spring Batch supports both remote (master and remote workers) and local partitioning (using threads for the workers). Two key differences between remote partitioning and remote chunking are that with remote partitioning, you don't need a durable method of communication, and the master serves only as a controller for a collection of worker steps. In this case, each of your worker steps is self-contained and configured the same as if it was locally deployed. The only difference is that the worker steps receive their work from the master node instead of the job itself. When all the workers have completed their work, the master step is considered complete. This configuration doesn't require durable communication with guaranteed delivery because the JobRepository guarantees that no work is duplicated and all work is completed—unlike the remote-chunking approach, in which the JobRepository has no knowledge of the state of the distributed work.

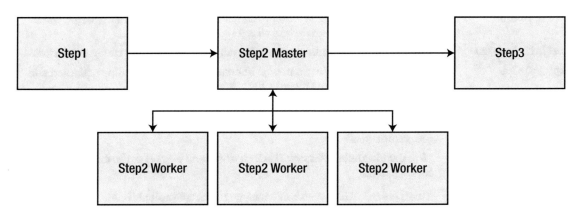

Figure 2-6. *Partitioning work*

Documentation

While I appreciate your reading this book, an open source project is only as good as its documentation. We've tried hard to create not only a comprehensive set of docs, but also a full suite of samples that demonstrate how to execute a Spring Batch job using many of the concepts within the framework. Table 2-2 provides a list of the samples included within the framework and what they do.

Table 2-2. *Sample Batch Jobs*

Batch Job	Description
adhocLoopJob	An infinite loop used to demonstrate the exposing of elements via JMX and the running of the job in a background thread (instead of the main JobLauncher thread).
amqpExampleJob	A job that demonstrates using AMQP as both the input and output data for the job.
beanWrapperMapper SampleJob	A job with two steps that is used to demonstrate the mapping of file fields to domain objects as well as validation of file-based input.
compositeItemWriter SampleJob	A step can have only one reader and writer. The CompositeWriter is the way around this. This sample job demonstrates how.
customerFilterJob	Uses an ItemProcessor to filter out customers that aren't valid. This job also updates the filter count field of the step execution.
delegatingJob	Using the ItemReaderAdapter, delegates the reading of input to a configured method of a POJO.
footballJob	A football statistics job. After loading two input files, one with player data and one with game data, the job generates a selection of summary statistics for the players and games and writes them to the log file.
groovyJob	Uses Groovy (a dynamic JVM language) to script the unzipping and zipping of a file.
headerFooterSample	Using callbacks, adds the ability to render a header and footer on the output.
hibernateJob	Spring Batch readers and writers don't use Hibernate by default. This job shows how to integrate Hibernate into your job.
infiniteLoopJob	Just a job with an infinite loop, used to demonstrate stop and restart scenarios.
ioSampleJob	Provides examples of a number of different I/O options including delimited and fix-width files, multiline records, XML, and JDBC integration.
jobSampleJob	Demonstrates the execution of a job from another job.
loopFlowSample	Using the decision tag, demonstrates how to control execution flow programmatically.
mailJob	Uses the SimpleMailMessageItemWriter to send e-mails as the form of output for each item.
multilineJob	Treats groups of file records as a list that represents a single item.
multilineOrder	As an expansion of the multiline input concept, reads in a file with multiline nested records using a custom reader. The output is also multiline, using standard writers.
parallelJob	Reads records into a staging table, where a multithreaded step processes them.
partitionFileJob	Uses the MultiResourcePartitioner to process a collection of files in parallel.

(continued)

Table 2-2. (*continued*)

Batch Job	Description
partitionJdbcJob	Instead of looking for multiple files and processing each one in parallel, divides the number of records in the database for parallel processing.
restartSampleJob	Throws a fake exception when processing has begun, to demonstrate the ability to restart a job that has errored and have it begin again where it left off.
retrySample	Using some interesting logic shows how Spring Batch can attempt to process an item multiple times before giving up and throwing an error.
skipSampleJob	Based on the tradeJob example. In this job, however, one of the records fails validation and is skipped.
taskletJob	The most basic use of Spring Batch is the Tasklet. This example shows how any existing method can be used as tasklets via the MethodInvokingTaskletAdapter.
tradeJob	Models a real-world scenario. This three-step job imports trade information into a database, updates customer accounts, and generates reports.

Project Setup

Up to this point, you've looked at why you'd use Spring Batch and examined the components of the framework. However, looking at diagrams and learning new lingo will only take you so far. At some point, you need to dig into the code: so, grab an editor, and let's start digging.

In this section, you build your first batch job. You walk through the setup of a Spring Batch project, including obtaining the required files from Spring. You then configure a job and code the "Hello, World!" version of Spring Batch. Finally, you learn how to launch a batch job from the command line.

Obtaining Spring Batch

Before you begin writing batch processes, you need to obtain the Spring Batch framework. There are a number of ways to get it, including grabbing the code from Github, using Maven or Gradle, and so on. However, for the purposes of this book, since we're going to focus on Spring Boot–based batch jobs, we'll start with Spring Initailizr. Spring Initailizr is a service provided by the Spring team that allows you to generate the shell of a project with a set of validated dependencies.

In order to use Spring Initailizr, you need one of two things: either an IDE that supports integration with Initializr directly (Spring Tool Suite and IntelliJ are two examples of this) or a web browser. Let's walk through each of those options. Going forward, each of the examples within this book will assume you have a clean Spring Boot–based project.

The Web Site

To use the Spring Initializr web site, we begin by navigating to https://start.spring.io as seen in Figure 2-7. There you are presented with a UI that allows you to define some basic parameters about your project including

- *Build system*: Maven or Gradle are both supported at the time of this book.

- *Language*: Spring has robust support for Java, Groovy, and Kotlin so Spring Initializr allows you to choose from these.

- *Spring Boot version*: Some features differ from version to version so Spring Initializr allows you to select which version you use.

- *Artifact coordinates*: This allows your POM or gradle.build file to be generated with the correct coordinates prepopulated.

- *Dependencies*: This allows you to specify the Spring Boot starters you'll want included in your project.

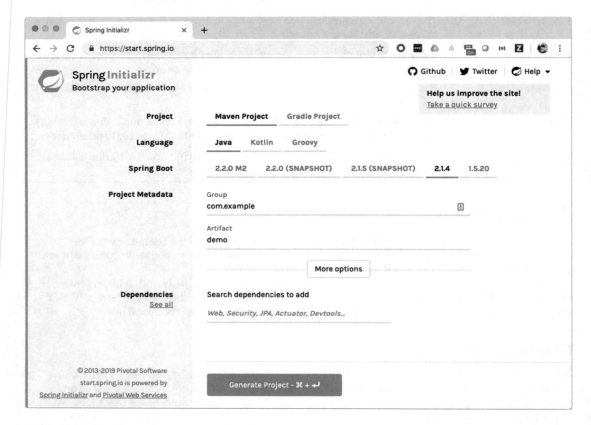

Figure 2-7. *The defaults for start.spring.io*

The UI provides two options for adding dependencies. If you know what you want, you can just type them into the search box to the right. If you don't know what you want, you can click the "Switch to the full version" link, which exposes a checkbox for each of the options available in Spring Initializr. In every project we create in this book, you'll be choosing the Batch Spring Boot starter as a minimum.

Once you have entered all the relevant data, click the "Generate Project" button. A zip file will be downloaded with a full project shell ready to add your code to. This project can then be imported directly into your favorite IDE for the rest of your development.

Spring Tool Suite

Spring Tool Suite (STS) is an Eclipse-based IDE maintained by the Spring team that provides added features around the Spring Framework and developing of microservices. It can be obtained either as an independent download or as a plug-in to add to your existing Eclipse installation. It can be obtained from the Spring website for free at `https://spring.io/tools`.

With STS installed, go to File ➤ New ➤ Spring Starter Project. You'll be presented with a window that looks like Figure 2-8.

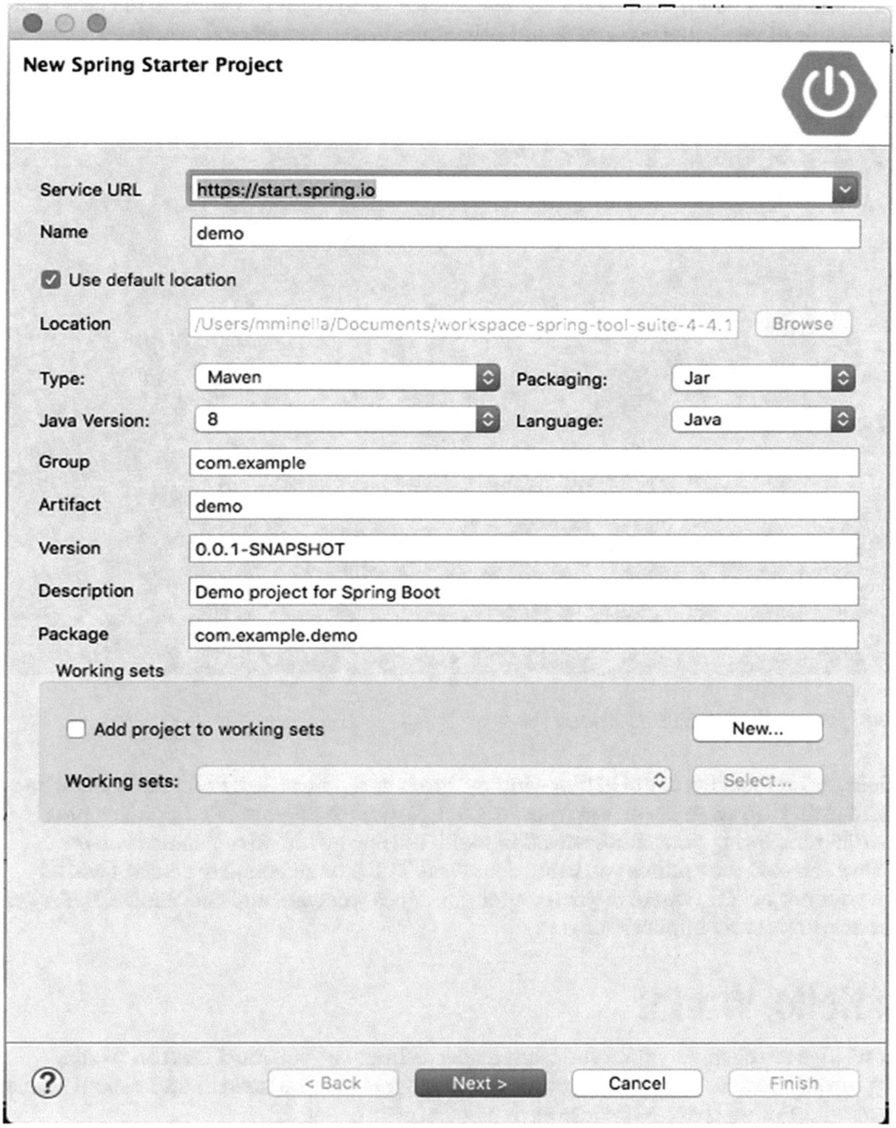

Figure 2-8. *The defaults for the Spring Tool Suite project setup*

The only differences between the website and this window are the location of the service (we'll still use https://start.spring.io), a location field for where to place the project locally, and the ability to add it to an Eclipse working set. Once the values for this window are configured, click the next button at the bottom. This will take you to a window that allows you to chose the Spring Boot starters you want to include in your project similar to how the checkboxes are made available on the website. Once the dependencies you want have been selected, you can click Finish and STS will download and import the new project for you.

IntelliJ IDEA

IntelliJ IDEA is another popular IDE for Java development that provides excellent integration with the Spring functionality including a similar experience using the Spring Initializr. With IDEA installed, go to File ➤ New ➤ Project... Along the left side of the dialog box you'll see Spring Initializr. Once selected, you'll see a screen like in Figure 2-9.

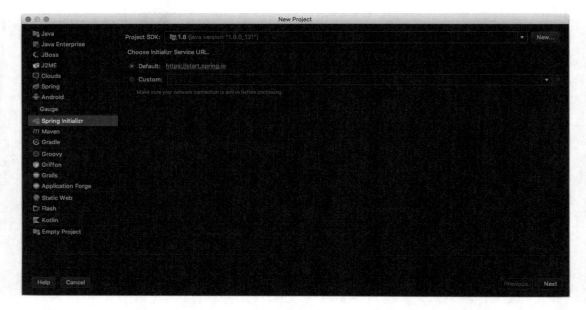

Figure 2-9. *The defaults for IntelliJ IDEA project setup*

We can use the defaults for this discussion in that window. From here, click Next. From here, you'll be presented with the options that are available on the website. Click Next once they are filled out. The next screen will display the different Spring Boot starters available with the Spring Intiailizr instance you are looking at. Once you have selected the options you want, click Next. The final screen in the wizard will let you chose the name of your project and where to put it. Enter the values you want and click Finish. IDEA will download the appropriate project and import it for you.

It's the Law: Hello, World!

The laws of computer science are clear. Any time you learn a new technology, you must create a "Hello, World!" program using the said technology, so let's get started. Don't feel like you need to understand all the moving parts of this example. Future chapters go into each piece in greater detail.

To get started with our Hello, World Spring Batch job, we'll need to create a new project. Use whichever method you prefer from the previous section to create a new project, naming it hello-world. As you go through the wizard in Spring Initializr (via whatever means you choose), you'll want to make the following selections:

- Group ID: io.spring.batch

- Artifact ID: hello-world

- Build System: Maven

- Language: Java 8+[2]

- Packaging: Jar

- Version: 0.0.1-SNAPSHOT

- Spring Boot Version: 2.1.2[3]

- Dependencies: Batch, H2, and JDBC

For the dependences, batch should be obvious. H2 is used as an in-memory database for our job repository. Finally JDBC is added for the database support (`DataSource`, etc.). Once you've imported your project into your IDE, you should be left with a project structure that looks like what is found in Figure 2-10.[4]

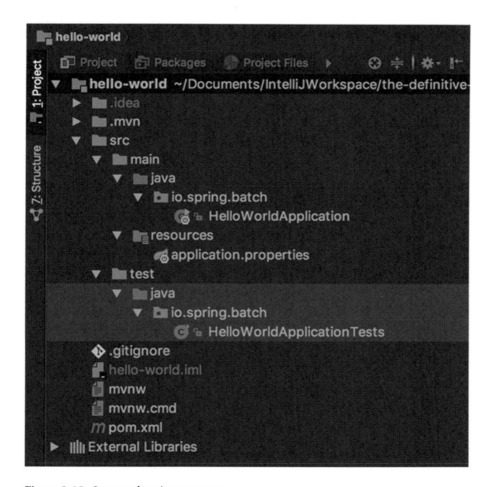

Figure 2-10. *Imported project structure*

---

[2]Any version Java 8 or higher should work with all examples in this book.
[3]As of the writing of this book, Spring Boot 2.1.2 is the latest version. However, any version greater than 2.1 should work for the examples in this book.
[4]This project uses Maven as the build system. Your layout will differ slightly if you use Gradle.

With our project imported, we can take a look at the different pieces. Really, a Spring Initializr project only consists of four main pieces:

1. A build system (I'll be using Maven throughout the book, but you can use whichever you choose.)

2. A source file to bootstrap Spring Boot (located in /src/main/java/your/package/ name/HelloWorldApplication.java)

3. A test file that simply bootstraps your context to see if it works (located in /src/ test/java/your/package/name/HelloWorldApplicationTests.java)

4. An application.properties for configuring your application (located in /src/main/ resources)

For this hello-world project, we'll focus on the HelloWorldApplication class. In Listing 2-1, is all the code needed to run your first batch job (minus imports).

Listing 2-1. The "Hello, World!" Job

```
@EnableBatchProcessing
@SpringBootApplication
public class HelloWorldApplication {

    @Autowired
    private JobBuilderFactory jobBuilderFactory;

    @Autowired
    private StepBuilderFactory stepBuilderFactory;

    @Bean
    public Step step() {
        return this.stepBuilderFactory.get("step1")
                        .tasklet(new Tasklet() {
                                @Override
                                public RepeatStatus execute(StepContribution contribution,
                                                                                ChunkContext
                                                                                chunkContext) {

                                System.out.println("Hello, World!");
                                return RepeatStatus.FINISHED;
                            }
                }).build();
    }

    @Bean
    public Job job() {
        return this.jobBuilderFactory.get("job")
                        .start(step())
                        .build();
    }

    public static void main(String[] args) {
        SpringApplication.run(HelloWorldApplication.class, args);
    }
}
```

If you read through the code in Listing 2-1, it should sound vaguely familiar from a high level. It's really just the Java representation of what is discussed in Examining Jobs and Steps section earlier in this chapter. However, if we break it down, it really consists of three main pieces that we added to the generated `HelloWorldApplication` class: the `@EnableBatchProcessing` annotation, the injection of the `JobBuilderFactory` and `StepBuilderFactory`, the definition of the step, and the definition of our job.

Working from the top down, our job begins with the `@EnableBatchPocessing` annotation. This annotation is provided by Spring Batch and is used to bootstrap the batch infrastructure. It provides Spring bean definitions for most of the batch infrastructure so you don't have to, including:

- A `JobRepository`: Used to record the state of the job as it runs
- A `JobLauncher`: Used to launch the job
- A `JobExplorer`: Used to perform read only operations with the `JobRepository`
- A `JobRegistry`: Used to find jobs when using certain launcher implementations
- A `PlatformTransactionManager`: Handles transactionality over the course of your job
- A `JobBuilderFactory`: A fluent builder for creating jobs
- A `StepBuilderFactory`: A fluent builder for creating steps

As you can see, that one annotation does a lot for us. It does have one requirement though, which is to provide it with a `DataSource`. The `JobRepository` and `PlatformTransactionManager` use it as they each require. Spring Boot handles that for us by having HSQLDB on the classpath. It detects it and creates an embedded `DataSource` for us on startup.

As we move down the class, the next thing is the `@SpringBootApplication` annotation. This is really a meta-annotation that combines `@ComponentScan` with `@EnableAutoConfiguration`. It triggers the autoconfiguration that will create our `DataSource` as well as any other Spring Boot–based autoconfigs that make sense.

After the class definition we autowire in two builders provided by Spring Batch: one for building `Jobs` and one for building `Steps`. Each one is automatically provided via the `@EnableBatchProcessing` annotation so all we need to do is have Spring inject them into our configuration class.

Next we create our `Step`. This job consists of only a single step so we simply name it `step`. It's configured as a Spring Bean and in this simple example, only requires two elements, a name and a `Tasklet`. The `Tasklet`, implemented inline here[5], is what does the actual work in our job. In this example, we have it just call `System.out.println("Hello, World!")` and then returning `RepeatStatus.FINISHED`.

Returning `RepeatStatus.FINISHED` indicates to Spring Batch that you are done with this `Tasklet`. The other option to return is `RepeatStatus.CONTINUABLE`. In this case, Spring Batch will call your `Tasklet` again. If you were to return `RepeatStatus.CONTINUABLE` here, the result would be an infinite loop.

Once our `Step` is configured, we can use it to create our `Job`. As mentioned earlier, a `Job` is composed of one or more `Step` instances. In this case, it's just our one `Step`. We configure our `Job` using the same paradigm we used for our `Step` via the `JobBuilderFactory`. We configure the name of the `Job`, and the `Step` to start at. In this simple example, that's all there is to it.

With all of that configured, you have defined your first Spring Batch job! Now, let's see it in action.

Running Your Job

That's really it. Let's try building and running the job. To compile it (assuming you're using Maven), run `mvn clean package` from the root of the project. When the build is successful, run the job. Spring Boot, by default, will run any `Job` it finds within the configured `ApplicationContext` on startup. You can configure

---

[5]The `Tasklet` here could also be defined as a lambda; however, this approach was selected here for clarity.

this behavior via properties as needed. Since the default behavior is actually what we want, all you need to do is go to the target directory of the project and execute java -jar hello-world-0.0.1-SNAPSHOT.jar. Spring Boot will do the rest to launch your job.

After you've run the job, notice that in traditional Spring Boot style, there is quite a bit of output (including some pretty awesome ASCII art) for a simple "Hello, World!" But if you look closely (around line 23 of the output), there it is:

```
2010-12-01 23:15:42,442 DEBUG 2017-05-22 21:15:04.274  INFO 2829 --- [          main]
o.s.batch.core.job.SimpleStepHandler    : Executing step: [step1]
Hello, World!
2017-05-22 21:15:04.293  INFO 2829 --- [          main] o.s.b.c.l.support.
SimpleJobLauncher      : Job: [SimpleJob: [name=job]] completed with the following
parameters: [{}] and the following status: [COMPLETED]
```

Congratulations! You just ran your first Spring Batch job. So, what actually happened? Spring Boot has a component called the JobLauncherCommandLineRunner. This component is loaded at startup when Spring Batch is found on the classpath and it uses our JobLauncher to run any Job definitions found in the ApplicationContext. So when we bootstrapped Spring Boot in our main method, the ApplicationContext was created, the JobLauncherCommandLineRunner was executed, and our Job was run.

Our Job executed the first Step which, in turn, started a transaction, executed our Tasklet, and updated the JobRepository with the results. We will look in more detail at what the results that ended up the JobRepository were later in this book.

Summary

In this chapter, you got your feet wet with Spring Batch. You walked through the batch domain covering what a Job and Step are and how they interact through the job repository. You learned about the different features of the framework, including the ability to map batch concepts in Java, robust parallelization options, and the formal documentation (including a list of the available sample jobs).

From there, you wrote the Spring Batch version of "Hello, World!" You learned the various ways to get started using Spring Initializr. When you had your project set up, you created your Job in Java and executed your Job.

I want to point out that you've barely taken a peek into what Spring Batch can do. The next chapter walks through the design of a sample application that you'll build later in this book and outlines how Spring Batch addresses issues that you'd have to deal with yourself without it.

■ ■ ■

CHAPTER 3

Sample Job

This book is designed to not only explain how the many features of Spring Batch work but also demonstrate them in detail. Each chapter includes a number of examples that show how each feature works. However, examples designed to communicate individual concepts and techniques may not be the best for demonstrating how those techniques work together in a real-world example. So, in Chapter 10, you create a sample application that is intended to emulate a real-world scenario.

The scenario I chose is simplified: a domain you can easily understand but that provides sufficient complexity so that using Spring Batch makes sense. Bank statements are an example of common batch processing. Run nightly, these processes generate statements based on the previous month's transactions. The batch process we'll create will apply transactions to an existing set of accounts and then generate a statement for each account. It will illustrate a number of important batch concepts:

- *Various input and output options*: Among the most significant features of Spring Batch are the well-abstracted options for reading and writing from a variety of sources. The bank statements job will obtain input from flat files and a database. On the output side, you write to databases as well as flat files. A variety of readers and writers are utilized.

- *Error handling*: The worst part about maintaining batch processes is that when they break, it's typically at 2:00 a.m., and you're the one getting the phone call to fix the problem. Because of this, robust error handling is a must. The example statement generation process covers a number of different scenarios including logging, skipping records with errors, and retry logic.

- *Scalability*: In the real world, batch processes need to be able to accommodate large amounts of data. Later in this book, you use the scalability features of Spring Batch to tune the batch process so it can process literally millions of customers.

In order to build our batch job we will want a set of requirements to work from. Since we will be using user stories to define our requirements, we will take a look at the agile development process as a whole in the next section.

Understanding Agile Development

Before this chapter digs into the individual requirements of the batch process you develop in Chapter 10, let's spend a little time going over the approach you use to do so. A lot has been said in our industry about various agile processes; so instead of banking on any previous knowledge you may have of the subject, let's start by establishing a base of what *agile* and the development process will mean for this book.

The agile process has 12 tenets that virtually all of its variants prescribe. They are as follows:

- Customer satisfaction comes from quick delivery of working software.

- Change is welcome regardless of the stage of development.

- Deliver working software frequently.

- Business and development must work hand in hand daily.

- Build projects with motivated teams. Give them the tools and trust them to get the job done.

- Face-to-face communication is the most effective form.

- Working software is the number one measure of success.

- Strive for sustainable development. All members of the team should be able to maintain the pace of development indefinitely.

- Continue to strive for technical excellence and good design.

- Minimizing waste by eliminating unnecessary work.

- Self-organizing teams generate the best requirements, architectures, and designs.

- At regular intervals, have the team reflect to determine how to improve.

It doesn't matter if you're using Extreme Programming (XP), Scrum, or any other currently hip variant. The point is that these dozen tenets still apply.

Notice that not all of them will necessarily apply in your case. It's pretty hard to work face to face with a book. You'll probably be working by yourself through the examples, so the aspects of team motivation don't exactly apply either. However, there are pieces that do apply. An example is quick delivery of working software. This will drive you through out the book. You'll accomplish it by building small pieces of the application, validating that they work with unit tests, and then adding onto them.

Even with the exceptions, the tenets of agile provide a solid framework for any development project, and this book applies as many of them as possible. Let's get started looking at how they're applied by examining the way you document the requirements for the sample job: user stories.

Capturing Requirements with User Stories

User stories are the agile method for documenting requirements. Written as a customer's take on what the application should do, a story's goal is to communicate the how a user will interact with the system and document testable results of that interaction. A user story has three main parts:

- *The title*: The title should be a simple and concise statement of what the story is about. *Load transaction file. Generate print file.* All of these are good examples of story titles. You notice that these titles aren't GUI specific. Just because you don't have a GUI doesn't mean you can't have interactions between users. In this case, the user is the batch process you're documenting or any external system you interface with.

- *The narrative*: This is a short description of the interaction you're documenting, written from the perspective of the user. Typically, the format is something like "Given the situation Y, X does something, and something else happens." You see in the upcoming sections how to approach stories for batch processes (given that they're purely technical in nature).

- *Acceptance criteria*: The acceptance criteria are testable requirements that can be used to identify when a story is complete. The important word in the previous statement is *testable*. In order for an acceptance criterion to be useful, it must be able to be verified in some way. These aren't subjective requirements but hard items that the developer can use to say, "Yes it does do that" or "No it doesn't."

Let's look at a user story for a universal remote control as an example:

- *Title*: Turn on Television.

- *Narrative*: As a user, with the television, receiver, and cable box off, I will be able to press the power button on my universal remote. The remote will then power on the television, receiver, and cable box and configure them to view a television show.

- *Acceptance criteria*:

  - Have a power button on the universal remote.

  - When the user presses the power button, the following will occur:

    1. The television will power on.

    2. The AV receiver will power on.

    3. The cable box will power on.

    4. The cable box will be set to channel 187.

    5. The AV receiver will be set to the SAT input.

    6. The television will be set to the Video 1 input.

The Turn on Television user story begins with a title—Turn on Television—that is short and descriptive. It continues with a narrative. In this case, the narrative provides a description of what happens when the user presses the power button. Finally, the acceptance criteria list the testable requirements for the developers and QA. Notice that each criterion is something the developers can easily check: they can look at their developed product and say yes or no, what they wrote does or doesn't do what the criteria state.

USER STORIES VS. USE CASES

Use cases are another familiar form of requirements documentation. Similar to user stories, they're actor centric. Use cases were the documentation form of choice for the Rational Unified Process (RUP). They're intended to document every aspect of the interaction between an actor and a system. Because of this, their overly documentation-centric focus (writing documents for the sake of documents), and their bloated format, use cases have fallen out of favor and been replaced with user stories in agile development.

User stories mark the beginning of the development cycle. Let's continue by looking at a few of the other tools used over the rest of the cycle.

Capturing Design with Test-Driven Development

Test-driven development (TDD) is another agile practice. When using TDD, a developer first writes a test that fails and then implements the code to make the test pass. Designed to require that developers think about what they're trying to code before they code it, TDD (also called *test-first development*) has been proven to make developers more productive, use their debuggers less, and end up with cleaner code.

Another advantage of TDD is that tests serve as executable documentation. Unlike user stories or other forms of documentation that become stale due to lack of maintenance, automated tests are always updated as part of the ongoing maintenance of the code. If you want to understand how a piece of code is intended to work, you can look at the unit tests for a complete picture of the scenarios in which the developers intended their code to be used.

Although TDD has a number of positives, you won't use it much in this book. It's a great tool for development, but it isn't the best for explaining how things work. However, Chapter 13 looks at testing of all types, from unit testing to functional testing, using open source tools including JUnit, Mockito, and the testing additions in Spring.

Using a Version-Control System

Although it isn't a requirement by any means, you're strongly encouraged to use a source-control system for all your development. Whether you choose to use git and Github or some other form of version control system, the features that source control provides are essential for productive programming.

You're probably thinking, "Why would I use source control for code that I'm going to throw away while I'm learning?" That is the strongest reason I can think of to use it. By using a version-control system, you give yourself a safety net to try things. Commit your working code; try something that may not work. If it does, commit the new revision. If not, roll back to the previous revision with no harm done. Think about the last time you learned a new technology and did so without version control. I'm sure there were times when you coded your way down a path that didn't pan out and were then stuck to debug your way out of it because you didn't have a previously working copy. Save yourself the headache and allow yourself to make mistakes in a controlled environment by using version control.

Working with a True Development Environment

There are many other pieces to development in an agile environment. Get yourself a good IDE. Because this book is purposely written to be IDE agnostic, it won't go into pros and cons of each. However, be sure you have a good one, and learn it well, including the keyboard shortcuts.

Although spending a lot of time setting up a continuous integration environment may not make sense for you while you learn a given technology, it may be worth setting one up to use in general for your personal development. You never know when that widget you're developing on the side will be the next big thing, and you'd hate to have to go back and set up source control and continuous integration (CI) when things are starting to get exciting. There are many options for setting up a CI environment. One option when using a service like Github is to use another cloud service called Travis CI (https://travis-ci.org/). Travis provides seamless integration with Github via their webhooks API so you just need to provide some simple configuration to enable CI on your projects.

Understanding the Requirements of the Statement Job

Now that you've seen the pieces of the development process you're encouraged to use as you learn Spring Batch, let's look at what you'll develop in this book. Figure 3-1 shows what you expect to see online from your bank each month as your bank statement. While many people receive their statements online, we'll use a printed statement for our example batch job.

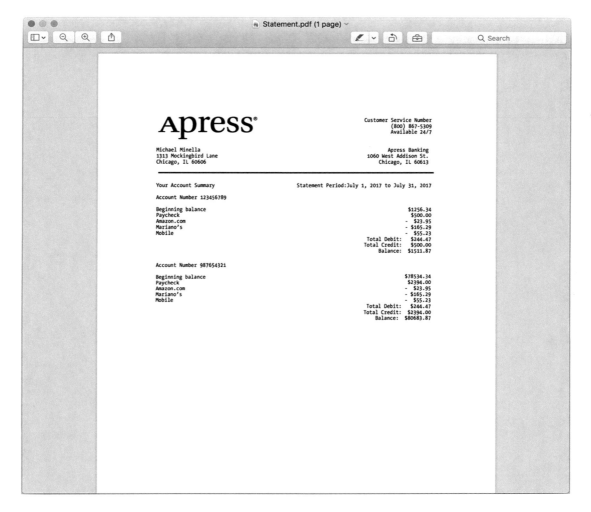

Figure 3-1. *Bank statement, formatted and printed on letterhead*

If you break down how the statement is created, there are really two pieces to it. The first is nothing more than a pretty piece of paper on which the second piece is printed. It's the second piece, shown in Figure 3-2, that you create in this book.

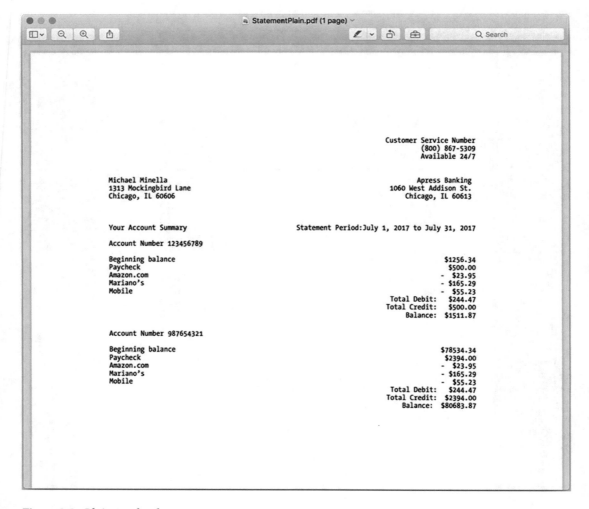

Figure 3-2. *Plain-text bank statement*

Typically, statements are created as follows. A batch process creates a print file consisting of a little more than text.[1] That print file is then sent to a printer that prints the text onto the decorated paper, producing the final statement. The print file is the piece you create using Spring Batch. Your batch process will perform the following functions:

1. Update customer information based on a file provided.

2. Import transactions for all customers in the database.

3. Update the account table with the account balance.

4. Print the file for the bank account for the past month.

---

[1]Another approach is to have the batch process generate PDF files for printing which is out of the scope of this book.

Let's look at what each of these features entails. Your job is provided with a customer flat file that consists of information about existing customers that needs to be updated. For example, a customer's address may have changed. This file will provide details about that address update. Our batch job will read this file and apply the updates to the existing customer data in our database.

The next feature our batch job needs to have is to import transactions for all customers. This data is made available via an XML file that we'll import into the existing database.

Once the transactions have been imported, we can update the account table that keeps a record of what a customer's current balance is. This is the value used to determine what their balance is without needing to re-evaluate all the transactions from the beginning of time every time we need to know the current balance.

With all the database updates complete, we can extract the print file consisting of the customer's information, a list of transactions, and a summary of their account.

This list of features is intended to provide a complete view into how Spring Batch is used in a real-world problem. Throughout the book, you learn about the features Spring Batch provides to help you develop batch processes like the one required for this scenario. In Chapter 10, you implement the batch job to meet the requirements outlined in the following user stories:

> *Update Customer Information*: As the batch process, I will import the customer information and use it to update existing customer records. Acceptance criteria:

- The batch job will read the CSV based customer update file.

- The updates will be applied to the customer records based on the type of update (each type will have its own record format).

  a. A record type of 1 will indicate a name change.

  b. A record type of 2 will indicate a mailing address change.

  c. A record type of 3 will indicate a contact information change.

The record formats are as follows:

Record type 1

Name	Required	Format
Record Type ID	True	\d
Customer ID	True	\d{9}
Customer First Name	False	\w+
Customer Middle Name	False	\w+
Customer Last Name	False	\w+

A record type 1 record will look like the following:

```
1,123456789,John,Middle,Doe
```

Record type 2

Name	Required	Format
Record Type Id	True	\d
Customer Id	True	\d{9}
Address1	False	\w+
Address2	False	\w+
City	False	\w+
State	False	\w{2}
Postal Code	False	\d{5}

A record type 2 record will look like the following:

```
2,123456789,123 4ᵗʰ Street,Unit 5,Chicago,IL,60606
```

Record type 3

Name	Required	Format
Record Type ID	True	\d
Customer ID	True	\d{9}
Email Address	False	\w+
Home Phone	False	\d{3}-\d{3}-\d{4}
Cell Phone	False	\d{3}-\d{3}-\d{4}
Work Phone	False	\d{3}-\d{3}-\d{4}
Notification Preference	False	\d

A record type 3 record will look like the following:

```
3,123456789,foo@bar.com,123-456-7890,123-456-7890,123-456-7890,2
```

Records with validation errors should be written to an error file for future validation and reprocessing.

> *Import transactions*: As the batch process, I will import all new transactions provided via an XML input file. Acceptance criteria:

- The process will read in the XML file of transactions.
- Each transaction will create a new record in the transaction table.
- Each record in the file will have the following fields:

Name	Required	Format
Transaction ID	True	\d{9}
Account Id	True	\d{9}
Credit	False	\d+\.\d{2}
Debit	False	\d+\.\d{2}
Timestamp	True	yyyy-MM-dd HH:mm:ss.ssss

An example of the data the transaction file consists of is as follows (either credit or debit will be filled out in each record):

```
<transactions>
    <transaction>
        <transactionId>123456789</ transactionId>
        <accountId>987654321</accountId>
        <description>Paycheck</description>
        <credit>500.00</credit>
        <debit/>
        <timestamp>2017-07-20 15:38:57.480</timestamp>
    </transaction>
    ...
</transactions>
```

Apply transaction updates to account table: As the batch process, I will update the account table with the latest balance. Acceptance criteria:

- The account table will have a balance field that will be updated with all transactions inserted in the most recent import.

Print Statement Header: As the batch process, at the top of each page I will print a header. This will provide generic information about the customer and the bank. Acceptance criteria:

- The header is all static text except for the customer's address.

- Following is an example of the header, where the Michael Minella name and address are the customer's name and address:

```
                                      Customer Service Number
                                        (800) 867-5309
                                        Available 24/7

Michael Minella                       Apress Banking
1313 Mockingbird Lane                 1060 West Addison St.
Chicago, IL 60606
```

Print Account Summary: As the batch process, after all calculations have been completed, I will print out a summary for each customer. This summary will provide an overview of the customer's account and a breakdown of what makes up the total value of their accounts. Acceptance criteria:

- The process will generate a single file for each customer.

- The summary will begin with a line that states the following, fully justified

```
Your Account Summary                    Statement Period:<BEGIN_DATE> to <END_DATE>
```

where `BEGIN_DATE` is the first calendar date after the last statement date in the account table and `END_DATE` is the date the job is being run.After the summary title, there will be a header line for each account the customer has.

- After the account header, there will be a list of transactions made against the account.

- After the list of transactions, there will be a line for the total amount of credits and total amount of debits within the statement period.

- Finally there will be a line with the current balance for the account.

- The account header, list of transactions, and balance line all will repeat for each account associated with the customer.

- Here is an example with one account:

```
Your Account Summary                    Statement Period: 07/20/2017 to 08/20/2017

Account Number 123456789

Beginning balance                                              $1256.34
Paycheck                                                        $500.00
Amazon.com                                                    -  $23.95
Mariano's                                                     - $165.29
Mobile                                                        -  $55.23
                                               Total Debit:    $244.47
                                              Total Credit:    $500.00
                                                   Balance:   $1511.87
```

That does it for the requirements. If your head is spinning about now, that's okay. In the next section, you begin to outline how to tackle this statement process with Spring Batch. Then, over the rest of this book, you learn how to implement the various pieces required to make it work.

Designing a Batch Job

As stated before, the goal of this project is to take a real-world example and work through it using the features that Spring Batch provides to create a robust, scalable, and maintainable solution. In order to accomplish this goal, the example includes elements that may seem a bit complex right now, such as headers, multiple file format imports, and complex output including subheadings. The reason is that Spring Batch provides facilities exactly for these features. Let's dig into how you structure this batch process by outlining the job and describing its steps.

Job Description

In order to implement the statement-generation process, you build a single job with four steps. Figure 3-3 shows the flow of the batch job for this process, and the following sections describe the steps.

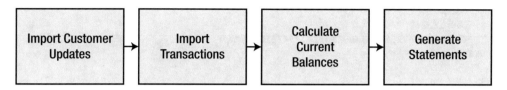

Figure 3-3. *Bank statement jobflow*

Importing Customer Data

To start the job, you begin by importing the customer updates. Contained in a flat file, this data has a complex format consisting of three record types as mentioned earlier. Spring Batch provides the ability to handle multiple record formats in a single file which we'll take advantage of in this step. Once we can read the data, we'll validate it using an ItemProcessor so that we can minimize errors in the write stage. From there, we'll use the appropriate `ItemWriter` implementation to do the correct update based on the record type. A sample of the input we'll be working with in this step can be found in Listing 3-1.

Listing 3-1. Customer Input File

```
2,3,"P.O. Box 554, 6423 Integer Street",,Provo,UT,10886
2,65,"2374 Aliquet, Street", ,Bellevue,WA,83841
3,73,Nullam@fames.net,,1-611-704-0026,1-119-888-1484,4
2,26,985 Malesuada. Avenue,P.O. Box 585,Aurora,IL,73863
2,23,686-1088 Porttitor Avenue,,Stamford,CT,89593
1,36,Zia,,Strong
2,60,313-8010 Commodo St.,,West Jordan,UT,26634
2,17,"P.O. Box 519, 3778 Vel Rd.",,Birmingham,AL,36907
```

Importing Transaction Data

Once the customer data has been imported, the transaction data import is next. Again, Spring Batch provides a robust set of `ItemReader` and `ItemWriter` implementations so we'll be able to use the implementations for reading XML and writing to a database within this step. Again for this step, we will validate the input and do an insert into the database for each record parsed. An example of the input for the transaction XML file is listed in Listing 3-2.

Listing 3-2. Transaction Input File

```xml
<?xml version="1.0" encoding="UTF-8" ?>
<transactions>
        <transaction>
                <transactionId>1</transactionId>
                <accountId>15</accountId>
                <credit>5.62</credit>
                <debit>1.95</debit>
```

```
                    <timestamp>2017-07-12 12:05:21</timestamp>
            </transaction>
            <transaction>
                    <transactionId>2</transactionId>
                    <accountId>68</accountId>
                    <credit>5.27</credit>
                    <debit>6.26</debit>
                    <timestamp>2017-07-23 16:28:37</timestamp>
            </transaction>
    ...
</transactions>
```

Calculating Current Balance

Once the data has been imported, we'll need to update the balance in the account table. This is precalculated for online accounts but is also what we'll use for statement generation. In order to do this, we'll use the driving query pattern (to be discussed later) to iterate over each account and calculate the impact of the current transactions are to the current balance. We'll then update the balance accordingly in the account table.

Generating Customer Monthly Statements

The last step is the most complex. In this step we'll generate one print file per account that contains the customer's statement. Similar to how we calculate the current balance via the driving query pattern, this step will work the same way. By using an `ItemReader` that reads the customers from the database, sending those customers to an `ItemProcessor` for enrichment where we'll add all the data required for each statement, and sending those items to a file based `ItemWriter`, we'll be able to address all the requirements for the statement generation using minimal custom code.

All of this sounds great in theory but leaves a lot of questions to be answered. That's good. You'll spend the rest of the book working through how these features are implemented in the processes as well as examining things like exception handling, restart/retry logic, and addressing scalability concerns. One final item you should be familiar with before you move on, though, is the data model. That will help clear the air regarding how this system is structured. Let's take a look.

Understanding the Data Model

Data being the lifeblood of any application, exploring the data model you'll be working with is a great way to begin to understand how a system works. This section looks at the data model used for the sample application.

Figure 3-4 outlines the application-specific tables for this batch process. To be clear, this diagram doesn't encompass all the tables required for this batch job to run. We'll take a look at those in a later chapter; however, all of those tables will exist in addition to these in your database.

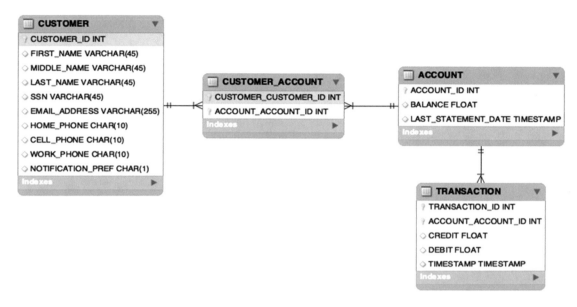

Figure 3-4. Sample application data model

For the batch application, you have four tables: CUSTOMER, ACCOUNT, TRANSACTION, and CUSTOMER_ ACCOUNT. When you look at the data in the tables, notice that you aren't storing all the required fields to generate the statement. There are fields (such as the total debits and total credits) that you calculate during processing. Other than that, the data model should appear relatively straightforward:

- *Customer*: This record contains all the customer-specific information, including name and contact information.

- *Account*: For every customer, an account is maintained. For your purposes, each account has a number and the current balance.

- *CustomerAccount*: This is a join table. Since an account can have many customers associated with it, and a customer can have many accounts, this addresses the many to many relationship between the two.

- *Transaction*. This contains all of the transactions that have occurred against a given account.

Summary

This chapter discussed the agile development process and how you can apply it to batch development. The chapter continued along those lines by defining requirements via user stories for the sample application you build in this book. From this point, the book switches from the "what" and "why" of Spring Batch to the "how."

In the next chapter, you take a deep dive into Spring Batch's concepts of jobs and steps and look at a number of other specific examples.

CHAPTER 4

■ ■ ■

Understanding Jobs and Steps

In Chapter 2, you created your first job. You walked through the configuration of a job and steps, executed the job, and configured a database to store your job repository. In that "Hello, World!" example, you began to scratch the surface of what jobs and steps are in Spring Batch. This chapter continues to explore jobs and steps at a much deeper level. You begin by learning what a job and a step are in relation to the Spring Batch framework.

From there, you dive into great detail about what happens when jobs or steps are executed, from loading them and validating that they're valid to running all the way through their completion. Then, you dig into some code, see the various parts of jobs and steps that you can configure, and learn best practices along the way. Finally, you see how different pieces of the batch puzzle can pass data to each other via the various scopes involved in a Spring Batch process.

Although you dive deep into steps in this chapter, the largest parts of a step are their readers and writers, which aren't covered here. Chapters 7 and 9 explore the input and output functionality available in Spring Batch. This chapter keeps the I/O aspects of each step as simple as possible so you can focus on the intricacies of steps in a job.

Introducing a Job

There are many different types of workloads a modern developer will see within an enterprise. Web applications, integration applications, big data, and others, all have some common paradigms. One of them is the idea that there is a logic flow that needs to occur to provide the business value your application is intended to provide. For example, in a web application, you may have a shopping cart flow where a user adds items to a shopping cart, provides a shipping address, provides payment information, and finally confirms their order. In an integration style application, a message may go through a number of transformers, filters, and so on, all before reaching the end of its flow.

A job is similar to those flows. We'll define a job as a unique, ordered, list of steps that can be executed from start to finish independently. Let's break down this definition, so you can get a better understanding of what you're working with:

- *Unique*: Jobs in Spring Batch are configured via Java or XML in the same way as to how beans are configured using the core Spring framework and are reusable as a result. You can execute a job as many times as you need to with the same configuration. Because of this there is no reason to define the same job multiple times.

- *Ordered list of steps*:[1] Going back to the checkout flow example, the order of the steps matter. You can't confirm your order if you haven't provided a shipping address. You can't execute the checkout process if your shopping cart is empty. The order of

---

[1]Although most jobs consist of an ordered list of steps, Spring Batch does support the ability to execute steps in parallel and conditionally. These features are discussed later.

M. T. Minella, *The Definitive Guide to Spring Batch*, https://doi.org/10.1007/978-1-4842-3724-3_4

steps in your job is important. You can't generate a customer's statement until their transactions have been imported into your system. You can't calculate the balance of an account until you've applied the transactions to the balance. You structure jobs in a sequence that allows all steps to be executed in a logical order.

- *Can be executed from start to finish*: Chapter 1 defined a batch process as a process that can run without additional interaction to some form of completion. A job is a series of steps that can be executed without external interactions. You don't structure a job so that the third step is to wait until a file is sent to a directory to be processed. Instead, you have a job begin when the file has arrived.

- *Independently*: Each batch job should be able to execute without external dependencies affecting it. This doesn't mean a job can't have dependencies. On the contrary, there are not many practical jobs (except "Hello, World") that don't have external dependencies. However, the job should be able to manage those dependencies. If a file isn't there, it handles the error gracefully. It doesn't wait for a file to be delivered (that's the responsibility of a scheduler, and so on). A job can handle all elements of the process it's defined to do.

As you can see in Figure 4-1, a batch process is executed with all of the input available for it as it runs. There are no user interactions. Each step is executed to completion before the next step is executed. Before you dig deeply into how to configure the various features of a job in Spring Batch, let's talk about a job's execution lifecycle.

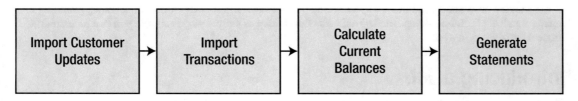

Figure 4-1. *Steps in a batch job*

Tracing a Job's Lifecycle

When a job is executed, it goes through a lifecycle. Knowledge of this lifecycle is important as you structure your jobs and understand what is happening as they run. When you define a job, what you're really doing is providing the blueprint for a job. Just like writing the code for a Java class is like defining a blueprint for the JVM from which to create an instance, your definition of a job is a blueprint for Spring Batch to create an instance of your job.

The execution of a job begins with a job runner. The job runner is intended to execute the job requested by name with the parameters passed. Spring Batch provides two job runners:

- CommandLineJobRunner: This job runner is intended to be used from a script or directly from the command line. When used, the CommandLineJobRunner bootstraps Spring and executes the job requested with the parameters passed.

- JobRegistryBackgroundJobRunner: When using a scheduler like Quartz or a JMX hook to execute a job, typically Spring is bootstrapped and the Java process is live before the job is to be executed. In this case, a JobRegistry is created when Spring is bootstrapped containing the jobs available to run. The JobRegistryBackgroundJobRunner is used to create the JobRegistry.

CommandLineJobRunner and JobRegistryBackgroundJobRunner (both located in the org. springframework.batch.core.launch.support package) are the two job runners provided by Spring Batch. Spring Boot provides yet another way to launch your jobs via the JobLauncherCommandLineRunner. This CommandLineRunner implementation looks for all beans of type Job defined in your ApplicaitonContext and executes them on startup (unless configured otherwise). We will be using this mechanism to launch all of our jobs throughout this book.

Although the job runner is what you use to interface with Spring Batch, it's not a standard piece of the framework. There is no JobRunner interface because each scenario would require a different implementation (although both of the two job runners provided by Spring Batch use main methods to start). Instead, the true entrance into the framework's execute is an implementation of the org.springframework. batch.core.launch.JobLauncher interface.

Spring Batch provides a single JobLauncher, the org.springframework.batch.core.launch. support.SimpleJobLauncher. This class, used internally by the CommandLineJobRunner and the JobLauncherCommandLineRunner, uses the TaskExecutor interface from Core Spring to execute the requested job. You see in a bit at how this is configured, but it's important to note that there are multiple ways to configure the org.springframework.core.task.TaskExecutor in Spring. If an org. springframework.core.task.SyncTaskExecutor is used, the job is executed in the same thread as the JobLauncher. Any other option executes the job in its own thread.

When a batch job is run, an org.springframework.batch.core.JobInstance is created. A JobInstance represents a logical run of the job and is identified by the job name and the identifying parameters passed to the job for this run. A run of the job is different than an attempt at executing the job. If you have a job that is expected to run daily, you would have it configured once. Each day you would have a new run or JobInstance because you pass a new set of parameters into the job (one being the date, for example). Each JobInstance would be considered complete when it has an attempt or JobExecution that has successfully completed. Figure 4-2 illustrates the relationship between the Job, JobInstance and JobExecution.

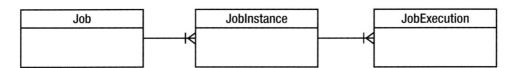

Figure 4-2. *The relationship between a Job, JobInstance, and JobExecution*

■ **Note** A JobInstance can only be executed once to a successful completion. Because a JobInstance is identified by the job name and identifying parameters passed in, this means you can only run a job once with the same identifying parameters.

You are probably wondering how Spring Batch knows the state of a JobInstance from attempt to attempt. In Chapter 4 we'll take a deeper look; however, in the database used by the JobRepository, there is a BATCH_JOB_INSTANCE table. This table is the base from which all other tables are derived. It's the BATCH_JOB_INSTANCE and BATCH_JOB_EXECUTION_PARAMS that identify a JobInstance (the BATCH_JOB_INSTANCE. JOB_KEY is actually a hash of the name and identifying parameters).

A JobExecution is an actual attempt to run the job. If a job runs from start to finish the first time, there is only one JobExecution related to a given JobInstance. If a job ends in an error state after the first run, a new JobExecution is created each time an attempt is made to run the JobInstance (by passing in the same identifying parameters to the same job). For each JobExecution that Spring Batch creates for your job, a record in the BATCH_JOB_EXECUTION table is created. As the JobExecution executes, its state is maintained in the BATCH_JOB_ EXECUTION_CONTEXT as well. This allows Spring Batch to restart a job at the correct point if an error occurs.

Configuring a Job

Enough about theory. Let's get into some code. This section digs into the various ways to configure a job. As mentioned in Chapter 2, as with all of Spring, Spring Batch configurations are done via XML or Java. We'll be using Java configuration for our examples.

Basic Job Configuration

To start, let's take a simple Spring Batch job. Functionally, this job will be the same as the HelloWorld job in Chapter 2. However, we'll simplify things by putting all the code in a single class. We also won't use an in-memory database for our job repository. The reason for that is that we are going to look at some features that span executions and to do that, we need our data to persist. I'll be using MySQL for these examples, but any JDBC-supported database will do.

Let's begin by creating a new project from Spring Initializr using all of the defaults and selecting the following dependencies: batch, jdbc, MySQL. We can name the project Chatper04 for the time being. Once you've imported the new project into your IDE, we will need to make two updates. The first is to configure our database connection properties in the application.properties that is provided by Spring Boot. Listing 4-1 illustrates the properties we'll need.

Listing 4-1. application.properties

```
spring.datasource.driverClassName=com.mysql.cj.jdbc.Driver
spring.datasource.url=jdbc:mysql://localhost:3306/spring_batch
spring.datasource.username=root
spring.datasource.password=p@ssw0rd
spring.batch.initialize-schema=always
```

In Listing 4-1, we've configured Spring Boot to create a DataSource that uses the MySQL driver, and pointed it at our local MySQL instance with the appropriate credentials. We've also configured the application to automatically create the batch schema if it isn't there.[2]

Once we have our database configured, we can create our application code. Listing 4-2 illustrates a simple HelloWorld job for us to discuss.

Listing 4-2. HelloWorld.java

```
...
@EnableBatchProcessing
@SpringBootApplication
public class HelloWorldJob {

    @Autowired
    private JobBuilderFactory jobBuilderFactory;

    @Autowired
    private StepBuilderFactory stepBuilderFactory;
```

---

[2]Spring Boot will always try to create the batch schema when configured this way. If it is already created, the script will fail. By default, this failure is ignored on subsequent runs.

```
@Bean
public Job job() {
        return this.jobBuilderFactory.get("basicJob")
                        .start(step1())
                        .build();
}

@Bean
public Step step1() {
        return this.stepBuilderFactory.get("step1")
                        .tasklet((contribution, chunkContext) -> {
                                System.out.println("Hello, world!");
                                return RepeatStatus.FINISHED;
                        }).build();
}

public static void main(String[] args) {
        SpringApplication.run(HelloWorldJob.class, args);
}
}
```

The first piece of the configuration is adding the @EnableBatchProcessing annotation. This annotation is only needed once within your application and provides the infrastructure needed to run batch jobs as discussed in Chapter 2. After that we have the @SpringBootApplication annotation bootstrapping the Spring Boot magic. By default, Spring Boot considers this a configuration class as well so we do not need to add the @Configuration annotation to it explicitly. Using Spring Boot, this class will be picked up via classpath scanning in most cases.

After the class declaration, we autowire in a JobBuilderFactory and a StepBuilderFactory. These factories are for creating JobBuilder and StepBuilder instances, which we will use to create our actual Spring Batch jobs and steps. With our factory wired in, we define a bean-named job of type Job. This factory method will return our fully configured Spring Batch Job. Constructing the job itself is done via the builders provided by Spring Batch. Passing the name of our job to the JobBuilderFactory.get(String name) call, we get a JobBuilder back, which we can use to configure the Job. We specify the step to start with and since this is a single step job, we call JobBuilder.build() to generate the actual Job.

The last bean definition in our configuration is the Step. We create a method that returns type Step and use the StepBuilderFactory to construct it. Calling the get() method and passing it a name returns a StepBuilder that we can then use to define our step. This step will use a Tasklet so we pass in a lambda to represent our Tasklet implementation. In this case, we just do a simple System.out.println, then return that our Tasklet is complete. We finish by calling build(). The last bit of code in this class is the same as in any Spring Boot application, the main method used to bootstrap Spring.

With regard to configuration, 90% of the configuration of a job is the configuration of the steps and the transitions from one to the next, which is covered later in this chapter.

Job Parameters

You've read a few times that a JobInstance is identified by the job name and the identifying parameters passed into the job. You also know that because of that, you can't run the same job more than once with the same identifying parameters. If you do, you receive an org.springframework.batch.core.launch. JobInstanceAlreadyCompleteException telling you that if you'd like to run the job again, you need to change the parameters. We can see this in action by executing our job twice with the same parameters. If we build our application then execute it once with the command java -jar target/Chapter04-0.0.1-SNAPSHOT.jar foo=bar, we see the normal log files as shown in Listing 4-3.

Listing 4-3. The First Execution with Parameters

```
...
2019-01-16 11:09:29.562  INFO 74578 --- [           main] o.s.b.a.b.JobLauncherCommandLine
Runner    : Running default command line with: [foo=bar]
2019-01-16 11:09:29.669  INFO 74578 --- [           main] o.s.b.c.l.support.SimpleJob
Launcher     : Job: [SimpleJob: [name=basicJob]] launched with the following parameters:
[{foo=bar}]
2019-01-16 11:09:29.714  INFO 74578 --- [           main] o.s.batch.core.job.SimpleStep
Handler     : Executing step: [step1]
Hello, world!
2019-01-16 11:09:29.793  INFO 74578 --- [           main] o.s.b.c.l.support.SimpleJob
Launcher     : Job: [SimpleJob: [name=basicJob]] completed with the following parameters:
[{foo=bar}] and the following status: [COMPLETED]
...
```

Now, if we execute the same command again we get a very different outcome as shown in Listing 4-4.

Listing 4-4. What Happens When You Try to Run a Job Twice with the Same Parameters

```
...
2019-01-16 11:09:34.250  INFO 74588 --- [           main] o.s.b.a.b.JobLauncherCommandLine
Runner    : Running default command line with: [foo=bar]
2019-01-16 11:09:34.436  INFO 74588 --- [           main] ConditionEvaluationReport
LoggingListener :

Error starting ApplicationContext. To display the conditions report re-run your application
with 'debug' enabled.
2019-01-16 11:09:34.447 ERROR 74588 --- [           main] o.s.boot.SpringApplication        :
Application run failed

java.lang.IllegalStateException: Failed to execute CommandLineRunner
        at org.springframework.boot.SpringApplication.callRunner(SpringApplication.java:816)
        [spring-boot-2.1.2.RELEASE.jar!/:2.1.2.RELEASE]
        ...
Caused by: org.springframework.batch.core.repository.JobInstanceAlreadyCompleteException:
A job instance already exists and is complete for parameters={foo=bar}.  If you want to run
this job again, change the parameters.
        at org.springframework.batch.core.repository.support.SimpleJobRepository.createJob
        Execution(SimpleJobRepository.java:132) ~[spring-batch-core-4.1.1.RELEASE.jar!/:
        4.1.1.RELEASE]
...
```

So how do you pass parameters to your jobs? Spring Batch allows you not only to pass parameters to your jobs but also to automatically increment them[3] or validate them before your job runs. You start by looking at how to pass parameters to your jobs.

Passing parameters to your job depends on how you're calling your job. One of the functions of the job runner is to create an instance of org.springframework.batch.core.JobParameters and pass it to the JobLauncher for execution. This makes sense because the way you pass parameters is different if you launch a job from a command line than if you launch your job from a Quartz scheduler. Because you've been using Spring Boot's JobLauncherCommandLineRunner up to now, let's start there.

Passing parameters to `JobLauncherCommandLineRunner` is as simple as passing `key=value` pairs on the command line as we did in the previous example. Listing 4-5 shows how to pass parameters to a job using the way you've been calling jobs up to this point.

Listing 4-5. Passing Parameters to the `CommandLineJobRunner`

```
java –jar demo.jar name=Michael
```

In Listing 4-5, you pass one parameter, `name`. When you pass a parameter into your batch job, your job runner creates an instance of `JobParameters`, which serves as a container for all the parameters the job received.

■ **Note** Spring Batch's `JobParameters` are different than configuring properties via Spring Boot's command line capabilities. As such, job parameters are not passed in with a -- prefix.

Spring Batch's `JobParameters` are different than system properties and should not be passed to the batch application via `-D` arguments on the command line.

`JobParameters` isn't much more than a wrapper for a `java.util.Map<String, JobParameter>` object. Notice that although you're passing in `Strings` in this example, the value of the `Map` is an `org.springframework.batch.core.JobParameter` instance. The reason for this is type. Spring Batch provides for type conversion of parameters, and with that, type-specific accessors on the `JobParameter` class. If you specify the type of parameter to be a `long`, it's available as a `java.lang.Long`. `String`, `Double`, and `java.util.Date` are all available out of the box for conversion. In order to utilize the conversions, you tell Spring Batch the parameter type in parentheses after the parameter name, as shown in Listing 4-6. Notice that Spring Batch requires that the name of each be all lowercase.

Listing 4-6. Specifying the Type of a Parameter

```
java –jar demo.jar executionDate(date)=2017/11/28
```

To view what parameters have been passed into your job, you can look in the job repository. In the database schema for the job repository, there is a table `BATCH_JOB_EXECUTION_PARAMS`. If you explore this table after executing the examples in Listing 4-6, you should see what is shown in Table 4-1 (columns `LONG_VAL` and `DOUBLE_VAL` are removed for brevity).

Table 4-1. *Contents of* `BATCH_JOB_EXECUTION_PARAMS`

JOB_EXECUTION_ID	TYPE_CD	KEY_NAME	STRING_VAL	DATE_VAL	IDENTIFYING
1	DATE	executionDate		2017-11-28 00:00:00	Y

---

[3]It may make sense to have a parameter that is incremented for each `JobInstance`. For example, if the date the job is run is one of its parameters, this can be addressed automatically via a parameter incrementer.

Up to this point, I've been repeatedly specifying that *identifying* parameters are what contribute to a job instance's identity. That must imply that there are nonidentifying parameters as well. And there are. Since Spring Batch 2.2, users were given the option to indicate if a job parameter will contribute to that identity. This is useful, for example, when you have a job that uses an execution date that is consistent for the job instance (and therefore should be identifying), but you want to be able to modify at runtime other parameters based on various conditions that can change from job execution to job execution (an input file name for example). To indicate that a job parameter is non-identifying, you provide a "–" as a prefix as shown in Listing 4-7.

Listing 4-7. Identifying a Job Parameter as Nonidentifying

```
java –jar demo.jar executionDate(date)=2017/11/28 -name=Michael
```

In Listing 4-7, the job parameter `executionDate` is identifying so it's what will be used to determine if the run is part of an existing job instance of if it triggers creating a new one. However, name is not identifying. If the first job execution for the execution Date 2017/11/28 fails, we can relaunch the job using the same `executionDate` but changing the name to John and Spring Batch will create a new job execution under the existing job instance.

Accessing Job Parameters

Now that you know how to get parameters into your batch jobs, how do you access them once you have them? If you take a quick look at the `ItemReader`, `ItemProcessor`, `ItemWriter`, and `Tasklet` interfaces, you quickly notice that all the methods of interest don't receive a `JobParameters` instance as one of their parameters. There are a few different options depending on where you're attempting to access the parameter:

- `ChunkContext`: If you look at the `HelloWorld Tasklet`, you see that the `execute` method receives two parameters. The first parameter is `org.springframework.batch.core.StepContribution`, which contains information about the current transaction that has not been committed yet (write count, read count, and so on). The second parameter is an instance of `ChunkContext`. It provides the state of the job at the point of execution. If you're in a `Tasklet`, it contains any information about the chunk you're processing. Information about that chunk includes information about the step and job. As you might guess, `ChunkContext` has a reference to `org.springframework.batch.core.scope.context.StepContext`, which contains your `JobParameters`.

- *Late binding*: For any piece of the framework that isn't a step or a job, the easiest way to get a handle on a parameter is to inject it via the Spring configuration. Given that `JobParameters` are immutable, binding them during bootstrapping makes perfect sense.

Listing 4-8 shows an updated `HelloWorld` job that utilizes a name parameter in the output as an example of how to access parameters from `ChunkContext`.

Listing 4-8. Accessing JobParameters in a Spring Configuration

```
...
@EnableBatchProcessing
@SpringBootApplication
public class HelloWorldJob {

    @Autowired
    private JobBuilderFactory jobBuilderFactory;

    @Autowired
    private StepBuilderFactory stepBuilderFactory;

    @Bean
    public Job job() {
        return this.jobBuilderFactory.get("basicJob")
                    .start(step1())
                    .build();
    }

    @Bean
    public Step step1() {
        return this.stepBuilderFactory.get("step1")
                    .tasklet(helloWorldTasklet())
                    .build();
    }

    @Bean
    public Tasklet helloWorldTasklet() {

        return (contribution, chunkContext) -> {
                    String name = (String) chunkContext.getStepContext()
                        .getJobParameters()
                        .get("name");

                    System.out.println(String.format("Hello, %s!", name));
                    return RepeatStatus.FINISHED;
            };
    }

    public static void main(String[] args) {
        SpringApplication.run(HelloWorldJob.class, args);
    }
}
```

Although Spring Batch stores the job parameters in an instance of the JobParameter class, when you obtain the parameters this way getJobParameters() returns a Map<String, Object>. Because of this, the previous cast is required.

Listing 4-9 shows how to use Spring's late binding to inject job parameters into components without having to reference any of the JobParameters code. Besides the use of Spring's EL (Expression Language) to pass in the value, any bean that is going to be configured with late binding is required to have the scope set to step or job.

Listing 4-9. Obtaining Job Parameters via Late Binding

```
...
@EnableBatchProcessing
@SpringBootApplication
public class HelloWorldJob {

    @Autowired
    private JobBuilderFactory jobBuilderFactory;

    @Autowired
    private StepBuilderFactory stepBuilderFactory;

    @Bean
    public Job job() {
        return this.jobBuilderFactory.get("basicJob")
                    .start(step1())
                    .build();
    }

    @Bean
    public Step step1() {
        return this.stepBuilderFactory.get("step1")
                    .tasklet(helloWorldTasklet(null))
                    .build();
    }

    @StepScope
    @Bean
    public Tasklet helloWorldTasklet(
                @Value("#{jobParameters['name']}") String name) {

        return (contribution, chunkContext) -> {
                    System.out.println(String.format("Hello, %s!", name));
                    return RepeatStatus.FINISHED;
            };
    }

    public static void main(String[] args) {
        SpringApplication.run(HelloWorldJob.class, args);
    }
}
```

The configuration for the step-scoped version of this bean (which is what allows for the late binding) is shown in Listing 4-10.

Listing 4-10. Step Scoped Bean Configuration

```
...
@StepScope
@Bean
public Tasklet helloWorldTasklet(
            @Value("#{jobParameters['name']}") String name) {

    return (contribution, chunkContext) -> {
                System.out.println(String.format("Hello, %s!", name));
                return RepeatStatus.FINISHED;
            };
}
...
```

The custom step and job scopes included in Spring Batch are what facilitate the late binding capabilities. What each of these does is delay the creation of the bean until you are in the scope of an executing step (for step scope) or job (for job scope). This allows job parameters to be ingested from the command line or other sources and be available for injection when the bean is created.

With the ability to pass parameters into your jobs as well as put them to use, two parameter-specific pieces of functionality are built into the Spring Batch framework that the chapter discusses next: parameter validation and the ability to increment a given parameter with each run. Let's start with parameter validation because it's been alluded to in previous examples.

Validating Job Parameters

Whenever a piece of software obtains outside input, it's a good idea to be sure the input is valid for what you're expecting. The web world uses client-side JavaScript as well as various server-side frameworks to validate user input, and the validation of batch parameters is no different. Fortunately, Spring has made it very easy to validate job parameters. To do so, you just need to implement the org.springframework.batch.core.JobParametersValidator interface and configure your implementation in your job. Listing 4-11 shows an example of a job parameter validator in Spring Batch.

Listing 4-11. A Parameter Validator That Validates the File Name Is a .csv

```
...
public class ParameterValidator implements JobParametersValidator {

    @Override
    public void validate(JobParameters parameters) throws JobParametersInvalidException {
        String fileName = parameters.getString("fileName");

        if(!StringUtils.hasText(fileName)) {
            throw new JobParametersInvalidException("fileName parameter is missing");
        }
        else if(!StringUtils.endsWithIgnoreCase(fileName, "csv")) {
            throw new JobParametersInvalidException("fileName parameter does " +
                                    "not use the csv file extension");
        }
    }
}
```

As you can see, the method of consequence is the validation method. Because this method is void, the validation is considered passing as long as a JobParametersInvalidException isn't thrown. In this example, if the fileName parameter is missing or does not end in a .csv, the exception is thrown and the job is never executed.

In addition to implementing your own custom parameter validator as you did earlier, Spring Batch offers a validator to confirm that all the required parameters have been passed: org.springframework.batch.core.job.DefaultJobParametersValidator. To use it, you configure it the same way you would your custom validator. DefaultJobParametersValidator has two optional dependencies: requiredKeys and optionalKeys. Both are String arrays that take in a list of parameter names that are either required or are the only optional parameters allowed. Listing 4-12 shows a sample configuration for the DefaultJobParametersValidator.

Listing 4-12. DefaultJobParametersValidator Configuration in BatchConfiguration.java

```
...
@Bean
public JobParametersValidator validator() {
    DefaultJobParametersValidator validator = new DefaultJobParametersValidator();

    validator.setRequiredKeys(new String[] {"fileName"});
    validator.setOptionalKeys(new String[] {"name"});

    return validator;
}
...
```

In Listing 4-12, the DefaultJobParametersValidator has the executionDate configured as a required parameter. As such, if the job is attempted to be executed without the fileName being passed as a job parameter, the validation will fail. We have also configured an optional key, name. By doing so, the only two parameters that can be passed to this job are executionDate and name. If any other parameters are passed, validation will also fail. If no optional keys are configured (only required keys are configured), you can pass any combination of keys in addition to the required keys and pass validation. It's important to note that no validation other than the parameter's existence is done via the DefaultJobParametersValidator. Any more robust logic must be done via a custom implementation of the JobParametersValidator.

In order to actually put these two validators in place, we need to configure our job to use them. Going back to the HelloWorld job we've been working with in this chapter, we can add our JobParametersValidators to the job and Spring Batch will execute them at the start of our job. However, there is a small problem. We have two validators we want to use and the method on the JobBuilder that we use to configure a validator only takes one JobParameterValidator instance. Fortunately, Spring Batch provides a CompositeJobParametersValidator for just this use case. Listing 4-13 shows our job configuration updated to use our validators.

Listing 4-13. Configured Job with JobParameters Validation

```
...
@EnableBatchProcessing
@SpringBootApplication
public class HelloWorldJob {

        @Autowired
        private JobBuilderFactory jobBuilderFactory;
```

```
@Autowired
private StepBuilderFactory stepBuilderFactory;

@Bean
public CompositeJobParametersValidator validator() {
        CompositeJobParametersValidator validator =
                    new CompositeJobParametersValidator();

        DefaultJobParametersValidator defaultJobParametersValidator =
                    new DefaultJobParametersValidator(
                                new String[] {"fileName"},
                                new String[] {"name"});

        defaultJobParametersValidator.afterPropertiesSet();

        validator.setValidators(
                    Arrays.asList(new ParameterValidator(),
                            defaultJobParametersValidator));

        return validator;
}

@Bean
public Job job() {
        return this.jobBuilderFactory.get("basicJob")
                    .start(step1())
                    .validator(validator())
                    .build();
}

@Bean
public Step step1() {
        return this.stepBuilderFactory.get("step1")
                    .tasklet(helloWorldTasklet(null, null))
                    .build();
}

@StepScope
@Bean
public Tasklet helloWorldTasklet(
                @Value("#{jobParameters['name']}") String name,
                @Value("#{jobParameters['fileName']}") String fileName) {

        return (contribution, chunkContext) -> {

                    System.out.println(
                                String.format("Hello, %s!", name));
                    System.out.println(
                                String.format("fileName = %s", fileName));
```

```
                        return RepeatStatus.FINISHED;
                };
    }

    public static void main(String[] args) {
            SpringApplication.run(HelloWorldJob.class, args);
    }
}
```

After building this application, if we execute it without the required parameter fileName or if the fileName parameter is misformatted (leaving off the csv at the end of the file name), an exception will be thrown and the job won't be executed. We can see the output of executing the command java -jar target/Chapter04-0.0.1-SNAPSHOT.jar in Listing 4-14.

Listing 4-14. JobParameters Validation Failure Output

```
...
2019-01-16 15:48:20.638  INFO 4023 --- [            main] o.s.b.a.b.JobLauncherCommandLine
Runner   : Running default command line with: []
2019-01-16 15:48:20.689  INFO 4023 --- [            main]
ConditionEvaluationReportLoggingListener :

Error starting ApplicationContext. To display the conditions report re-run your application
with 'debug' enabled.
2019-01-16 15:48:20.696 ERROR 4023 --- [            main] o.s.boot.SpringApplication       :
Application run failed

java.lang.IllegalStateException: Failed to execute CommandLineRunner
        at org.springframework.boot.SpringApplication.callRunner(SpringApplication.java:816)
        [spring-boot-2.1.2.RELEASE.jar!/:2.1.2.RELEASE]
        at
...
Caused by: org.springframework.batch.core.JobParametersInvalidException: fileName parameter
is missing
        at com.example.Chapter04.batch.ParameterValidator.validate(ParameterValidator.java:33)
        ~[classes!/:0.0.1-SNAPSHOT]
        at org.springframework.batch.core.job.CompositeJobParametersValidator.validate(Composite
        JobParametersValidator.java:49) ~[spring-batch-core-4.1.1.RELEASE.jar!/:4.1.1.RELEASE]
        at
...
```

If we take the same code and execute it by passing just the required parameter, the job will still run, but the System.out we have in our Tasklet that says Hello, will output null. We need to provide both parameters via the command java -jar target/Chapter04-0.0.1-SNAPSHOT.jar fileName=foo.csv name=Michael to have everything work as expected. Listing 4-15 illustrates the final output with all job parameters provided.

Listing 4-15. Output with all parameters provided

```
...
2019-01-16 15:48:41.124  INFO 4044 --- [            main] o.s.b.a.b.JobLauncherCommandLine
Runner   : Running default command line with: [fileName=foo.csv, name=bar]
```

```
2019-01-16 15:48:41.216  INFO 4044 --- [            main] o.s.b.c.l.support.SimpleJob
Launcher        : Job: [SimpleJob: [name=basicJob]] launched with the following parameters:
[{name=bar, fileName=foo.csv}]
2019-01-16 15:48:41.249  INFO 4044 --- [            main] o.s.batch.core.job.SimpleStep
Handler         : Executing step: [step1]
Hello, bar!
fileName = foo.csv!
2019-01-16 15:48:41.320  INFO 4044 --- [            main] o.s.b.c.l.support.
SimpleJobLauncher       : Job: [SimpleJob: [name=basicJob]] completed with the following
parameters: [{name=bar, fileName=foo.csv}] and the following status: [COMPLETED]
...
```

Incrementing Job Parameters

Up to now, you've been running under the limitation that a job can only be run once with a given set of identifying parameters. If you've been following along with the examples, you've probably hit what happens if you attempt to run the same job twice with the same parameters as shown in Listing 4-4. However, there is a small loophole: using JobParametersIncrementer.

org.springframework.batch.core.JobParametersIncrementer is an interface that Spring Batch provides to allow you to uniquely generate parameters for a given job. You can add a timestamp to each run. You may have some other business logic that requires a parameter to be incremented with each run. The framework provides a single implementation of the interface, which increments a single long parameter with the default name run.id.

Listing 4-16 shows how to configure a JobParametersIncrementer for your job by adding the reference to the job we have been working on in this chapter.

Listing 4-16. Using a JobParametersIncrementer in a Job

```
...
@Bean
public CompositeJobParametersValidator validator() {
        CompositeJobParametersValidator validator =
                    new CompositeJobParametersValidator();

        DefaultJobParametersValidator defaultJobParametersValidator =
                new DefaultJobParametersValidator(
                            new String[] {"fileName"},
                            new String[] {"name", "run.id"});

        defaultJobParametersValidator.afterPropertiesSet();

        validator.setValidators(
                    Arrays.asList(new ParameterValidator(),
                        defaultJobParametersValidator));

        return validator;
}
```

```
@Bean
public Job job() {
      return this.jobBuilderFactory.get("basicJob")
                    .start(step1())
                    .validator(validator())
                    .incrementer(new RunIdIncrementer())
                    .build();
}
...
```

You'll notice that in our example job we had to configure more than just our job to accept the RunIdIncrementer. We also had to update our JobParametersValidator to allow for the new parameter it would introduce.

Once you've configured JobParametersIncrementer (the framework provides org.springframework. batch.core.launch.support.RunIdIncrementer in this case), you can run your job as many times as you want with the same parameters passed in as shown in Listing 4-17.

Listing 4-17. Command to Run a Job and Increment Parameters

```
java -jar target/Chapter04-0.0.1-SNAPSHOT.jar fileName=foo.csv name=Michael
```

In fact, go ahead and give it a try. When you've run the job three or four times, look in the BATCH_JOB_ EXECUTION_PARAMS table and see how Spring Batch is executing your job with three parameters: one String named name with the value *Michael*, one String named fileName with the value foo.csv, and one long named run.id. run.id's value changes each time, increasing by one with each execution as shown in Listing 4-18.

Listing 4-18. The Results of the RunIdIncrementer After Three Executions

```
mysql> select job_execution_id as id, type_cd as type, key_name as name, string_val,
long_val, identifying from SPRING_BATCH.BATCH_JOB_EXECUTION_PARAMS;
+----+--------+----------+------------+----------+-------------+
| id | type   | name     | string_val | long_val | identifying |
+----+--------+----------+------------+----------+-------------+
|  1 | STRING | name     | Michael    |        0 | Y           |
|  1 | LONG   | run.id   |            |        1 | Y           |
|  1 | STRING | fileName | foo.csv    |        0 | Y           |
|  2 | STRING | name     | Michael    |        0 | Y           |
|  2 | STRING | fileName | foo.csv    |        0 | Y           |
|  2 | LONG   | run.id   |            |        2 | Y           |
|  3 | STRING | name     | Michael    |        0 | Y           |
|  3 | STRING | fileName | foo.csv    |        0 | Y           |
|  3 | LONG   | run.id   |            |        3 | Y           |
+----+--------+----------+------------+----------+-------------+
9 rows in set (0.00 sec)
```

You saw earlier that you may want to have a parameter be a timestamp with each run of the job. This is common in jobs that run once a day. To do so, you need to create your own implementation of JobParametersIncrementer. The configuration and execution are the same as before. However, instead of using RunIdIncrementer, you use DailyJobTimestamper, the code for which is in Listing 4-19.

Listing 4-19. `DailyJobTimestamper.java`

```
...
public class DailyJobTimestamper implements JobParametersIncrementer {
    @Override
    public JobParameters getNext(JobParameters parameters) {

        return new JobParametersBuilder(parameters)
            .addDate("currentDate", new Date())
            .toJobParameters();
    }
}
```

Once the incrementer has been created, we need to add it to our job. We also need to update our parameter validation to handle the removal of the RunIdIncrementer and the addition of the new currentDate parameter our incrementer will introduce. Listing 4-20 illustrates the updated job configuration.

Listing 4-20. Updated Job to Use `DailyJobTimestamper`

```
@EnableBatchProcessing
@SpringBootApplication
public class HelloWorldJob {

    @Autowired
    private JobBuilderFactory jobBuilderFactory;

    @Autowired
    private StepBuilderFactory stepBuilderFactory;

    @Bean
    public CompositeJobParametersValidator validator() {
        CompositeJobParametersValidator validator =
                    new CompositeJobParametersValidator();

        DefaultJobParametersValidator defaultJobParametersValidator =
                    new DefaultJobParametersValidator(
                            new String[] {"fileName"},
                            new String[] {"name", "currentDate"});

        defaultJobParametersValidator.afterPropertiesSet();

        validator.setValidators(
                    Arrays.asList(new ParameterValidator(),
                        defaultJobParametersValidator));

        return validator;
    }

    @Bean
    public Job job() {
        return this.jobBuilderFactory.get("basicJob")
                    .start(step1())
```

```
                         .validator(validator())
                         .incrementer(new DailyJobTimestamper())
                         .build();
    }

    @Bean
    public Step step1() {
            return this.stepBuilderFactory.get("step1")
                         .tasklet(helloWorldTasklet(null, null))
                         .build();
    }

    @StepScope
    @Bean
    public Tasklet helloWorldTasklet(
                    @Value("#{jobParameters['name']}") String name,
                    @Value("#{jobParameters['fileName']}") String fileName) {

            return (contribution, chunkContext) -> {

                    System.out.println(
                            String.format("Hello, %s!", name));
                    System.out.println(
                            String.format("fileName = %s", fileName));

                    return RepeatStatus.FINISHED;
            };
    }

    public static void main(String[] args) {
            SpringApplication.run(HelloWorldJob.class, args);
    }
}
```

Once built, we can execute this job with the same command we used previously: java -jar target/
Chapter04-0.0.1-SNAPSHOT.jar fileName=foo.csv name=Michael, and in an empty database, see the
results of our new JobParameterIncrementer as shown in Listing 4-21.

Listing 4-21. BATCH_JOB_EXECUTION_PARAMS after using the DailyJobTimestamper

```
mysql> select job_execution_id as id, type_cd as type, key_name as name, string_val as s_
val, date_val as d_val, identifying from SPRING_BATCH.BATCH_JOB_EXECUTION_PARAMS;
+----+--------+-------------+---------+---------------------+-------------+
| id | type   | name        | s_val   | d_val               | identifying |
+----+--------+-------------+---------+---------------------+-------------+
|  1 | STRING | name        | Michael | 1969-12-31 18:00:00 | Y           |
|  1 | DATE   | currentDate |         | 2019-01-16 16:40:55 | Y           |
|  1 | STRING | fileName    | foo.csv | 1969-12-31 18:00:00 | Y           |
+----+--------+-------------+---------+---------------------+-------------+
3 rows in set (0.00 sec)
```

It's pretty obvious that job parameters are an important part of the framework. They allow you to specify values at runtime for your job. They also are used to uniquely identify a run of your job. You use them more throughout the book for things like configuring the dates for which to run the job and reprocessing error files. For now, let's look at another powerful feature at the job level: job listeners.

Working with Job Listeners

Every job has a lifecycle. In fact, just about every aspect of Spring Batch has a well-defined lifecycle. This allows us to provide hooks into the different phases of that lifecycle to add additional logic. In the case of a job execution, we have the `JobExecutionListener`. This interface provides two methods, `beforeJob(JobExecution jobExecution)` and `afterJob(JobExecution jobExecution)`. Each of these callbacks are executed as early and late, respectively, as possible in the lifecycle of your jobs. This allows you to utilize these callbacks for a number of different use cases:

- *Notifications*: Spring Cloud Task[4] provides a `JobExecutionListener` that emits messages over a message queue notifying other systems that a job has started or ended.

- *Initialization*: If there are some preparations that need to occur prior to the execution of the job, the `beforeJob` is a good place to execute that logic.

- *Cleanup*: Many jobs have cleanup that needs to occur after it has run (delete or archive files, etc.). This cleanup shouldn't impact the success/failure indications of the job, but still need to be executed. The `afterJob` is a perfect place to handle these use cases.

There are two ways to create a job listener. The first is by implementing the `org.springframework.batch.core.JobExecutionListener` interface. This interface has two methods of consequence: `beforeJob` and `afterJob`. Each takes `JobExecution` as a parameter, and they're executed—you guessed it, before the job executes and after the job executes, respectively. One important thing to note about the `afterJob` method is that it's called regardless of the status the job finishes in. Because of this, you may need to evaluate the status in which the job ended to determine what to do. Listing 4-22 has an example of a simple listener that prints out some information about the job being run before and after as well as the status of the job when it completed.

Listing 4-22. `JobLoggerListener.java`

```
...
public class JobLoggerListener implements JobExecutionListener {

    private static String START_MESSAGE = "%s is beginning execution";
    private static String END_MESSAGE =
                "%s has completed with the status %s";

    @Override
    public void beforeJob(JobExecution jobExecution) {
        System.out.println(String.format(START_MESSAGE,
                    jobExecution.getJobInstance().getJobName()));
    }
```

---

[4]We'll discuss Spring Cloud Task in more detail later in this book.

```
    @Override
    public void afterJob(JobExecution jobExecution) {
        System.out.println(String.format(END_MESSAGE,
                    jobExecution.getJobInstance().getJobName(),
                    jobExecution.getStatus()));
    }
}
```

To configure your job to use this new listener, we simply call the `.listener` method on our `JobBuilder` as shown in Listing 4-23.

Listing 4-23. Job using the `JobLoggerListener`

```
...
@Bean
public Job job() {

    return this.jobBuilderFactory.get("basicJob")
                .start(step1())
                .validator(validator())
                .incrementer(new DailyJobTimestamper())
                .listener(new JobLoggerListener())
                .build();
}
...
```

When we execute the updated code, Spring Batch automatically calls the `beforeJob` method before any additional processing occurs in our job and once all other processing is complete within our job, the `afterJob` method is called. Listing 4-24 illustrates the updated output.

Listing 4-24. `JobExecutionListener` Output

```
...
019-01-16 21:22:25.094  INFO 9006 --- [           main] o.s.b.a.b.JobLauncherCommandLine
Runner     : Running default command line with: [fileName=foo.csv, name=Michael]
2019-01-16 21:22:25.186  INFO 9006 --- [           main] o.s.b.c.l.support.SimpleJob
Launcher      : Job: [SimpleJob: [name=basicJob]] launched with the following parameters:
[{name=Michael, currentDate=1547695345140, fileName=foo.csv}]
basicJob is beginning execution
2019-01-16 21:22:25.217  INFO 9006 --- [           main] o.s.batch.core.job.SimpleStep
Handler     : Executing step: [step1]
Hello, Michael!
fileName = foo.csv
basicJob has completed with the status COMPLETED
2019-01-16 21:22:25.281  INFO 9006 --- [           main] o.s.b.c.l.support.SimpleJob
Launcher      : Job: [SimpleJob: [name=basicJob]] completed with the following parameters:
[{name=Michael, currentDate=1547695345140, fileName=foo.csv}] and the following status:
[COMPLETED]
...
```

As with just about everything in Spring these days, if you can implement an interface for something, there are probably annotations that will make your life easier. Creating listeners is no exception to that. Spring Batch provides the @BeforeJob and @AfterJob annotations just for that use. When using the annotations, the only difference, as shown in Listing 4-25, is that you don't need to implement the JobExecutionListener interface.

Listing 4-25. JobLoggerListener.java

```
...
public class JobLoggerListener {

    private static String START_MESSAGE = "%s is beginning execution";

    private static String END_MESSAGE = "%s has completed with the status %s";

    @BeforeJob
    public void beforeJob(JobExecution jobExecution) {
        System.out.println(String.format(START_MESSAGE,
                        jobExecution.getJobInstance().getJobName()));
    }

    @AfterJob
    public void afterJob(JobExecution jobExecution) {
        System.out.println(String.format(END_MESSAGE,
                        jobExecution.getJobInstance().getJobName(),
                        jobExecution.getStatus()));
    }
}
```

The configuration of the annotation option is slightly different. Spring Batch needs to wrap the listener for us to inject it into our job. To do so, we use the JobListenerFactoryBean as shown in Listing 4-26, resulting in the same output as our previous execution.

Listing 4-26. Configuring Job Listeners in BatchConfiguration.java

```
...
@Bean
public Job job() {

    return this.jobBuilderFactory.get("basicJob")
                .start(step1())
                .validator(validator())
                .incrementer(new DailyJobTimestamper())
                .listener(JobListenerFactoryBean.getListener(
                        new JobLoggerListener()))
                .build();
}
...
```

Listeners are a useful tool to be able to execute logic at certain points of your job. Listeners are also available for many other pieces of the batch puzzle, such as steps, readers, writers, and so on. You see each of those as you cover their respective components later in the book. For now, there is just one more piece to cover that pertains to jobs: ExecutionContext.

ExecutionContext

Batch processes are stateful by their nature. They need to know what step they're on. They need to know how many records they have processed within that step. These and other stateful elements are vital to not only the ongoing processing for any batch process but also restarting it if the process failed before. For example, suppose a batch process that processes a million transactions a night goes down after processing 900,000 of those records. Even with periodic commits along the way, how do you know where to pick back up when you restart? The idea of reestablishing that execution state can be daunting, which is why Spring Batch handles it for you.

You read earlier about how a JobExecution represents an actual attempt at executing the job. This is one level where state needs to be maintained. As a JobExecution progresses through a job or step, the state changes. This state for a job is maintained in the job execution's ExecutionContext.

If you think about how web applications store state, typically it's through the HttpSession.[5] ExecutionContext is essentially the session for your batch job. Holding nothing more than simple key-value pairs, ExecutionContext provides a way to store state within your job in a safe way. One difference between a web application's session and ExecutionContext is that you actually have multiple ExecutionContexts over the course of your job. JobExecution has an ExecutionContext, as does each StepExecution (which you'll see later in this chapter). This allows data to be scoped at the appropriate level (either data-specific for the step or global data for the entire job). Figure 4-3 shows how these elements are related.

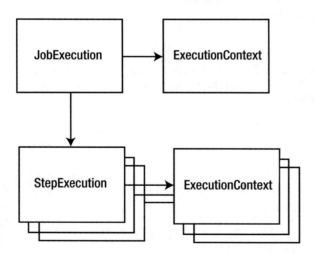

Figure 4-3. *The relationship between* ExecutionContexts

ExecutionContext provides a "safe" way to store data. The storage is safe because everything that goes into an ExecutionContext is persisted in the job repository. Let's look at how to add data to and retrieve data from the ExecutionContext and what it looks like in the database when you do.

---

[5]We ignore web frameworks that maintain state in some form of client form (cookies, thick client, and so on).

Manipulating the ExecutionContext

The ExecutionContext is part of the JobExecution or StepExecution as mentioned earlier. Because of this, to get a handle on the ExecutionContext, you obtain it from the JobExecution or StepExecution based on which you want to use. Listing 4-27 shows how to get a handle on ExecutionContext in the HelloWorld Tasklet and add to the context the name of the person you're saying hello to.

Listing 4-27. Adding a Name to the Job's ExecutionContext

```
...
public class HelloWorld implements Tasklet {
    private static final String HELLO_WORLD = "Hello, %s";

    public RepeatStatus execute( StepContribution step,
                                 ChunkContext context ) throws Exception {
        String name = (String) context.getStepContext()
                            .getJobParameters()
                            .get("name");

        ExecutionContext jobContext = context.getStepContext()
                                        .getStepExecution()
                                        .getJobExecution()
                                        .getExecutionContext();
        jobContext.put("user.name", name);

        System.out.println( String.format(HELLO_WORLD, name) );
        return RepeatStatus.FINISHED;
    }
}
```

Notice that you have to do a bit of traversal to get to the job's ExecutionContext. All you're doing in this case is going from the chunk to the step to the job, working your way up the tree of scopes. If you look at the API for StepContext, you see that there is a getJobExecutionContext() method. This method returns a Map<String, Object> that represents the current state of the job's ExecutionContext. Although this is a handy way to get access to the current values, it has one limiting factor in its use: updates made to the Map returned by the StepContext.getJobExecutionContext() method aren't persisted to the actual ExecutionContext. Thus any changes you make to that Map that aren't also made to the real ExecutionContext are lost in the event of an error.

Listing 4-27's example showed using the job's ExecutionContext, but the ability to obtain and manipulate the step's ExecutionContext works the same way. In that case, you get the ExecutionContext directly from the StepExecution instead of the JobExecution. Listing 4-28 shows the code updated to use the step's ExecutionContext instead of the job's.

Listing 4-28. Adding a Name to the Job's ExecutionContext

```
...
public class HelloWorld implements Tasklet {
    private static final String HELLO_WORLD = "Hello, %s";
```

```
    public RepeatStatus execute( StepContribution step,
                                  ChunkContext context ) throws Exception {
        String name =
            (String) context.getStepContext()
                            .getJobParameters()
                            .get("name");

        ExecutionContext jobContext = context.getStepContext()
                                             .getStepExecution()
                                             .getExecutionContext();
        jobContext.put("user.name", name);

        System.out.println( String.format(HELLO_WORLD, name) );
        return RepeatStatus.FINISHED;
    }
}
```

Another way to manipulate the job execution's ExecutionContext is via promoting keys from the step execution's ExecutionContext to the job execution's ExecutionContext. This can be useful if there is data you want to share between steps but don't want it shared unless the first step is successful. The mechanism to do this promotion is via the ExecutionContextPromotionListener. Listing 4-29 shows the configuration of this listener in a batch job to promote the name key assuming it was put in the step's ExecutionContext.

Listing 4-29. Adding a Name to the Job's ExecutionContext

```
...
public class BatchConfiguration {

        @Autowired
        public JobBuilderFactory jobBuilderFactory;

        @Autowired
        public StepBuilderFactory stepBuilderFactory;

        @Bean
        public Job job() {
                return this.jobBuilderFactory.get("job")
                                .start(step1())
                                .next(step2())
                                .build();
        }

        @Bean
        public Step step1() {
                this.stepBuilderFactcry.get("step1")
                                .tasklet(new HelloTasklet())
                                .listener(promotionListener())
                                .build();
        }
```

```
        @Bean
        public Step step2() {
                this.stepBuilderFactory.get("step2")
                                .tasklet(new GoodByeTasklet())
                                .build();
        }

        @Bean
        public StepExecutionListener promotionListener() {
                ExecutionContextPromotionListener listener = new
                                ExecutionContextPromotionListener();

                listener.setKeys(new String[] {"name"});

                return listener;
        }
}
```

The promotionListner configured in Listing 4-29 will look for the key "name" in the step's ExecutionContext and if it is found after the step has successfully completed, it will be copied into the job execution's ExecutionContext. By default if it's not found, nothing will happen but the listener can be configured to throw an exception if the key isn't found as well.

The final way to access the ExecutionContext is via the ItemStream interface. This will be covered later in this book.

ExecutionContext Persistence

As your jobs process, Spring Batch persists your state as part of committing each chunk. Part of that persistence is the saving of the job and current step's ExecutionContexts. Chapter 2 went over the layout of the tables. Let's take the job in Listing 4-30 and execute it to see what the values look like persisted in the database.

Listing 4-30. Adding a Name to the Job's ExecutionContext

```
...
@EnableBatchProcessing
@Configuration
public class BatchConfiguration {

        @Autowired
        private JobBuilderFactory jobBuilderFactory;

        @Autowired
        private StepBuilderFactory stepBuilderFactory;

        @Bean
        public Job job() {
                return this.jobBuilderFactory.get("job")
                                .start(step1())
                                .build();
        }
```

```
@Bean
public Step step1() {
        return this.stepBuilderFactory.get("step1")
                        .tasklet(helloWorldTasklet())
                        .build();
}

@Bean
public Tasklet helloWorldTasklet() {
        return new HelloWorld();
}

public static class HelloWorld implements Tasklet {
        private static final String HELLO_WORLD = "Hello, %s";

        public RepeatStatus execute( StepContribution step,
                        ChunkContext context ) throws Exception {
                String name =
                                (String) context.getStepContext()
                                        .getJobParameters()
                                        .get("name");

                ExecutionContext jobContext = context.getStepContext()
                        .getStepExecution()
                        .getExecutionContext();
                jobContext.put("name", name);

                System.out.println( String.format(HELLO_WORLD, name) );

                return RepeatStatus.FINISHED;
        }
    }
}
```

Table 4-2 shows what the BATCH_JOB_EXECUTION_CONTEXT table has in it after a single run with the name parameter set as Michael.

Table 4-2. *Contents of BATCH_JOB_EXECUTION_CONTEXT*

JOB_EXECUTION_ID	SHORT_CONTEXT	SERIALIZED_CONTEXT
1	{"batch.taskletType":"io.spring.batch. demo.configuration.BatchConfiguration $HelloWorld","name":"Michael","batch. stepType":"org.springframework.batch. core.step.tasklet.TaskletStep"}	NULL

Table 4-2 consists of three columns. The first is a reference to the JobExecution that this ExecutionContext is related to. The second is a JSON representation of the Job's ExecutionContext. This field is updated as processing occurs. Finally, the SERIALIZED_CONTEXT field contains a serialized Java object. The SERIALIZED_CONTEXT is only populated while a job is running or when it has failed.

You'll notice in the SHORT_CONTEXT field there is the `"name" : "Michael"` String in there as well as other fields. The other fields (`"batch.taskletType"` and `"batch.stepType"`) are both values used by Spring Cloud Data Flow, an orchestration tool we'll look at later in this book.

This section of the chapter has gone through different pieces of what a job is in Spring Batch. In order for a job to be valid, however, it requires at least one step, which brings you to the next major piece of the Spring Batch framework: steps.

Working with Steps

If a job defines the entire process, a step is the building block of a job. It is an independent, sequential batch processor. I call it a batch processor for a reason. A step contains all of the pieces a unit of work requires. It handles its own input. It can have its own processor. It handles its own output. Transactions are self-contained within a step. It's by design that steps are as disjointed. This allows you as the developer to structure your job as freely as needed.

In this section you take the same style deep dive into steps that you did with jobs in the previous section. You cover the way Spring Batch breaks processing down in a step by chunks and how transactions are handled within that style of execution. You also look at a number of examples on how to configure steps within your job including how to control the flow from step to step and conditional step execution. Finally you configure the steps required for your statement job. With all of this in mind, let's start looking at steps by looking at how steps process data.

Tasklet vs. Chunk Processing

Batch processes in general are about processing data. Some work within a batch job just requires the execution of a single command. Maybe a shell script to cleanup a directory or single SQL statement to delete the contents of a staging table. Other work requires the iteration over a large collection of data, reading it a record or item at a time, performing some type of logic on it, then writing it out to a data store of some kind. Spring Batch supports both processing models.

The first model is what we've used so far in our batch jobs, the `Tasklet` model. The `Tasklet` interface that we have used up to this point allows a developer to create a block of code that is executed within the scope of a transaction repeatedly until the `Tasklet.execute` method returns `RepeatStatus.FINISHED`.[6]

The second model is chunk-based processing. A chunk-based step consists of at least two and up to three main components: an `ItemReader`, an optional `ItemProcessor`, and an `ItemWriter`. Using these components, Spring Batch processes records in chunks, or groups of records. Each chunk is executed within its own transaction allowing Spring Batch to restart after the last successful transaction after a failure.

Using these three components, the framework performs three loops. The first loop is with the `ItemReader`. It reads all the records to be processed within this chunk into memory. The second loop is with the optional `ItemProcessor`. If an `ItemProcessor` is configured, the items that were read into memory will be looped over, each one being passed through the `ItemProcessor`. Finally all of the items are passed in a single call to the `ItemWriter` where they can be written out at once. This single call to the `ItemWriter` allows for IO optimizations by batching the physical write. Figure 4-4 shows a sequence diagram of how chunk-based processing works.

---

[6]Each of our examples up to this point using a Tasklet have returned RepeatStatus.FINISHED after the first execution, so we haven't demonstrated the possibility of iterating with a Tasklet.

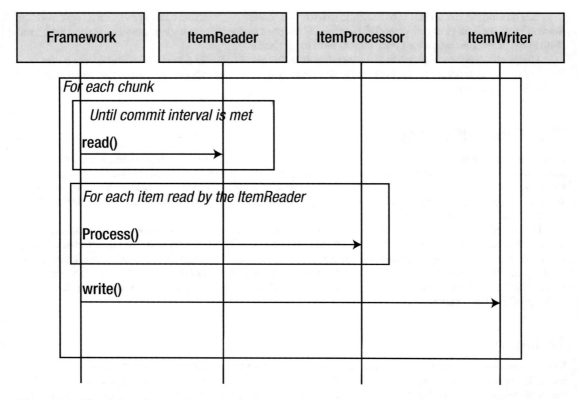

Figure 4-4. *Chunk-based processing*

As you learn more about steps, readers, writers, and scalability throughout the book, keep in mind the chunk-based processing that Spring Batch is based on. Let's move on by digging into how to configure the building blocks of your jobs: steps.

Step Configuration

By now, you've identified that a job is really not much more than container for steps that transition from one to another via configured transitions. Does that paradigm sound familiar? It's a state machine. Spring Batch fundamentally is a state machine where steps represent the states with a collection of transitions that can be made from one state to the next. Let's start looking at steps by taking a look at the most commonly used step type, the Tasklet step.

Tasklet Step

The tasklet step is one of the two main styles of step in Spring Batch. It should also be the most familiar to you, because it's what you used in virtually every job up to now. The way it's different is that in this case, you're writing your own code to be executed as the tasklet. Using MethodInvokingTaskletAdapter is one way to define a tasklet step. In that case, you allow Spring to forward the processing to your code. This lets you develop regular POJOs and use them as steps.

The other way to create a tasklet step is to implement the Tasklet interface as you did when you created the HelloWorld Tasklet in Chapter 2. There, you implement the execute method required in the interface and return a RepeatStatus object to tell Spring Batch what to do after you completed processing. Since the Tasklet interface is technically a functional interface, you also can implement your Tasklet as a lambda. Listing 4-31 illustrates the configuration of a lambda as a Tasklet.

Listing 4-31. HelloWorld Tasklet

```
@EnableBatchProcessing
@Configuration
public class BatchConfiguration {

    @Autowired
    private JobBuilderFactory jobBuilderFactory;

    @Autowired
    private StepBuilderFactory stepBuilderFactory;

    @Bean
    public Job job() {
        return this.jobBuilderFactory.get("job")
                            .start(step1())
                            .build();
    }

    @Bean
    public Step step1() {
        return this.stepBuilderFactory.get("step1")
                        .tasklet((stepContribution, chunkContext) -> {
                                System.out.println("Hello, World!");
                                return RepeatStatus.FINISHED;
                        })
                        .build();
    }
}
```

When processing is complete in your Tasklet implementation, you return an org.springframework. batch.repeat.RepeatStatus object. There are two options with this: RepeatStatus.CONTINUABLE and RepeatStatus.FINISHED. These two values can be confusing at first glance. If you return RepeatStatus. CONTINUABLE, you aren't saying that the job can continue. You're telling Spring Batch to run the Tasklet again. Say, for example, that you wanted to execute a particular Tasklet in a loop until a given condition was met, yet you still wanted to use Spring Batch to keep track of how many times the Tasklet was executed, transactions, and so on. Your Tasklet could return RepeatStatus.CONTINUABLE until the condition was met. If you return RepeatStatus.FINISHED, that means the processing for this Tasklet is complete (regardless of success) and to continue with the next piece of processing.

Understanding the Other Types of Tasklets

While each of our samples up to this point have used custom implementations of the Tasklet interface, custom implementations of the Tasklet interface are not the only way to use the Tasklet step. Spring Batch provides three other implementations of Tasklet: CallableTaskletAdapter, MethodInvokingTaskletAdapter, and SystemCommandTasklet. Let's look at CallableTaskletAdapter first.

CallableTaskletAdapter

org.springframework.batch.core.step.tasklet.CallableTaskletAdapter is an adapter that allows you to configure an implementation of the java.util.concurrent.Callable<RepeatStatus> interface. If you're unfamiliar with this interface, the Callable<V> interface is similar to the java.lang.Runnable interface in that it's intended to be run in a new thread. However, unlike the Runnable interface, which doesn't return a value and can't throw checked exceptions, the Callable interface can return a value (a RepeatStatus, in this case) and can throw checked exceptions.

The adapter is actually extremely simple in its implementation. It calls the call() method on your Callable object and returns the value that the call() method returns. That's it. Obviously you would use this if you wanted to execute the logic of your step in another thread than the thread in which the step is being executed. If you look at Listing 4-32, you can see that to use this adapter, you configure the CallableTaskletAdapter as a normal Spring Bean, then register it as the Tasklet in your step. The CallableTaskletAdapter does have a single dependency, the callable object itself.

Listing 4-32. Using CallableTaskletAdapter

```
...
@EnableBatchProcessing
@SpringBootApplication
public class CallableTaskletConfiguration {

    @Autowired
    private JobBuilderFactory jobBuilderFactory;

    @Autowired
    private StepBuilderFactory stepBuilderFactory;

    @Bean
    public Job callableJob() {
        return this.jobBuilderFactory.get("callableJob")
                    .start(callableStep())
                    .build();
    }

    @Bean
    public Step callableStep() {
        return this.stepBuilderFactory.get("callableStep")
                    .tasklet(tasklet())
                    .build();
    }

    @Bean
    public Callable<RepeatStatus> callableObject() {
        return () -> {
                System.out.println("This was executed in another thread");

                return RepeatStatus.FINISHED;
        };
    }
```

```
    @Bean
    public CallableTaskletAdapter tasklet() {
        CallableTaskletAdapter callableTaskletAdapter =
                    new CallableTaskletAdapter();

        callableTaskletAdapter.setCallable(callableObject());

        return callableTaskletAdapter;
    }

    public static void main(String[] args) {
        SpringApplication.run(CallableTaskletConfiguration.class, args);
    }
}
```

One thing to note with CallableTaskletAdapter is that although the tasklet is executed in a different thread than the step itself, this doesn't parallelize your step execution. The execution of this step won't be considered complete until the Callable object returns a valid RepeatStatus object. Until this step is considered complete, no other steps in the flow in which this step is configured will execute. You see how to parallelize processing in a number of ways, including executing steps in parallel, later in this book.

MethodInvokingTaskletAdapter

The next Tasklet implementation is org.springframework.batch.core.step.tasklet. MethodInvokingTaskletAdapter. This class is similar to a number of utility classes available in the Spring framework. It allows you to execute a preexisting method on another class as a tasklet in your job. Say for example you already have a service that does a piece of logic that you want to run once in your batch job. Instead of writing an implementation of the Tasklet interface that really just wraps that method call, you can use MethodInvokingTaskletAdapter to call the method. Listing 4-33 shows an example of the configuration for MethodInvokingTaskletAdapter.

Listing 4-33. Using MethodInvokingTaskletAdapter

```
...
@EnableBatchProcessing
@SpringBootApplication
public class MethodInvokingTaskletConfiguration {

    @Autowired
    private JobBuilderFactory jobBuilderFactory;

    @Autowired
    private StepBuilderFactory stepBuilderFactory;

    @Bean
    public Job methodInvokingJob() {
        return this.jobBuilderFactory.get("methodInvokingJob")
                    .start(methodInvokingStep())
                    .build();
    }
```

```
@Bean
public Step methodInvokingStep() {
        return this.stepBuilderFactory.get("methodInvokingStep")
                        .tasklet(methodInvokingTasklet())
                        .build();
}

@Bean
public MethodInvokingTaskletAdapter methodInvokingTasklet() {
        MethodInvokingTaskletAdapter methodInvokingTaskletAdapter =
                        new MethodInvokingTaskletAdapter();

        methodInvokingTaskletAdapter.setTargetObject(service());
        methodInvokingTaskletAdapter.setTargetMethod("serviceMethod");

        return methodInvokingTaskletAdapter;
}

@Bean
public CustomService service() {
        return new CustomService();
}

public static void main(String[] args) {
        SpringApplication.run(MethodInvokingTaskletConfiguration.class, args);
}
}
```

The CustomService referenced in Listing 4-33 is nothing more than a simple POJO that does a System.out.println stating that it was called as shown in Listing 4-34.

Listing 4-34. CustomService

```
...
public class CustomService {

    public void serviceMethod() {
            System.out.println("Service method was called");
    }
}
```

The example shown in Listing 4-33 specifies an object and a method. With this configuration, the adapter calls the method with no parameters and returns an ExitStatus.COMPLETED result unless the method specified also returns the type org.springframework.batch.core.ExitStatus. If it does return an ExitStatus, the value returned by the method is returned from the Tasklet. If you want to configure a static set of parameters, you can use the late-binding method of passing job parameters that you read about earlier in this chapter, as shown in Listing 4-35.

Listing 4-35. Using MethodInvokingTaskletAdapter with Parameters

```
...
@EnableBatchProcessing
@SpringBootApplication
public class MethodInvokingTaskletConfiguration {

    @Autowired
    private JobBuilderFactory jobBuilderFactory;

    @Autowired
    private StepBuilderFactory stepBuilderFactory;

    @Bean
    public Job methodInvokingJob() {
        return this.jobBuilderFactory.get("methodInvokingJob")
                    .start(methodInvokingStep())
                    .build();
    }

    @Bean
    public Step methodInvokingStep() {
        return this.stepBuilderFactory.get("methodInvokingStep")
                    .tasklet(methodInvokingTasklet(null))
                    .build();
    }

    @StepScope
    @Bean
    public MethodInvokingTaskletAdapter methodInvokingTasklet(
                @Value("#{jobParameters['message']}") String message) {

        MethodInvokingTaskletAdapter methodInvokingTaskletAdapter =
                    new MethodInvokingTaskletAdapter();

        methodInvokingTaskletAdapter.setTargetObject(service());
        methodInvokingTaskletAdapter.setTargetMethod("serviceMethod");
        methodInvokingTaskletAdapter.setArguments(new String[] {message});

        return methodInvokingTaskletAdapter;
    }

    @Bean
    public CustomService service() {
        return new CustomService();
    }

    public static void main(String[] args) {
        SpringApplication.run(MethodInvokingTaskletConfiguration.class, args);
    }
}
```

For the code in Listing 4-35 to work, we need to update the `CustomService` as well to accept and print out a message. Listing 4-36 illustrates those updates.

Listing 4-36. CustomService with a Parameter

```
...
public class CustomService {

    public void serviceMethod(String message) {
        System.out.println(message);
    }
}
```

SystemCommandTasklet

The last type of `Tasklet` implementation that Spring Batch provides is `org.springframework.batch.core.step.tasklet.SystemCommandTasklet`. This `Tasklet` is used to—you guessed it—execute a system command! The system command specified is executed asynchronously. Because of this, the timeout value (in milliseconds) as shown in Listing 4-37 is important. The `interruptOnCancel` attribute in the listing is optional but indicates to Spring Batch whether to kill the thread the system process is associated with if the job exits abnormally.

Listing 4-37. Using SystemCommandTasklet

```
...
@EnableBatchProcessing
@SpringBootApplication
public class SystemCommandJob {

    @Autowired
    private JobBuilderFactory jobBuilderFactory;

    @Autowired
    private StepBuilderFactory stepBuilderFactory;

    @Bean
    public Job job() {
        return this.jobBuilderFactory.get("systemCommandJob")
                    .start(systemCommandStep())
                    .build();
    }

    @Bean
    public Step systemCommandStep() {
        return this.stepBuilderFactory.get("systemCommandStep")
                    .tasklet(systemCommandTasklet())
                    .build();
    }
```

```
    @Bean
    public SystemCommandTasklet systemCommandTasklet() {
            SystemCommandTasklet systemCommandTasklet = new SystemCommandTasklet();

            systemCommandTasklet.setCommand("rm -rf /tmp.txt");
            systemCommandTasklet.setTimeout(5000);
            systemCommandTasklet.setInterruptOnCancel(true);

            return systemCommandTasklet;
    }

    public static void main(String[] args) {
            SpringApplication.run(SystemCommandJob.class, args);
    }
}
```

SystemCommandTasklet allows you to configure a number of parameters that can have an effect on how a system command executes. Listing 4-38 shows a more robust example.

Listing 4-38. Using SystemCommandTasklet with Full Environment Configuration

```
...
@EnableBatchProcessing
@SpringBootApplication
public class AdvancedSystemCommandJob {

    @Autowired
    private JobBuilderFactory jobBuilderFactory;

    @Autowired
    private StepBuilderFactory stepBuilderFactory;

    @Bean
    public Job job() {
            return this.jobBuilderFactory.get("systemCommandJob")
                        .start(systemCommandStep())
                        .build();
    }

    @Bean
    public Step systemCommandStep() {
            return this.stepBuilderFactory.get("systemCommandStep")
                        .tasklet(systemCommandTasklet())
                        .build();
    }

    @Bean
    public SystemCommandTasklet systemCommandTasklet() {
            SystemCommandTasklet tasklet = new SystemCommandTasklet();
```

```
        tasklet.setCommand("touch tmp.txt");
        tasklet.setTimeout(5000);
        tasklet.setInterruptOnCancel(true);

        // Change this directory to something appropriate for your environment
        tasklet.setWorkingDirectory("/Users/mminella/spring-batch");

        tasklet.setSystemProcessExitCodeMapper(touchCodeMapper());
        tasklet.setTerminationCheckInterval(5000);
        tasklet.setTaskExecutor(new SimpleAsyncTaskExecutor());
        tasklet.setEnvironmentParams(new String[] {
                    "JAVA_HOME=/java",
                    "BATCH_HOME=/Users/batch"});

        return tasklet;
    }

    @Bean
    public SimpleSystemProcessExitCodeMapper touchCodeMapper() {
        return new SimpleSystemProcessExitCodeMapper();
    }

    public static void main(String[] args) {
        SpringApplication.run(AdvancedSystemCommandJob.class, args) ;
    }
}
```

Listing 4-38 includes five more optional parameters in the configuration:

- workingDirectory: This is the directory from which to execute the command. In this example, it's the equivalent of executing cd ~/spring-batch before executing the actual command.

- systemProcessExitCodeMapper: System codes may mean different things depending on the command you're executing. This property allows you to use an implementation of the org.springframework.batch.core.step.tasklet. SystemProcessExitCodeMapper interface to map what system-return codes go with what Spring Batch status values. Spring provides two implementations of this interface by default: org.springframework.batch.core.step.tasklet. ConfigurableSystemProcessExitCodeMapper, which allows you to configure the mapping in your configuration, and org.springframework.batch.core.step. tasklet.SimpleSystemProcessExitCodeMapper, which returns ExitStatus. FINISHED if the return code was 0 and ExitStatus.FAILED if it was anything else.

- terminationCheckInterval: Because the system command is executed in an asynchronous way by default, the tasklet checks periodically to see if it has completed. By default, this value is set to 1 second, but you can configure it to any value you wish in milliseconds.

- taskExecutor: This allows you to configure your own TaskExecutor to execute the system command. You're highly discouraged from configuring a synchronous task executor due to the potential of locking up your job if the system command causes problems.

- *environmentParams:* This is a list of environment parameters you can set prior to the execution of your command.

You've seen over the previous section that many different tasklet types are available in Spring Batch. Now let's take a look at the other most commonly used step type, the chunk-based step.

Chunk-Based Step

As you saw earlier, chunks are defined by their commit intervals. If the commit interval is set to 50 items, then your job reads in 50 items, processes 50 items, and then writes out 50 items at once. Listing 4-39 shows how to configure a basic step for chunk-oriented processing.

Listing 4-39. `BatchConfiguration.java`

```
...
@EnableBatchProcessing
@Configuration
public class BatchConfiguration {

    @Autowired
    private JobBuilderFactory jobBuilderFactory;

    @Autowired
    private StepBuilderFactory stepBuilderFactory;

    @Bean
    public Job job() {
        return this.jobBuilderFactory.get("job")
                    .start(step1())
                    .build();
    }

    @Bean
    public Step step1() {
        return this.stepBuilderFactory.get("step1")
                    .<String, String>chunk(10)
                    .reader(itemReader(null))
                    .writer(itemWriter(null))
                    .build();
    }

    @Bean
    @StepScope
    public FlatFileItemReader<String> itemReader(
                @Value("#{jobParameters['inputFile']}") Resource inputFile) {

        return new FlatFileItemReaderBuilder<String>()
                    .name("itemReader")
                    .resource(inputFile)
                    .lineMapper(new PassThroughLineMapper())
                    .build();
    }
```

```
@Bean
@StepScope
public FlatFileItemWriter<String> itemWriter(
            @Value("#{jobParameters['outputFile']}") Resource outputFile) {

    return new FlatFileItemWriterBuilder<String>()
                    .name("itemWtiter")
                    .resource(outputFile)
                    .lineAggregator(new PassThroughLineAggregator<>())
                    .build();
    }
}
```

Listing 4-39 may look intimidating, but let's focus on the job and step configuration at the top. The rest of the file is the configuration of a basic `ItemReader` and `ItemWriter`, which are covered in Chapters 7 and 9, respectively. When you look through the job in Listing 4-39, you see that the step begins with getting a `StepBuilder` from the `StepBuilderFactory`. We then identify it as a chunk-based step via the chunk method. The 10 we pass to it is the configuration for the commit interval so in this example, the job will commit after processing 10 records. The chunk-based step takes a reader (an implementation of the `ItemReader` interface) and a writer (an implementation of the `ItemWriter` interface) before the build method is called.

It's important to note the `commit interval`. It's set at 10 in the example. This means no records will be written until 10 records are read and processed. If an error occurs after processing nine items, Spring Batch will roll back the current chunk (transaction) and mark the job as failed. If you were to set the commit interval value to 1, your job would read in a single item, process that item, and then write that item. Essentially, you would be going back to item-based processing. The issue with this is that there is more than just that single item being persisted at the commit interval. The state of the job is being updated in the job repository as well. You experiment with the commit interval later in this book but you needed to know now that it's important to set `commit interval` as high as reasonably possible to get the best performance on the write side.

We will look at the components of a chunk-based step in detail next.

Chunk-Size Configuration

Because chunk-based processing is the foundation of Spring Batch, it's important to understand how to configure its various options to take full advantage of this important feature. This section covers the two options for configuring the size of a chunk: a static commit count and a `CompletionPolicy` implementation. All other chunk configuration options relate to error handling and are discussed in that section.

To start looking at chunk configuration, Listing 4-40 has a basic example of nothing more than a reader, writer, and commit-interval configured. The reader is an implementation of the `ItemReader` interface, and the writer an implementation of `ItemWriter`. Each of these interfaces has its own dedicated chapter later in the book, so this section doesn't go into detail about them. All you need to know is that they supply input and output, respectively, for the step. The commit interval defines how many items make up a chunk (10 items, in this case).

Listing 4-40. A Basic Chunk Configuration

```
...
@EnableBatchProcessing
@SpringBootApplication
public class ChunkJob {
```

```java
    @Autowired
    private JobBuilderFactory jobBuilderFactory;

    @Autowired
    private StepBuilderFactory stepBuilderFactory;

    @Bean
    public Job chunkBasedJob() {
        return this.jobBuilderFactory.get("chunkBasedJob")
                    .start(chunkStep())
                    .build();
    }

    @Bean
    public Step chunkStep() {
        return this.stepBuilderFactory.get("chunkStep")
                    .<String, String>chunk(1000)
                    .reader(itemReader())
                    .writer(itemWriter())
                    .build();
    }

    @Bean
    public ListItemReader<String> itemReader() {
        List<String> items = new ArrayList<>(100000);

        for (int i = 0; i < 100000; i++) {
            items.add(UUID.randomUUID().toString());
        }

        return new ListItemReader<>(items);
    }

    @Bean
    public ItemWriter<String> itemWriter() {
        return items -> {
            for (String item : items) {
                System.out.println(">> current item = " + item);
            }
        };
    }

    public static void main(String[] args) {
        SpringApplication.run(ChunkJob.class, args);
    }
}
```

Although typically you define the size of a chunk based on a hard number configured via a commit interval as configured in Listing 4-40, that isn't always a robust enough option. Say that you have a job that needs to process chunks that aren't all the same size (processing all transactions for an account in a single transaction, for example). Spring Batch provides the ability to programmatically define when a chunk is complete via an implementation of the org.springframework.batch.repeat.CompletionPolicy interface.

The CompletionPolicy interface allows the implementation of decision logic to decide if a given chunk is complete. Spring Batch comes with a number of implementations of this interface. By default it uses org.springframework.batch.repeat.policy.SimpleCompletionPolicy, which counts the number of items processed and flags a chunk complete when the configured threshold is reached. Another out-of-the-box implementation is org.springframework.batch.repeat.policy.TimeoutTerminationPolicy. This allows you to configure a timeout on a chunk so that it may exit gracefully after a given amount of time. What does "exit gracefully" mean in this context? It means that the chunk is considered complete and all transaction processing continues normally.

As you can undoubtedly deduce, there are few times when a timeout by itself is enough to determine when a chunk of processing will be complete. TimeoutTerminationPolicy is more likely to be used as part of org.springframework.batch.repeat.policy.CompositeCompletionPolicy. This policy lets you configure multiple policies that determine whether a chunk has completed. When you use CompositeCompletionPolicy, if any of the policies consider a chunk complete, then the chunk is flagged as complete. Listing 4-41 shows an example of using a timeout of 3 milliseconds along with the normal commit count of 200 items to determine if a chunk is complete.

Listing 4-41. Using a Timeout Along with a Regular Commit Count

```
...
@EnableBatchProcessing
@SpringBootApplication
public class ChunkJob {

    @Autowired
    private JobBuilderFactory jobBuilderFactory;

    @Autowired
    private StepBuilderFactory stepBuilderFactory;

    @Bean
    public Job chunkBasedJob() {
        return this.jobBuilderFactory.get("chunkBasedJob")
                    .start(chunkStep())
                    .build();
    }

    @Bean
    public Step chunkStep() {
        return this.stepBuilderFactory.get("chunkStep")
                    .<String, String>chunk(completionPolicy())
                    .reader(itemReader())
                    .writer(itemWriter())
                    .build();
    }

    @Bean
    public ListItemReader<String> itemReader() {
        List<String> items = new ArrayList<>(100000);

        for (int i = 0; i < 100000; i++) {
                items.add(UUID.randomUUID().toString());
        }
```

```
            return new ListItemReader<>(items);
    }

    @Bean
    public ItemWriter<String> itemWriter() {
            return items -> {
                    for (String item : items) {
                            System.out.println(">> current item = " + item);
                    }
            };
    }

    @Bean
    public CompletionPolicy completionPolicy() {
            CompositeCompletionPolicy policy =
new CompositeCompletionPolicy();

            policy.setPolicies(
                            new CompletionPolicy[] {
                                    new TimeoutTerminationPolicy(3),
                                    new SimpleCompletionPolicy(1000)});

            return policy;
    }

    public static void main(String[] args) {
            SpringApplication.run(ChunkJob.class, args);
    }
}
```

You will notice that if you execute both of the previous two examples, by the looks of things, everything occurred exactly the same. However, that isn't the case. The first example (the one listed in Listing 4-40), you get 101 commits (100000/1000 + 1 for the empty transaction at the end). However, in the second example (from Listing 4-41), you get something around 191 commits,[7] showing the impact of the TimeoutTerminationPolicy being added to the mix.

Using the implementations of the CompletionPolicy interface isn't your only option to determine how large a chunk is. You can also implement it yourself. Before you look at an implementation, let's go over the interface.

The CompletionPolicy interface requires four methods: two versions of isComplete, start, and update. If you look at this through the lifecycle of the class, first the start method is called first. This method initializes the policy so that it knows the chunk is starting. It's important to note that an implementation of the CompletionPolicy interface is intended to be stateful and should be able to determine if a chunk has been completed by its own internal state. The start method resets this internal state to whatever is required by the implementation at the beginning of the chunk. Using SimpleCompletionPolicy as an example, the start method resets an internal counter to 0 at the beginning of a chunk. The update method is called once for each item that has been processed to update the internal state. Going back to the SimpleCompletionPolicy example, update increments the internal counter by one after each item. Finally, there are two isComplete methods. The first isComplete method signature accepts a RepeatContext

---

[7]Due to the time-based nature of the TimeoutTerminationPolicy, the number of commits here will vary based on the environment the job is run in.

as its parameter. This implementation is intended to use its internal state to determine if the chunk has completed. The second signature takes the RepeatContext and also the RepeatStatus as parameters. This implementation is expected to determine based on the status whether a chunk has completed. Listing 4-42 shows an example of a CompletionPolicy implementation that considers a chunk complete once a random number of items fewer than 20 have been processed; Listing 4-42 showing the configuration.

Listing 4-42. Random Chunk Size CompletionPolicy Implementation

```
...
public class RandomChunkSizePolicy implements CompletionPolicy {

    private int chunksize;
    private int totalProcessed;
    private Random random = new Random();

    @Override
    public boolean isComplete(RepeatContext context,
                RepeatStatus result) {

        if(RepeatStatus.FINISHED == result) {
            return true;
        }
        else {
            return isComplete(context);
        }
    }

    @Override
    public boolean isComplete(RepeatContext context) {
        return this.totalProcessed >= chunksize;
    }

    @Override
    public RepeatContext start(RepeatContext parent) {
        this.chunksize = random.nextInt(20);
        this.totalProcessed = 0;

        System.out.println("The chunk size has been set to " +
                    this.chunksize);

        return parent;
    }

    @Override
    public void update(RepeatContext context) {
        this.totalProcessed++;
    }
}
```

Listing 4-43. Configuring RandomChunkSizePolicy

```
...
@EnableBatchProcessing
@SpringBootApplication
public class ChunkJob {

    @Autowired
    private JobBuilderFactory jobBuilderFactory;

    @Autowired
    private StepBuilderFactory stepBuilderFactory;

    @Bean
    public Job chunkBasedJob() {
        return this.jobBuilderFactory.get("chunkBasedJob")
                    .start(chunkStep())
                    .build();
    }

    @Bean
    public Step chunkStep() {
        return this.stepBuilderFactory.get("chunkStep")
                    .<String, String>chunk(randomCompletionPolicy())
                    .reader(itemReader())
                    .writer(itemWriter())
                    .build();
    }

    @Bean
    public ListItemReader<String> itemReader() {
        List<String> items = new ArrayList<>(100000);

        for (int i = 0; i < 100000; i++) {
            items.add(UUID.randomUUID().toString());
        }

        return new ListItemReader<>(items);
    }

    @Bean
    public ItemWriter<String> itemWriter() {
        return items -> {
            for (String item : items) {
                System.out.println(">> current item = " + item);
            }
        };
    }
```

```
@Bean
public CompletionPolicy randomCompletionPolicy() {
        return new RandomChunkSizePolicy();
}

public static void main(String[] args) {
        SpringApplication.run(ChunkJob.class, args);
}
}
```

Executing the job in Listing 4-43, you will see scattered throughout your output the chunk size printed out as each new chunk starts as well as you can count the number of items between each of those output lines to see the impact the CompletionPolicy has on the chunk size.

You explore the rest of chunk configuration when you get to error handling. That section covers retry and skip logic, which the majority of the remaining options center around. The next pieces of a step this chapter looks at also carry over from a job: listeners.

Step Listeners

When you looked at job listeners, earlier this chapter, you saw the two events they can fire on: the start and end of a job. Step listeners cover the same types of events (start and end), but for individual steps instead of an entire job. This section covers the org.springframework.batch.core.StepExecutionListener and org.springframework.batch.core.ChunkListener interfaces, both of which allow the processing of logic at the beginning and end of a step and chunk respectively. Notice that the Step's listener is named the StepExecutionListener and not just StepListener. There actually is a StepListener interface; however. it's just a marker interface that all step-related listeners extend.

Both the StepExecutionListener and ChunkListener provide methods that are similar to the ones in the JobExecutionListener interface. StepExecutionListener has a beforeStep and an afterStep, and ChunkListener has a beforeChunk and an afterChunk, as you would expect. All of these methods are void except afterStep. afterStep returns an ExitStatus because the listener is allowed to modify the ExitStatus that was returned by the step itself prior to it being returned to the job. This feature can be useful when a job requires more than just knowing whether an operation was successful to determine if the processing was successful. An example would be doing some basic integrity checks after importing a file (whether the correct number of records were written to the database, and so on). The ability to configure listeners via annotations also continues to be consistent, with Spring Batch providing @BeforeStep, @AfterStep, @BeforeChunk, and @AfterChunk annotations to simplify the implementation. Listing 4-44 shows a StepExecutionListener that uses annotations to identify the methods.

Listing 4-44. Logging Step Start and Stop Listeners

```
...
public class LoggingStepStartStopListener {

    @BeforeStep
    public void beforeStep(StepExecution stepExecution) {
        System.out.println(stepExecution.getStepName() + " has begun!");
    }
```

```
    @AfterStep
    public ExitStatus afterStep(StepExecution stepExecution) {
            System.out.println(stepExecution.getStepName() + " has ended!");

            return stepExecution.getExitStatus();
    }
}
```

The configuration for all the step listeners is combined into a single list in the step configuration. Listing 4-45 configures the LoggingStepStartStopListener that you coded earlier.

Listing 4-45. Configuring LoggingStepStartStopListener

```
...
@EnableBatchProcessing
@SpringBootApplication
public class ChunkJob {

    @Autowired
    private JobBuilderFactory jobBuilderFactory;

    @Autowired
    private StepBuilderFactory stepBuilderFactory;

    @Bean
    public Job chunkBasedJob() {
            return this.jobBuilderFactory.get("chunkBasedJob")
                        .start(chunkStep())
                        .build();
    }

    @Bean
    public Step chunkStep() {
            return this.stepBuilderFactory.get("chunkStep")
                        .<String, String>chunk(1000)
                        .reader(itemReader())
                        .writer(itemWriter())
                        .listener(new LoggingStepStartStopListener())
                        .build();
    }

    @Bean
    public ListItemReader<String> itemReader() {
            List<String> items = new ArrayList<>(100000);

            for (int i = 0; i < 100000; i++) {
                    items.add(UUID.randomUUID().toString());
            }

            return new ListItemReader<>(items);
    }
```

```
    @Bean
    public ItemWriter<String> itemWriter() {
        return items -> {
            for (String item : items) {
                System.out.println(">> current item = " + item);
            }
        };
    }

    @Bean
    public CompletionPolicy randomCompletionPolicy() {
        return new RandomChunkSizePolicy();
    }

    public static void main(String[] args) {
        SpringApplication.run(ChunkJob.class, args);
    }
}
```

As you can see, listeners are available at just about every level of the Spring Batch framework to allow you to hang processing off your batch jobs. They're commonly used not only to perform some form of preprocessing before a component or evaluate the result of a component but also in error handling, as you see in a bit.

The next section covers the flow of steps. Although all your steps up to this point have been processed sequentially, that isn't a requirement in Spring Batch. You learn how to perform simple logic to determine what step to execute next and how to externalize flows for reuse.

Step Flow

A single file line: that is what your jobs have looked like up to this point. You've lined up the steps and allowed them to execute one after another. However, if that were the only way you could execute steps, Spring Batch would be very limited. Instead, the authors of the framework provided a robust collection of options for customizing the flow of your jobs.

To start, let's look at how you can decide what step to execute next or even if you execute a given step at all. This occurs using Spring Batch's conditional logic.

Conditional Logic

Within a job in Spring Batch, steps are executed in the order you specify using the next method on the StepBuilder. If you want to execute steps in a different order, it's quite easy: you configure transitions. As Listing 4-46 shows, you can use the builders to direct a job to go from firstStep to successStep if things go okay or to failureStep if step1 returns an ExitStatus of FAILED.

Listing 4-46. If/Else Logic in Step Execution

```
...
@EnableBatchProcessing
@SpringBootApplication
public class ConditionalJob {
```

```
@Autowired
private JobBuilderFactory jobBuilderFactory;

@Autowired
private StepBuilderFactory stepBuilderFactory;

@Bean
public Tasklet passTasklet() {
    return (contribution, chunkContext) -> {
        return RepeatStatus.FINISHED;
        //                      throw new RuntimeException("This is a failure");
    };
}

@Bean
public Tasklet successTasklet() {
    return (contribution, context) -> {
        System.out.println("Success!");
        return RepeatStatus.FINISHED;
    };
}

@Bean
public Tasklet failTasklet() {
    return (contribution, context) -> {
        System.out.println("Failure!");
        return RepeatStatus.FINISHED;
    };
}

@Bean
public Job job() {
    return this.jobBuilderFactory.get("conditionalJob")
                .start(firstStep())
                .on("FAILED").to(failureStep())
                .from(firstStep()).on("*").to(successStep())
                .end()
                .build();
}

@Bean
public Step firstStep() {
    return this.stepBuilderFactory.get("firstStep")
                .tasklet(passTasklet())
                .build();
}
```

```
@Bean
public Step successStep() {
        return this.stepBuilderFactory.get("successStep")
                        .tasklet(successTasklet())
                        .build();
}

@Bean
public Step failureStep() {
        return this.stepBuilderFactory.get("failureStep")
                        .tasklet(failTasklet())
                        .build();
}

public static void main(String[] args) {
        SpringApplication.run(ConditionalJob.class, args);
}
}
```

The on method configures Spring Batch to evaluate the ExitStatus of the step and determine what to do. It's important to note that you've seen both org.springframework.batch.core.ExitStatus and org.springframework.batch.core.BatchStatus over the course of this chapter. BatchStatus is an attribute of the JobExecution or StepExecution that identifies the current state of the job or step. ExitStatus is the value returned to Spring Batch at the end of a job or step. Spring Batch evaluates the ExitStatus for its transitions. So, the example in Listing 4-46 is the equivalent of saying, "If the exit code of firstStep doesn't equal FAILED, go to successStep, else go to failureStep."

Because the values of the ExitStatus are really just Strings, the ability to use wildcards can make things interesting. Spring Batch allows for two wildcards in on criteria:

- * matches zero or more characters. For example, C* matches *C*, *COMPLETE*, and *CORRECT*.

- ? matches a single character. In this case, ?AT matches *CAT* or *KAT* but not *THAT*.

Although evaluating the ExitStatus gets you started in determining what to do next, it may not take you all the way. For example, what if you didn't want to execute a step if you skipped any records in the current step? You wouldn't know that from the ExitStatus alone.

■ **Note** Spring Batch helps you when it comes to configuring transitions. It automatically orders the transitions from most to least restrictive and applies them in that order.

Spring Batch has provided a programmatic way to determine what to do next. You do this by creating an implementation of the org.springframework.batch.core.job.flow.JobExecutionDecider interface. This interface has a single method, decide, that takes both the JobExecution and the StepExecution and returns a FlowExecutionStatus (a wrapper for a BatchStatus/ExitStatus pair). With both the JobExecution and StepExecution available for evaluation, all information should be available to you to make the appropriate decision about what your job should do next. Listing 4-47 shows an implementation of the JobExecutionDecider that randomly decides what the next step should be.

Listing 4-47. RandomDecider

```
...
public class RandomDecider implements JobExecutionDecider {

    private Random random = new Random();

    public FlowExecutionStatus decide(JobExecution jobExecution,
            StepExecution stepExecution) {

        if (random.nextBoolean()) {
            return new
                FlowExecutionStatus(FlowExecutionStatus.COMPLETED.getName());
        } else {
            return new
                FlowExecutionStatus(FlowExecutionStatus.FAILED.getName());
        }
    }
}
```

To use RandomDecider, you configure an extra attribute on your step called decider. This attribute refers to the Spring bean that implements JobExecutionDecider. Listing 4-48 shows RandomDecider configured. You can see that the configuration maps the values you return in the decider to steps available to execute.

Listing 4-48. If/Else Logic in Step Execution

```
...
@EnableBatchProcessing
@SpringBootApplication
public class ConditionalJob {

    @Autowired
    private JobBuilderFactory jobBuilderFactory;

    @Autowired
    private StepBuilderFactory stepBuilderFactory;

    @Bean
    public Tasklet passTasklet() {
        return (contribution, chunkContext) -> RepeatStatus.FINISHED;
    }

    @Bean
    public Tasklet successTasklet() {
        return (contribution, context) -> {
            System.out.println("Success!");
            return RepeatStatus.FINISHED;
        };
    }
```

```java
    @Bean
    public Tasklet failTasklet() {
        return (contribution, context) -> {
            System.out.println("Failure!");
            return RepeatStatus.FINISHED;
        };
    }

    @Bean
    public Job job() {
        return this.jobBuilderFactory.get("conditionalJob")
                    .start(firstStep())
                    .next(decider())
                    .from(decider())
                    .on("FAILED").to(failureStep())
                    .from(decider())
                    .on("*").to(successStep())
                    .end()
                    .build();
    }

    @Bean
    public Step firstStep() {
        return this.stepBuilderFactory.get("firstStep")
                    .tasklet(passTasklet())
                    .build();
    }

    @Bean
    public Step successStep() {
        return this.stepBuilderFactory.get("successStep")
                    .tasklet(successTasklet())
                    .build();
    }

    @Bean
    public Step failureStep() {
        return this.stepBuilderFactory.get("failureStep")
                    .tasklet(failTasklet())
                    .build();
    }

    @Bean
    public JobExecutionDecider decider() {
        return new RandomDecider();
    }

    public static void main(String[] args) {
        SpringApplication.run(ConditionalJob.class, args);
    }
}
```

Because you now know how to direct your processing from step to step either sequentially or via logic, you won't always want to just go to another step. You may want to end or pause the job. The next section covers how to handle those scenarios.

Ending a Job

You learned earlier that a JobInstance can't be executed more than once to a successful completion and that a JobInstance is identified by the job name and the parameters passed into it. Because of this, you need to be aware of the state in which you end your job if you do it programmatically. In reality, there are three states in which you can programmatically end a job in Spring Batch:

- *Completed:* This end state tells Spring Batch that processing has ended in a successful way. When a JobInstance is completed, it isn't allowed to be rerun with the same parameters.

- *Failed:* In this case, the job has not run successfully to completion. Spring Batch allows a job in the failed state to be rerun with the same parameters.

- *Stopped:* In the stopped state, the job can be restarted. The interesting part about a job that is stopped is that the job can be restarted from where it left off, although no error has occurred. This state is very useful in scenarios when human intervention or some other check or handling is required between steps.

It's important to note that these states are identified by Spring Batch evaluating the ExitStatus of the step to determine what BatchStatus to persist in the JobRepository. ExitStatus can be returned from a step, chunk, or job. BatchStatus is maintained in StepExecution or JobExecution and persisted in the JobRepository. Let's begin looking at how to end the job in each state with the completed state.

To configure a job to end in the completed state based on the exit status of a step, you use the builder's end method. In this state, you can't execute the same job again with the same parameters. Listing 4-49 shows that the end tag has a single attribute that declares the ExitStatus value that triggers the job to end.

Listing 4-49. Ending a Job in the Completed State

```
...
@EnableBatchProcessing
@SpringBootApplication
public class ConditionalJob {

    @Autowired
    private JobBuilderFactory jobBuilderFactory;

    @Autowired
    private StepBuilderFactory stepBuilderFactory;

    @Bean
    public Tasklet passTasklet() {
        return (contribution, chunkContext) -> {
            return RepeatStatus.FINISHED;
//            throw new RuntimeException("Causing a failure");
        };
    }
```

```java
@Bean
public Tasklet successTasklet() {
        return (contribution, context) -> {
                System.out.println("Success!");
                return RepeatStatus.FINISHED;
        };
}

@Bean
public Tasklet failTasklet() {
        return (contribution, context) -> {
                System.out.println("Failure!");
                return RepeatStatus.FINISHED;
        };
}

@Bean
public Job job() {
        return this.jobBuilderFactory.get("conditionalJob")
                        .start(firstStep())
                        .on("FAILED").end()
                        .from(firstStep()).on("*").to(successStep())
                        .end()
                        .build();
}

@Bean
public Step firstStep() {
        return this.stepBuilderFactory.get("firstStep")
                        .tasklet(passTasklet())
                        .build();
}

@Bean
public Step successStep() {
        return this.stepBuilderFactory.get("successStep")
                        .tasklet(successTasklet())
                        .build();
}

@Bean
public Step failureStep() {
        return this.stepBuilderFactory.get("failureStep")
                        .tasklet(failTasklet())
                        .build();
}

@Bean
public JobExecutionDecider decider() {
        return new RandomDecider();
}
```

```
        public static void main(String[] args) {
                SpringApplication.run(ConditionalJob.class, args);
        }
}
```

Once you run conditionalJob, as you would expect, the BATCH_STEP_EXECUTION table contains the ExitStatus returned by the step, and BATCH_JOB_EXECUTION contains COMPLETED regardless of the path taken.

For the failed state, which allows you to rerun the job with the same parameters, the configuration looks similar. Instead of using the end method, you use the fail method. Listing 4-50 shows this configuration.

Listing 4-50. Ending a Job in the Failed State

```
...
@EnableBatchProcessing
@SpringBootApplication
public class ConditionalJob {

        @Autowired
        private JobBuilderFactory jobBuilderFactory;

        @Autowired
        private StepBuilderFactory stepBuilderFactory;

        @Bean
        public Tasklet passTasklet() {
                return (contribution, chunkContext) -> {
//                      return RepeatStatus.FINISHED;
                        throw new RuntimeException("Causing a failure");
                };
        }

        @Bean
        public Tasklet successTasklet() {
                return (contribution, context) -> {
                        System.out.println("Success!");
                        return RepeatStatus.FINISHED;
                };
        }

        @Bean
        public Tasklet failTasklet() {
                return (contribution, context) -> {
                        System.out.println("Failure!");
                        return RepeatStatus.FINISHED;
                };
        }

        @Bean
        public Job job() {
                return this.jobBuilderFactory.get("conditionalJob")
                                .start(firstStep())
```

95

```
                        .on("FAILED").fail()
                        .from(firstStep()).on("*").to(successStep())
                        .end()
                        .build();
    }

    @Bean
    public Step firstStep() {
        return this.stepBuilderFactory.get("firstStep")
                    .tasklet(passTasklet())
                    .build();
    }

    @Bean
    public Step successStep() {
        return this.stepBuilderFactory.get("successStep")
                    .tasklet(successTasklet())
                    .build();
    }

    @Bean
    public Step failureStep() {
        return this.stepBuilderFactory.get("failureStep")
                    .tasklet(failTasklet())
                    .build();
    }

    @Bean
    public JobExecutionDecider decider() {
        return new RandomDecider();
    }

    public static void main(String[] args) {
        SpringApplication.run(ConditionalJob.class, args);
    }
}
```

When you rerun conditionalJob with the configuration in Listing 4-50, the results are a bit different. This time, if firstStep ends with the ExitStatus FAILURE, the job is identified in the jobRepository as failed, which allows it to be re-executed with the same parameters.

The last state you can leave a job in when you end it programmatically is the stopped state. In this case, you can restart the job; and when you do, it restarts at the step you configure. Listing 4-51 shows an example.

Listing 4-51. Ending a Job in the Stopped State

```
...
@EnableBatchProcessing
@SpringBootApplication
public class ConditionalJob {
```

```
     @Autowired
     private JobBuilderFactory jobBuilderFactory;

     @Autowired
     private StepBuilderFactory stepBuilderFactory;

     @Bean
     public Tasklet passTasklet() {
           return (contribution, chunkContext) -> {
//                 return RepeatStatus.FINISHED;
                   throw new RuntimeException("Causing a failure");
           };
     }

     @Bean
     public Tasklet successTasklet() {
           return (contribution, context) -> {
                   System.out.println("Success!");
                   return RepeatStatus.FINISHED;
           };
     }

     @Bean
     public Tasklet failTasklet() {
           return (contribution, context) -> {
                   System.out.println("Failure!");
                   return RepeatStatus.FINISHED;
           };
     }

     @Bean
     public Job job() {
          return this.jobBuilderFactory.get("conditionalJob")
                      .start(firstStep())
                      .on("FAILED").stopAndRestart(successStep())
                      .from(firstStep())
                      .on("*").to(successStep())
                      .end()
                      .build();
     }

     @Bean
     public Step firstStep() {
          return this.stepBuilderFactory.get("firstStep")
                      .tasklet(passTasklet())
                      .build();
     }
```

```
@Bean
public Step successStep() {
        return this.stepBuilderFactory.get("successStep")
                        .tasklet(successTasklet())
                        .build();
}

@Bean
public Step failureStep() {
        return this.stepBuilderFactory.get("failureStep")
                        .tasklet(failTasklet())
                        .build();
}

@Bean
public JobExecutionDecider decider() {
        return new RandomDecider();
}

public static void main(String[] args) {
        SpringApplication.run(ConditionalJob.class, args);
}
}
```

Executing conditionalJob with this final configuration, as in Listing 4-51, allows you to rerun the job with the same parameters. However, this time, if the FAILED path is chosen, when the job is restarted execution begins at successStep.

The flow from one step to the next isn't just another layer of configuration you're adding to potentially complex job configurations; it's also configurable in a reusable component. The next section discusses how to encapsulate flows of steps into reusable components.

Externalizing Flows

You've already identified that a step can be defined as a bean. This lets you extract the definition of your steps from a given job into reusable components. The same goes for the order of steps. In Spring Batch, there are three options for how to externalize the order of steps. The first is to create a flow, which is an independent sequence of steps. The second is to use the flow step; although the configuration is very similar, the state persistence in the JobRepository is slightly different. The last way is to actually call another job from within your job. This section covers how all three of these options work.

A flow looks a lot like a job. It's configured in a similar way. Listing 4-52 shows how to define a flow using the flow builder, giving it an id and then referencing it in your job.

Listing 4-52. Defining a Flow

```
...
@EnableBatchProcessing
@SpringBootApplication
public class FlowJob {

        @Autowired
        private JobBuilderFactory jobBuilderFactory;
```

```java
@Autowired
private StepBuilderFactory stepBuilderFactory;

@Bean
public Tasklet loadStockFile() {
      return (contribution, chunkContext) -> {
            System.out.println("The stock file has been loaded");
            return RepeatStatus.FINISHED;
      };
}

@Bean
public Tasklet loadCustomerFile() {
      return (contribution, chunkContext) -> {
            System.out.println("The customer file has been loaded");
            return RepeatStatus.FINISHED;
      };
}

@Bean
public Tasklet updateStart() {
      return (contribution, chunkContext) -> {
            System.out.println("The start has been updated");
            return RepeatStatus.FINISHED;
      };
}

@Bean
public Tasklet runBatchTasklet() {
      return (contribution, chunkContext) -> {
            System.out.println("The batch has been run");
            return RepeatStatus.FINISHED;
      };
}

@Bean
public Flow preProcessingFlow() {
      return new FlowBuilder<Flow>("preProcessingFlow").start(loadFileStep())
                  .next(loadCustomerStep())
                  .next(updateStartStep())
                  .build();
}

@Bean
public Job conditionalStepLogicJob() {
      return this.jobBuilderFactory.get("conditionalStepLogicJob")
                  .start(preProcessingFlow())
                  .next(runBatch())
                  .end()
                  .build();
}
```

```
@Bean
public Step loadFileStep() {
      return this.stepBuilderFactory.get("loadFileStep")
                    .tasklet(loadStockFile())
                    .build();
}

@Bean
public Step loadCustomerStep() {
      return this.stepBuilderFactory.get("loadCustomerStep")
                    .tasklet(loadCustomerFile())
                    .build();
}

@Bean
public Step updateStartStep() {
      return this.stepBuilderFactory.get("updateStartStep")
                    .tasklet(updateStart())
                    .build();
}

@Bean
public Step runBatch() {
      return this.stepBuilderFactory.get("runBatch")
                    .tasklet(runBatchTasklet())
                    .build();
}

public static void main(String[] args) {
      SpringApplication.run(HelloWorldJob.class, args);
}
}
```

When you execute a flow as part of a job and look at the JobRepository, you see the steps from the flow recorded as part of the job as if they were configured there in the first place. In the end, there is no difference between using a flow and configuring the steps within the job itself from a JobRepository perspective.

The next option for externalizing steps is to use the FlowStep. With this technique, the configuration of a flow is the same. But instead of configuring the execution of your flow by passing it to the JobBuilder, you wrap that flow in a step and pass the step to the JobBuilder. Listing 4-53 demonstrates how to use a FlowStep to configure the same example Listing 4-52 used.

Listing 4-53. Using a Flow Step

```
...
@EnableBatchProcessing
@SpringBootApplication
public class FlowJob {

      @Autowired
      private JobBuilderFactory jobBuilderFactory;
```

```
@Autowired
private StepBuilderFactory stepBuilderFactory;

@Bean
public Tasklet loadStockFile() {
      return (contribution, chunkContext) -> {
            System.out.println("The stock file has been loaded");
            return RepeatStatus.FINISHED;
      };
}

@Bean
public Tasklet loadCustomerFile() {
      return (contribution, chunkContext) -> {
            System.out.println("The customer file has been loaded");
            return RepeatStatus.FINISHED;
      };
}

@Bean
public Tasklet updateStart() {
      return (contribution, chunkContext) -> {
            System.out.println("The start has been updated");
            return RepeatStatus.FINISHED;
      };
}

@Bean
public Tasklet runBatchTasklet() {
      return (contribution, chunkContext) -> {
            System.out.println("The batch has been run");
            return RepeatStatus.FINISHED;
      };
}

@Bean
public Flow preProcessingFlow() {
      return new FlowBuilder<Flow>("preProcessingFlow").start(loadFileStep())
                  .next(loadCustomerStep())
                  .next(updateStartStep())
                  .build();
}

@Bean
public Job conditionalStepLogicJob() {
      return this.jobBuilderFactory.get("conditionalStepLogicJob")
                  .start(intializeBatch())
                  .next(runBatch())
                  .build();
}
```

```java
@Bean
public Step intializeBatch() {
        return this.stepBuilderFactory.get("initalizeBatch")
                        .flow(preProcessingFlow())
                        .build();
}

@Bean
public Step loadFileStep() {
        return this.stepBuilderFactory.get("loadFileStep")
                        .tasklet(loadStockFile())
                        .build();
}

@Bean
public Step loadCustomerStep() {
        return this.stepBuilderFactory.get("loadCustomerStep")
                        .tasklet(loadCustomerFile())
                        .build();
}

@Bean
public Step updateStartStep() {
        return this.stepBuilderFactory.get("updateStartStep")
                        .tasklet(updateStart())
                        .build();
}

@Bean
public Step runBatch() {
        return this.stepBuilderFactory.get("runBatch")
                        .tasklet(runBatchTasklet())
                        .build();
}

public static void main(String[] args) {
        SpringApplication.run(HelloWorldJob.class, args);
}
}
```

What is the difference between passing the Flow to the JobBuilder and the FlowStep? It comes down to what happens in the JobRepository. Using the flow method on the JobBuilder yields the same results as if you configured the steps in your job. Using a FlowStep adds an additional entry. When you use a FlowStep, Spring Batch records the step that includes the flow as a separate step. Why is this a good thing? The main benefit is for monitoring and reporting purposes. Using a FlowStep allows you to see the impact of the flow as a whole instead of having to aggregate the individual steps.

The last way to externalize the order in which steps occur is to not externalize them at all. In this case, instead of creating a flow, you call a job from within another job. Similar to the FlowStep, which creates a StepExecutionContext for the execution of the flow and each step within it, the JobStep creates a JobExecutionContext for the step that calls the external job. Listing 4-54 shows the configuration of a JobStep.

Listing 4-54. Using a Job Step

```
...
@EnableBatchProcessing
@SpringBootApplication
public class JobJob {

    @Autowired
    private JobBuilderFactory jobBuilderFactory;

    @Autowired
    private StepBuilderFactory stepBuilderFactory;

    @Bean
    public Tasklet loadStockFile() {
        return (contribution, chunkContext) -> {
            System.out.println("The stock file has been loaded");
            return RepeatStatus.FINISHED;
        };
    }

    @Bean
    public Tasklet loadCustomerFile() {
        return (contribution, chunkContext) -> {
            System.out.println("The customer file has been loaded");
            return RepeatStatus.FINISHED;
        };
    }

    @Bean
    public Tasklet updateStart() {
        return (contribution, chunkContext) -> {
            System.out.println("The start has been updated");
            return RepeatStatus.FINISHED;
        };
    }

    @Bean
    public Tasklet runBatchTasklet() {
        return (contribution, chunkContext) -> {
            System.out.println("The batch has been run");
            return RepeatStatus.FINISHED;
        };
    }

    @Bean
    public Job preProcessingJob() {
        return this.jobBuilderFactory.get("preProcessingJob")
                    .start(loadFileStep())
                    .next(loadCustomerStep())
```

```
                        .next(updateStartStep())
                        .build();
    }

    @Bean
    public Job conditionalStepLogicJob() {
        return this.jobBuilderFactory.get("conditionalStepLogicJob")
                        .start(intializeBatch())
                        .next(runBatch())
                        .build();
    }

    @Bean
    public Step intializeBatch() {
        return this.stepBuilderFactory.get("initalizeBatch")
                        .job(preProcessingJob())
                        .parametersExtractor(new DefaultJobParametersExtractor())
                        .build();
    }

    @Bean
    public Step loadFileStep() {
        return this.stepBuilderFactory.get("loadFileStep")
                        .tasklet(loadStockFile())
                        .build();
    }

    @Bean
    public Step loadCustomerStep() {
        return this.stepBuilderFactory.get("loadCustomerStep")
                        .tasklet(loadCustomerFile())
                        .build();
    }

    @Bean
    public Step updateStartStep() {
        return this.stepBuilderFactory.get("updateStartStep")
                        .tasklet(updateStart())
                        .build();
    }

    @Bean
    public Step runBatch() {
        return this.stepBuilderFactory.get("runBatch")
                        .tasklet(runBatchTasklet())
                        .build();
    }

    public static void main(String[] args) {
        SpringApplication.run(HelloWorldJob.class, args);
    }
}
```

You might be wondering about the `jobParametersExtractor` bean in Listing 4-54. When you launch a job, it's identified by the job name and the job parameters. In this case, you aren't passing the parameters to your sub-job, `preProcessingJob`, by hand. Instead, you define a class to extract the parameters from either the `JobParameters` of the parent job or the `ExecutionContext` (the `DefaultJobParameterExtractor` checks both places) and pass those parameters to the child job. Your extractor pulls the values from the `job.stockFile` and `job.customerFile` job parameters and passes those as parameters to `preProcessingJob`.

When `preProcessingJob` executes, it's identified in the `JobRepository` just like any other job. It has its own job instance, execution context, and related database records.

■ **Note** In order to run the example in Listing 4-54, the property `spring.batch.job.names=conditionalStepLogicJob` must be configured in your application.properties to prevent Spring Boot from executing `preProcessingJob` automatically on startup.

A word of caution about using the `JobStep` approach: this may seem like a good way to handle job dependencies. Creating individual jobs and being able to then string them together with a master job is a powerful feature. However, this can severely limit the control of the process as it executes. It isn't uncommon in the real world to need to pause a batch cycle or skip jobs based on external factors (another department can't get you a file in time to have the process finished in the required window, and so on). However, the ability to manage jobs exists at a single job level. Managing entire trees of jobs that could be created using this functionality is problematic and should be avoided. Linking jobs together in this manner and executing them as one master job severely limits the capability to handle these types of situations and should also be avoided.

Summary

This chapter covered a large amount of material. You learned what a job is and saw its lifecycle. You looked at how to configure a job and how to interact with it via job parameters. You wrote and configured listeners to execute logic at the beginning and end of a job, and you worked with the `ExecutionContext` for a job and step.

You began looking at the building blocks of a job: its steps. As you looked at steps, you explored one of the most important concepts in Spring Batch: chunk-based processing. You learned how to configure chunks and some of the more advanced ways to control them (through things like policies). You learned about listeners and how to use them to execute logic at the start and end of a step. Finally, you walked through how to order steps either using basic ordering or logic to determine what step to execute next.

The job and step are structural components of the Spring Batch framework. They're used to lay out a process. The majority of the book from here on covers all the different things that go into the structure laid out by these pieces.

CHAPTER 5

■ ■ ■

JobRepository and Metadata

When you look into writing a finite process, the ability to execute processes without a UI in a stand-alone manner isn't that hard. Spring Boot's CommandLineRunner allows a developer to write a Spring Boot application that executes a single function that implements any business logic you could think of and completes. You don't need Spring Batch to do that.

Where things get interesting, however, is where things go wrong. If your batch job is running and an error occurs, how do you recover? How does your job know where it was in processing when the error occurred, and what should happen when the job is restarted? State management is an important part of processing large volumes of data. This is one of the key features that Spring Batch brings to the table. Spring Batch, as discussed previously in this book, maintains the state of a job as it executes in a job repository. It then uses this information when a job is restarted or an item is retried to determine how to continue. The power of this feature can't be overstated.

Another aspect of batch processing in which the job repository is helpful is monitoring. The ability to see how far a job is in its processing as well as trend elements such as how long operations take or how many items were retried due to errors is vital in the enterprise environment. The fact that Spring Batch does the number gathering for you makes this type of trending much easier.

This chapter covers job repositories in detail. It goes over ways to configure a job repository for most environments by using either a database or an in-memory repository. After you have the job repository configured, you learn how to put the job information stored by the job repository to use using the JobExplorer and the JobOperator.

What Is the Job Repository?

When referring to the job repository within the context of Spring Batch, you can be referring to one of two things: the interface JobRepository or the data store that is used by the implementation of that interface to persist the data. Since the user should almost never need to interact with the interface itself beyond potentially configuring an instance of it, this section is going to focus on the data stores provided by Spring Batch used by a JobRepository implementation. The two data stores provided out of the box by Spring Batch to use within a batch job, in memory and a relational database. We will look at the relational database option first.

Using a Relational Database

A relational database is the default option for the job repository in Spring Batch. This option utilizes a set of database tables provided by Spring Batch to persist the batch meta data. Let's take a look at the schema in Figure 5-1.

© Michael T. Minella 2019
M. T. Minella, *The Definitive Guide to Spring Batch*, https://doi.org/10.1007/978-1-4842-3724-3_5

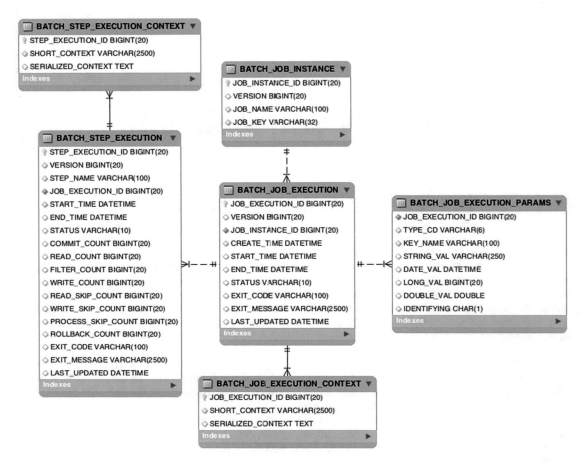

Figure 5-1. *The job repository schema*

As you can see in Figure 5-1, there are six tables in the job repository:

- BATCH_JOB_INSTANCE

- BATCH_JOB_EXECUTION

- BATCH_JOB_EXECUTION_PARAMS

- BATCH_JOB_EXECUTION_CONTEXT

- BATCH_STEP_EXECUTION

- BATCH_STEP_EXECUTION_CONTEXT

The schema really begins at the BATCH_JOB_INSTANCE table. As we saw earlier, a single job instance is created when the job is executed for the first time with a unique set of identifying job parameters. This record represents the logical run of a job. Table 5-1 illustrates what makes up the BATCH_JOB_INSTANCE table.

Table 5-1. *BATCH_JOB_INSTANCE Table*

Field	Description
JOB_EXECUTION_ID	Primary key of the table
VERSION	The version[1] for the record used for optimistic locking
JOB_NAME	The name of the job executed
JOB_KEY	A hash of the job name and identifying parameters used to uniquely identify a job instance

From there the BATCH_JOB_EXECUTION table represents each of the physical runs of a batch job. Every time a job is launched, a new record will be created here and be updated periodically as the job progresses. Table 5-2 walks through the columns of the BATCH_JOB_EXECUTION table.

Table 5-2. *BATCH_JOB_EXECUTION Table*

Field	Description
JOB_EXECUTION_ID	Primary key of the table
VERSION	The version[2] for the record used for optimistic locking
JOB_INSTANCE_ID	Foreign key to the BATCH_JOB_INSTANCE table
CREATE_TIME	The time the record was created
START_TIME	The time the job execution was started
END_TIME	The time the job execution finished
STATUS	The batch status of the job execution
EXIT_CODE	The exit code of the job execution
EXIT_MESSAGE	The message (potentially a stack trace) associated with the EXIT_CODE
LAST_UPDATED	The last time this record was updated

There are three tables that are associated with the BATCH_JOB_EXECUTION table. The first one we'll look at is the BATCH_JOB_EXECUTION_CONTEXT table. We looked at the idea of the ExecutionContext and how components use it to store state in it in the previous chapter. However, in order for it to be useful across executions (on restarts, for example), Spring Batch needs to persist it and the BATCH_JOB_EXECUTION_CONTEXT is where the JobExecution's ExecutionContext is persisted. Table 5-3 has the details about this table.

Table 5-3. *BATCH_JOB_EXECUTION_CONTEXT Table*

Field	Description
JOB_EXECUTION_ID	Primary key of the table
SHORT_CONTEXT	A trimmed version of the SERIALIZED_CONTEXT
SERIALIZED_CONTEXT	The actual serialized ExecutionContext

---

[1]To learn more about the versions and entities in domain-driven design, read *Domain Driven Design* by Eric Evans (Addison-Wesley, 2003).
[2]To learn more about the versions and entities in domain-driven design, read *Domain Driven Design* by Eric Evans (Addison-Wesley, 2003).

I should note that there are a few options for ways to serialize the ExecutionContext. Before Spring Batch 4, XStream's JSON facilities were the preferred method and what the framework used by default. However, as of when Spring Batch 4 came out, XStream's JSON support wasn't fully capable for what Spring Batch needed so the default was changed to use Jackson 2. We'll cover how to customize the configuration of the ExecutionContext's serialization later in this chapter.

The second table that associated with the BATCH_JOB_EXECUTION table is the BATCH_JOB_EXECUTION_PARAMS table. This table is where the job parameters are stored for each execution. We talked before about how identifying job parameters are used to determine if a run requires a new job instance or not. However, the table actually stores all parameters passed to the job. On a restart, only the identifying ones are passed in automatically. Table 5-4 describes the BATCH_JOB_EXECUTION_PARAMS table.

Table 5-4. *BATCH_JOB_EXECUTION_PARAMS Table*

Field	Description
JOB_EXECUTION_ID	Primary key of the table
TYPE_CODE	A String indicating the type the value of this parameter is
KEY_NAME	The name of the parameter
STRING_VAL	The value of the parameter if the type is a String
DATE_VAL	The value of the parameter if the type is a Date
LONG_VAL	The value of the parameter if the type is a long
DOUBLE_VAL	The value of the parameter if the type is a double
IDENTIFYING	A flag that indicates if a parameter is identifying

Beyond all the tables used to define the metadata for the job, two additional tables are used in the job repository. These tables are used to store the metadata for a step. The first is the BATCH_STEP_EXECUTION. This table is not only responsible for maintaining metadata about start, finish, and status of a step in the same way the BATCH_JOB_EXECUTION does for the job, but also is used to persist various counts to provide analytics of the step. Data points like read counts, process counts, write counts, skip counts, and more are all persisted as well. Table 5-5 identifies all of the data points stored in the BATCH_STEP_EXECUTION table.

Table 5-5. *BATCH_STEP_EXECUTION Table*

Field	Description
STEP_EXECUTION_ID	Primary key of the table
VERSION	The version[3] for the record used for optimistic locking
STEP_NAME	The name of the step
JOB_EXECUTION_ID	A foreign key to the BATCH_JOB_EXECUTION table
START_TIME	The time the step execution began
END_TIME	The time the step execution completed
STATUS	The batch status of the step

(*continued*)

---

[3]To learn more about the versions and entities in domain-driven design, read *Domain Driven Design* by Eric Evans (Addison-Wesley, 2003).

Table 5-5. (*continued*)

Field	Description
COMMIT_COUNT	The number of transactions that committed during the execution of the step
READ_COUNT	The number of items read
FILTER_COUNT	The number of items that were filtered out by the ItemProcessor returning null
WRITE_COUNT	The number of items written
READ_SKIP_COUNT	The number of items skipped due to an exception thrown in the ItemReader
PROCESS_SKIP_COUNT	The number of items skipped due to an exception thrown in the ItemProcessor
WRITE_SKIP_COUNT	The number of items skipped due to an exception thrown in the ItemWriter
ROLLBACK_COUNT	The number of transactions that were rolled back over the step execution
EXIT_CODE	The exit code for the step
EXIT_MESSAGE	The message or stack trace returned by the step execution
LAST_UPDATED	The last time the record was updated

The final table in the job repository is the BATCH_STEP_EXECUTION_CONTEXT. Just as the JobExecution has an ExecutionContext that is used to store the state of components, the StepExecution also has an ExecutionContext that is used for the same purpose. The StepExecution's ExecutionContext is used to store state of components at a step level. We'll look at its use in great detail as we dig deeper into components like ItemReaders and ItemWriters but for now, Table 5-6 outlines the columns of the BATCH_STEP_EXECUTION_CONTEXT table.

Table 5-6. *BATCH_STEP_EXECUTION_CONTEXT Table*

Field	Description
STEP_EXECUTION_ID	Primary key of the table
SHORT_CONTEXT	A trimmed version of the SERIALIZED_CONTEXT
SERIALIZED_CONTEXT	The actual serialized ExecutionContext

This section has reviewed what the components of the relational database implementation of a job repository consists of. However, Spring Batch does provide another option for use in development or testing use cases. That is an in-memory job repository. The next section will look at that.

The In-Memory Job Repository

When developing a Spring Batch job or running unit tests, configuring an external database may be more trouble than it's worth. Because of that, Spring Batch provides an implementation of the JobRepository that utilizes java.util.Map instances as the datastore. In the next section we'll look at how to configure this type of JobRepository as well as how to customize the configuration of a JobRepository.

■ **Note** The Map-based JobRepository is not intended for production use. If you wish to run a batch job without an external database, use an in-memory database like H2 or HSQLDB which has better support for things like multithreading and transactions.

Configuring the Batch Infrastructure

With the use of the @EnableBatchProcessing annotation, Spring Batch provides a JobRepository out of the box with no additional configuration required. However, there are plenty of times when customization of that JobRepository is needed. In this section we'll look at the customization of all of the Spring Batch infrastructure including the JobRepository via the BatchConfigurer interface.

The BatchConfigurer Interface

The BatchConfigurer interface is a strategy interface that provides the ability to customize the configuration of the Spring Batch infrastructure components. Spring Batch uses it to obtain instances of each of the infrastructure components used by the framework when using the @EnableBatchProcessing annotation. In essence, the addition of the beans that annotation provides is a two-step process. First they are created via the BatchConfigurer implementation, then added to the Spring ApplicationContext via the SimpleBatchConfiguration. In the vast majority of cases, you won't need to touch the SimpleBatchConfiguration. However, customizing the components it exposes is done via the BatchConfigurer and that is a common place to need customization. Let's start by looking at the interface itself in Listing 5-1.

Listing 5-1. The BatchConfigurer Interface

```
public interface BatchConfigurer {

    JobRepository getJobRepository() throws Exception;

    PlatformTransactionManager getTransactionManager() throws Exception;

    JobLauncher getJobLauncher() throws Exception;

    JobExplorer getJobExplorer() throws Exception;
}
```

Each of the methods provides one of the main components for the Spring Batch infrastructure. We've talked about what the JobRepository and JobLauncher are. The PlatformTransactionManager provided via this interface is used by Spring Batch in all of the transaction management the framework provides. Finally, the JobExplorer provides a read-only view into the data in the job repository.

Most of the time, you won't need to implement the entire interface yourself. Spring Batch uses a DefaultBatchConfigurer to provide all of the default options for these components. Typical use cases only require you to override the configuration of one or two of the components so extending the DefaultBatchConfigurer and overriding the appropriate methods is usually an easier option. Let's take a look at some common ways each of the components provided via the BatchConfigurer are customized and why.

Customizing the JobRepository

The JobRepository is created via a FactoryBean, not surprisingly named the JobRepositoryFactoryBean. This FactoryBean provides the ability to customize each of the attributes specified in Table 5-7.

Table 5-7. *JobRepositoryFactoryBean Customizations*

Setter Name	Description
setClobType(int type)	Takes a java.sql.Type value to indicate the type to be used for CLOB columns.
setDatabaseType(String dbType)	Configures the database type. Not typically needed to be set since Spring Batch attempts to identify the type automatically.
setDataSource(DataSource dataSource)	The DataSource to be used with the JobRepository.
setIncrementerFactory (DataFieldMaxValueIncrementerFactory incrementerFactory)	A factory for an incrementer used to increment the primary keys for most tables.
setIsolationLevelForCreate (String isoltationLevelForCreate)	Transaction serialization level used when JobExecution entities are created. Defaults to ISOLATION_SERIALIZABLE.
setJdbcOperations(JdbcOperations jdbcTemplate)	A setter for a JdbcOperations instance. If one is not provided, one will be created with the DataSource provided in the related setter.
setLobHandler(LobHandler lobHandler)	Really only needed for old versions of Oracle where special handling of LOBs was required.
setMaxVarCharLength(int maxLength)	Used to trim the length of the exit message (step and job) as well as the short execution context columns. Should not be set unless the schema has been modified from what Spring Batch provides.
setSerializer (ExecutionContextSerializer serializer)	Configures what implementation of the ExecutionContextSerializer to use to serialize and deserialize both the JobExecution's ExecutionContext and the StepExecution's ExecutionContext.
setTablePrefix(String tablePrefix)	Allows a user to configure a new prefix for all the tables besides the "BATCH_" used by default.
setTransactionManager (PlatformTransactionManager transactionManager)	When using multiple databases, a transaction manager that supports two phase commits is necessary to keep both databases in sync.
setValidateTransactionState(boolean validateTransactionState)	Flag that indicates if an existing transaction when a JobExecution is created. Defaults to true since this usually is a mistake.

The most common scenario where you'd need to extend DefaultBatchConfigurer and override its createJobRepository() method is when you have more than one DataSource in your ApplicationContext. For example, if you have one DataSource for your business data and one for your JobRepository, you'll need to explicitly configure which DataSource is used with the JobRepository. Listing 5-2 shows an example of a JobRepository being customized by extending the DefaultBatchConfigurer and overriding the createJobRepository() method.

Listing 5-2. Customizing the JobRepository

```
...
public class CustomBatchConfigurer extends DefaultBatchConfigurer {

        @Autowired
        @Qualifier("repositoryDataSource")
        private DataSource dataSource;

        @Override
        protected JobRepository createJobRepository() throws Exception {
                JobRepositoryFactoryBean factoryBean = new JobRepositoryFactoryBean();

                factoryBean.setDatabaseType(DatabaseType.MYSQL.getProductName());
                factoryBean.setTablePrefix("FOO_");
                factoryBean.setIsolationLevelForCreate("ISOLATION_REPEATABLE_READ");
                factoryBean.setDataSource(this.dataSource);

                factoryBean.afterPropertiesSet();

                return factoryBean.getObject();
        }
}
```

If we walk through Listing 5-2, we see that `CustomBatchConfigurer` extends `DefaultBatchConfigurer`, freeing us from the need to reimplement everything the interface requires. We autowire in a `DataSource` called `repositoryDataSource`. It is assumed in this listing that somewhere within the `ApplicationContext` there is a bean of type `DataSource` named `repositoryDataSource` that can be autowired in. From there, we can see that the `DefaultBatchConfigurer#createJobRepository()` method is overridden. This is the method used by the `DefaultBatchConfigurer` to actually create the `JobRepository`. In our implementation, we also create a `JobRepository` but we customize some of the default settings. Specifically we specify the database type, configure our table prefix to be "FOO_" instead of the default "BATCH_", set the transaction isolation level for create to be "ISOLATION_REPEATABLE_READ" instead of the default of "ISOLATION_ SERIALIZED", and finally set the `DataSource` we autowired earlier.

It's important to note that none of the `create*` methods found on the `DefaultBatchConfigurer` are called by the Spring container directly as bean definitions. Because of that, it's our responsibility to call `InitializingBean#afterPropertiesSet()` and `FactoryBean#getObject()` (two things that the Spring container would normally do for us).

The `JobRepository` isn't the only thing you can customize via the `BatchConfigurer` mechanism. Typically the `TransactionManager` used for your batch application is provided by Spring Boot so there isn't much to do with that. However if you do want to customize it or if your application has multiple `TransactionManagers`, you'll want to use the `BatchConfigurer` to specify which one to use. The next section will look at how that works.

Customizing the `TransactionManager`

Spring Batch is heavily transactional. It uses transactions as a core component of the framework, so the `TransactionManager` is a key component within the framework. Going through all the configuration options for various `TransactionManager` options is outside the scope of this book. However, Listing 5-3 illustrates how extending the `DefaultBatchConfigurer` to specify what `TransactionManager` is returned can be useful.

Listing 5-3. Customizing the TransactionManager

```
...
public class CustomBatchConfigurer extends DefaultBatchConfigurer {

    @Autowired
    @Qualifier("batchTransactionManager")
    private PlatformTransactionManager transactionManager;

    @Override
    public PlatformTransactionManager getTransactionManager() {
        return this.transactionManager;
    }
}
```

In Listing 5-3, a PlatformTransactionManager defined elsewhere that is intended to be used by our batch processing is explicitly returned for the call to BatchConfigurer#getTransactionManager(). You'll notice that this didn't override a protected method in the DefaultBatchConfigurer. This is because the DefaultBatchConfigurer by default creates a DataSourceTransactionManager in the setter of the DataSource if one has not already been created. This is the only component exposed by the BatchConfigurer that is handled like this.

The next component to look at customization options is the JobExplorer. The next section looks at its options.

Customizing the JobExplorer

The JobRepository provides an API for persisting and retrieving the data from the underlying data store used to store a batch job's state. However, there are use cases where you want to only expose a read-only view of that data. The JobExplorer provides that read only view into the batch metadata.

Since the JobExplorer is a read-only view into the same data that the JobRepository manipulates, the underlying data access layer is actually the same, a common set of DAOs shared between the JobRepository and the JobExplorer. Because of this, the customization options for the JobRepository and JobExplorer are the same for all attributes that are involved in reading from the database. Table 5-8 walks through the options on the JobExplorerFactoryBean.

Table 5-8. JobExplorerFactoryBean Customizations

Setter Name	Description
setDataSource(DataSource dataSource)	The DataSource to be used with the JobRepository.
setJdbcOperations(JdbcOperations jdbcTemplate)	A setter for a JdbcOperations instance. If one is not provided, one will be created with the DataSource provided in the related setter.
setLobHandler(LobHandler lobHandler)	Really only needed for old versions of Oracle where special handling of LOBs was required.
setSerializer (ExecutionContextSerializer serializer)	Configures what implementation of the ExecutionContextSerializer to use to serialize and deserialize both the JobExecution's ExecutionContext and the StepExecution's ExecutionContext.
setTablePrefix(String tablePrefix)	Allows a user to configure a new prefix for all the tables besides the "BATCH_" used by default.

Listing 5-4 shows an example of customizing a JobExplorer to match a JobRepository configured as in Listing 5-2.

Listing 5-4. Customizing the JobExplorer

```
...
public class CustomBatchConfigurer extends DefaultBatchConfigurer {

    @Autowired
    @Qualifier("batchTransactionManager")
    private DataSource dataSource;

    @Override
    protected JobExplorer createJobExplorer() throws Exception {
        JobExplorerFactoryBean factoryBean = new JobExplorerFactoryBean();

        factoryBean.setDataSource(this.dataSource);
        factoryBean.setTablePrefix("FOO_");

        factoryBean.afterPropertiesSet();

        return factoryBean.getObject();
    }
}
```

Just like when we customized the behavior in the JobRepository in Listing 5-2, Listing 5-4 illustrates that we configure the same things: DataSource, serializer, and table prefix. Again, since the BatchConfigurer's methods are not directly exposed to the Spring container, we need to call Initializer Bean#afterPropertiesSet() and FactoryBean#getObject().

■ **Note** Since the JobRepository and JobExplorer use the same underlying data store, it's good practice to customize both if you customize only one so that they are in sync.

The last piece of the Spring Batch infrastructure that is customizable via the BatchConfigurer mechanisms is a JobLauncher. The next section will walk through customizing it.

Customizing the JobLauncher

The JobLauncher is the entry point for launching a Spring Batch job. In most cases, you won't need to customize it when running jobs via Spring Boot's default mechanisms using the SimpleJobLauncher provided by Spring Batch. However, if you want to expose the ability to launch a job another way (say via a Controller as part of a Spring MVC application), you may want to tweak how the SimpleJobLauncher works. Table 5-9 walks through the various options for configuring a JobLauncher.

Table 5-9. *JobLauncher Customizations*

Setter Name	Description
setJobRepository(JobRepository jobRepository)	Configures the JobRepository to be used.
setTaskExecutor(TaskExecutor taskExecutor)	Sets the TaskExecutor to be used for this JobLauncher. Defaults to the SyncTaskExecutor.

With all of the components the BatchConfigurer can be used to customize (JobRepository, PlatformTransactionManager, JobLauncher, and JobExplorer), each is an interface that you can implement yourself as well. This chapter has covered how to customize the implementations that are provided out of the box by Spring Batch. However, there is one other aspect of the Spring Batch infrastructure that we need to learn to configure—the database. Both configuring the connection information as well as how to initialize the Spring Batch database schema are required. The next section will look into how Spring Boot can be used to accomplish this.

Database Configuration

Spring Boot makes doing simple things extra simple. Configuring a database is one example of this. In order to configure a database to use with Spring Boot, all you need to do is add your database driver to your classpath and configure the appropriate properties. In this book, we'll be using either HSQLDB for in-memory use cases and MySQL for use cases that require an external database.

Once you've added your database driver to your project, there are a list of properties you need to configure via one of the mechanisms Spring Boot supports: via the application.properties, application.yml, environment variables, or command line arguments. For this book, we're going to use application.yml for most of the configuration options. Listing 5-5 shows how to configure a MySQL database using Spring Boot properties.

Listing 5-5. Configuring a Database

```
spring:
  datasource:
    driverClassName: com.mysql.cj.jdbc.Driver
    url: jdbc:mysql://localhost:3306/spring_batch
    username: 'root'
    password: 'myPassword'
  batch:
    initialize-schema: always
```

The first four properties are pretty self-explanatory, driver class name, url, username, and password. These are standard configurations for any database. The last property, spring.batch.initalize-schema: always is used to tell Spring Boot to execute the Spring Batch schema script. This property has three possible values:

- always: This will run the script every time you run your application. Since errors are ignored if they occur and there are no drop statements in the Spring Batch SQL files, this is the easiest option for development environments.

- never: This will never run the script.

- embedded: This will only run the script when using an embedded database under the assumption that if you are, you are starting with a clean database instance on each start.

In this section we looked at how to configure the infrastructure of Spring Batch. Most of that infrastructure exists to manage and query batch metadata. However, all the metadata in the world is useless unless you have a way to use it. In the next section we'll take a look at an example for using a job's metadata.

Using Job Metadata

Although Spring Batch accesses the job repository tables through a collection of DAOs, they expose a much more practical API for the use of the framework and for you to use. This section you look at how Spring Batch exposes the data in the job repository. We've already learned how to configure it, but the main method for accessing the metadata provided by Spring Batch is the JobExplorer so let's dig into it.

The JobExplorer

The org.springframework.batch.core.explore.JobExplorer interface is the starting point for all access to historical and active data in the job repository. Figure 5-2 shows that although most of the framework accesses the information stored about job execution through the JobRepository, the JobExplorer accesses it directly from the database itself.

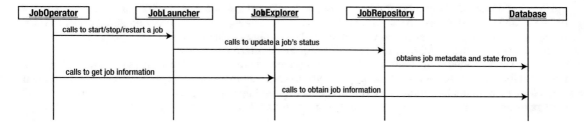

Figure 5-2. *The relationship between the job administration components*

The underlying purpose of the JobExplorer is to provide read-only access to the data in the job repository. The interface provides seven methods you can use to obtain information about job instances and executions. Table 5-10 lists the available methods and their use.

Table 5-10. *Methods of the JobExplorer*

Method	Description
java.util.Set<JobExecution>findRunningjob Executions(java.lang.String jobName)	Returns all JobExecutions without an end time.
List<JobInstance> findJobInstancesByName(java.lang.String name, int start, int count)	Returns a page of JobInstances with the name provided.
JobExecution getJobExecution(java.lang. Long executionId)	Returns the JobExecution identified by the supplied id and null if not found.
java.util.List<JobExecution> getJobExecutions(JobInstance instance)	Returns a list of all JobExecutions related to the JobInstance supplied.

(continued)

Table 5-10. (*continued*)

Method	Description
JobInstance getJobInstance(java.lang.Long instanceId)	Returns the JobInstance identified by the supplied id or null if none is found.
java.util.List<JobInstance> getJobInstances(java.lang.String jobName, int start, int count)	Returns a range of JobInstances starting with the index specified (the start parameter). The final parameter specifies the maximum number of JobInstances to return.
int getJobInstanceCount(String jobName)	Returns the number of JobInstances that have been created for a given job name.
java.util.List<java.lang.String> getJobNames()	Returns all unique job names from the job repository in alphabetical order.
StepExecution getStepExecution(java.lang.Long jobExecutionId, java.lang.Long stepExecutionId)	Returns the specified StepExecution based on the id of the StepExecution and the id of its parent JobExecution.

As you can see, the entire job repository is available from the methods exposed by the JobExplorer interface. To see how the JobExplorer works, you can inject it into a Tasklet and do some exploring with it. From there, you can see what you can use the JobExplorer for. In Listing 5-6, you configure the new Tasklet with the JobExplorer injected.

Listing 5-6. Configuration of an ExploringTasklet Tasklet and JobExplorer

```
...
@EnableBatchProcessing
@SpringBootApplication
public class DemoApplication {

    @Autowired
    private JobBuilderFactory jobBuilderFactory;

    @Autowired
    private StepBuilderFactory stepBuilderFactory;

    @Autowired
    private JobExplorer jobExplorer;

    @Bean
    public Tasklet explorerTasklet() {
        return new ExploringTasklet(this.jobExplorer);
    }

    @Bean
    public Step explorerStep() {
        return this.stepBuilderFactory.get("explorerStep")
                    .tasklet(explorerTasklet())
                    .build();
    }
```

```
    @Bean
    public Job explorerJob() {
        return this.jobBuilderFactory.get("explorerJob")
                    .start(explorerStep())
                    .build();
    }

    public static void main(String[] args) {
        SpringApplication.run(DemoApplication.class, args);
    }

}
```

With the JobExplorer configured, there are a number of things you can do with it. Within the Spring Batch framework, you can use the JobExplorer in the RunIdIncrementer you looked at in Chapter 4 to look up the previous run.id parameter value. Another place it's used is in the Spring Cloud Data Flow server to determine whether a job is currently running before launching a new instance. We will look at that in Chapter 12. In this example, we'll generate a simple report to standard out that illustrates how many job instances have been run for this job, how many executions they had, and what their results were.

Listing 5-7. ExploringTasklet

```
...
public class ExploringTasklet implements Tasklet {

    private JobExplorer explorer;

    public ExploringTasklet(JobExplorer explorer) {
        this.explorer = explorer;
    }

    public RepeatStatus execute(StepContribution stepContribution,
                ChunkContext chunkContext) {

        String jobName = chunkContext.getStepContext().getJobName();

        List<JobInstance> instances =
                    explorer.getJobInstances(jobName,
                            0,
                            Integer.MAX_VALUE);

        System.out.println(
                    String.format("There are %d job instances for the job %s",
                    instances.size(),
                    jobName));

        System.out.println("They have had the following results");
        System.out.println("***********************************");

        for (JobInstance instance : instances) {
            List<JobExecution> jobExecutions =
                        this.explorer.getJobExecutions(instance);
```

```
System.out.println(
                String.format("Instance %d had %d executions",
                        instance.getInstanceId(),
                        jobExecutions.size())));

        for (JobExecution jobExecution : jobExecutions) {
            System.out.println(
                    String.format("\tExecution %d resulted in Exit
                    Status %s",
                            jobExecution.getId(),
                            jobExecution.getExitStatus())));
        }
    }

    return RepeatStatus.FINISHED;
    }
}
```

The code in Listing 5-7 begins by getting the current job's name. From there, it looks up all the JobInstances that have been run. It is important to realize that the current JobInstance will be returned by this call. We then print out the number of JobInstances that were returned for this job. Then, for each JobInstance, we use the JobExplorer to find all the JobExecutions associated with it and display their results.

With the code and configuration in place, run the job a few times and you will begin to see output like what is displayed in Listing 5-8.

Listing 5-8. ExplorerJob Output

```
2019-01-18 00:01:27.392  INFO 35356 --- [           main] o.s.b.c.l.support.
SimpleJobLauncher      : Job: [SimpleJob: [name=explorerJob]] launched with the following
parameters: [{1=1}]
2019-01-18 00:01:27.423  INFO 35356 --- [           main] o.s.batch.core.job.
SimpleStepHandler      : Executing step: [explorerStep]
There are 2 job instances for the job explorerJob
They have had the following results
**********************************
Instance 2 had 1 executions
     Execution 2 resulted in Exit Status exitCode=UNKNOWN;exitDescription=
Instance 1 had 1 executions
     Execution 1 resulted in Exit Status exitCode=COMPLETED;exitDescription=
2019-01-18 00:01:27.517  INFO 35356 --- [           main] o.s.b.c.l.support.
SimpleJobLauncher      : Job: [SimpleJob: [name=explorerJob]] completed with the following
parameters: [{1=1}] and the following status: [COMPLETED]
2
```

This section looked at how to access data in the job repository via the JobExplorer. You use APIs like the JobExplorer to access the data to use it in a safe way.

Summary

Spring Batch's ability to manage metadata about a job as well as maintain the state of the job as it runs for error handling is one of the primary reasons, if not the primary reason, to use Spring Batch for enterprise batch processing. Not only does it provide the ability for robust error handling, but it also allows processes to make decisions about what to do based on what has happened elsewhere in the job. In the next chapter, you put this metadata to further use as you take a deep look at how to start, stop, and restart jobs in a variety of environments.

CHAPTER 6

Running a Job

It is pretty amazing how Spring Boot has changed how we look at running Java applications. Before Spring Boot, were you going to run your application in a servlet container or on an application server? Maybe you were going to deploy your application as an executable jar file. If you went that route, you either had to deal with building your classpath via some script or use something like the Shade Maven plug-in. With the introduction of Spring Boot, that decision has mostly been marginalized. Now most applications are bootstrapped via the launching of the executable jar file Spring Boot generates.

Spring Boot also provides facilities for running Spring Batch jobs. In this chapter we'll walk through how Spring Boot makes running Spring Batch jobs easy. However, the default facilities are not the only way to launch a Spring Batch job so we'll explore the components that go into running a batch job so you can develop your own mechanisms for launching a job.

Why would you want to write your own code to launch a job when Spring Boot can handle it for you? Well, Spring Boot only handles the most simple of use cases, executing your batch job on startup. However, there are a list of other use cases where that may not work out. For example, it's common for a batch job to need to execute at a given time based on a schedule. Integration with various schedulers would require custom code (we'll get to why later in this chapter). You may also want to launch a job based on some form of event as a reaction to an external system.

Starting your Spring Batch job isn't the only part of being able to run a job. Being able to stop a running job is another important piece of the execution puzzle. If your job fails, being able to restart it is also a critical piece of being able to execute production grade batch processes. You wouldn't want a job that is expected to take hours processing millions of records to have to start back at the beginning if a failure occurs. Spring Batch provides restart functionality that allows for it to pick up where it left off. We'll cover both stopping and restarting your batch jobs in this chapter.

Let's get started by looking at how Spring Boot handles launching your batch jobs for you.

Starting a Job with Spring Boot

We've been launching our jobs via the native functionality within Spring Boot exclusively up to this point in the book. However, we haven't really taken a look at how it actually works. In this section, we'll look at how Spring Boot launches your jobs for you on startup.

Spring Boot has two mechanisms for running logic at startup, a `CommandLineRunner` and an `ApplicationRunner`. Both interfaces provide a single method called after the `ApplicationContext` is refreshed and ready to run to allow for your application to execute code. When using Spring Boot with Spring Batch, a special `CommandLineRunner` is used, the `JobLauncherCommandLineRunner`.

The `JobLauncherCommandLineRunner` uses Spring Batch's `JobLauncher` to execute your job. We'll look at the `JobLauncher` interface in more depth later in this chapter. For now, just know that it knows how to launch a Spring Batch job. When Spring Boot executes all the `CommandLineRunners` configured in the `ApplicationContext`, if you have the spring-boot-starter-batch on your classpath, the

M. T. Minella, *The Definitive Guide to Spring Batch*, https://doi.org/10.1007/978-1-4842-3724-3_6

JobLauncherCommandLineRunner will run any Job definitions it finds in the context. This is the mechanism we've used for every example up to this point to run our batch jobs. However, Spring Boot does provide a few configuration options worth looking at.

The first is to define what jobs are launched at startup. A Spring Boot uber jar may contain more than one job. For example, if you plan on having your batch job executed as the result of something like a REST call or an event of some kind, you probably don't want that job to execute when the application is started up. To set this behavior, Spring Boot exposes the property spring.batch.job.enabled equal to false in your application.yml (this property is set to true by default). Listing 6-1 shows an example of this in action. When you execute that code with the spring.batch.job.enabled set to false, you'll see that no job is run. The context is created then immediately shut down.

Listing 6-1. A Job That Doesn't Run

```
@EnableBatchProcessing
@SpringBootApplication
public class NoRunJob {

        @Autowired
        private JobBuilderFactory jobBuilderFactory;

        @Autowired
        private StepBuilderFactory stepBuilderFactory;

        @Bean
        public Job job() {
                return this.jobBuilderFactory.get("job")
                                .start(step1())
                                .build();
        }

        @Bean
        public Step step1() {
                return this.stepBuilderFactory.get("step1")
                                .tasklet((stepContribution, chunkContext) -> {
                                        System.out.println("step1 ran!");
                                        return RepeatStatus.FINISHED;
                                }).build();
        }

        public static void main(String[] args) {
                SpringApplication application = new SpringApplication(NoRunJob.class);

                Properties properties = new Properties();
                properties.put("spring.batch.job.enabled", false);
                application.setDefaultProperties(properties);

                application.run(args);
        }
}
```

While Listing 6-1 has a job configured to run, Spring Boot hasn't run it because we have configured it not to run on startup.

Another use case that can occur when using Spring Boot is having multiple jobs defined within the context and only wanting specific ones executed on startup. An example of where this is useful is when you have a job that launches other jobs. In that case, you'd want only the parent or master job to be kicked off by Spring Boot since the others will be orchestrated by your parent job. Spring Boot supports this use case by allowing you to configure what jobs are run at startup by using the `spring.batch.job.names` to identify the names of the jobs to be executed on startup. Spring Boot will take this comma-delimited list and execute them in order.

Spring Boot isn't the only way to execute a job. You can write your own mechanisms to trigger the execution of your jobs. In the next section, we'll take a look at how to execute a batch job via a REST API.

Launching a Job via REST

REST APIs are the most popular way to expose functionality these days. They can easily be used to launch a batch job as well. However, there is no REST API available to launch batch jobs out of the box,[1] which means we will have to write our own. But how do we launch a Spring Batch job programmatically? The `JobLauncher` provides this capability so let's take a look.

The `JobLauncher` interface is an interface that, well, launches jobs. It has a single method, `run(Job job, JobParameters jobParameters)`, which takes two arguments: the `Job` to be executed and the `JobParameters` to be passed to the `Job`. Listing 6-2 illustrates the interface.

Listing 6-2. The `JobLauncher` Interface

```
public interface JobLauncher {

    public JobExecution run(Job job, JobParameters jobParameters) throws
                                    JobExecutionAlreadyRunningException,
                                    JobRestartException,
                                    JobInstanceAlreadyCompleteException,
                                    JobParametersInvalidException;

}
```

Out of the box, Spring Batch provides one `JobLauncher` implementation, the `SimpleJobLauncher`. In the vast majority of cases, this `JobLauncher` will address all of the launching requirements you have. It handles the determination of whether the run is part of an existing `JobInstance` or a new one and acts accordingly.

■ **Note** The `SimpleJobLauncher` does not manipulate the provided `JobParameters`. So if your job is using a `JobParametersIncrementer`, that needs to be applied before the parameters are passed to the `SimpleJobLauncher`.

The `JobLauncher` interface does not have any specific prescription as to whether the job it is running is synchronous or asynchronously nor does it have any strong opinion on the matter. It allows you to choose how the jobs it run are executed by allowing you to configure the `TaskExecutor` it uses. By default, it uses a synchronous `TaskExecutor` so the jobs are executed synchronously by the `SimpleJobLauncher` (in the same thread as the caller). However, if you'd like to free up the existing thread (e.g., if you wanted to return a REST call once the job has been started), then one of the asynchronous `TaskExecutor` implementations may be a better option.

---

[1]Spring Cloud Data Flow offers this feature, but it's a bit of a different animal. We'll look at Data Flow later in this book.

To get started with our REST API launching application, we need to create a new project. Our new project will come from Spring Initializr with the following dependencies selected:

- Batch

- MySQL

- JDBC

- Web

Once we have our project in place, we'll need to configure our application.yml to not run our batch jobs on startup (since we only want them to run when we call the REST API, as well as our database configuration). Listing 6-3 illustrates the contents of the application.yml required for our new application.

Listing 6-3. application.yml

```
spring:
  batch:
    job:
      enabled: false
    initialize-schema: always
  datasource:
    driverClassName: com.mysql.cj.jdbc.Driver
    url: jdbc:mysql://localhost:3306/spring_batch
    username: 'root'
    password: 'p@ssw0rd'
    platform: mysql
```

We'll use the `SimpleJobLauncher` when we create our REST API to launch our job. Conveniently, when using the `@EnableBatchProcessing` annotation, Spring Batch provides a `SimpleJobLauncher` for you out of the box so we don't need to do anything to get one. Given that's really all we need to launch a job, let's take a look at a controller that would accept a job name and job parameters as request parameters and launches the appropriate job. Listing 6-4 shows the complete application.

Listing 6-4. The `JobLaunchingController` Application

```
...
@EnableBatchProcessing
@SpringBootApplication
public class RestApplication {

    @Autowired
    private JobBuilderFactory jobBuilderFactory;

    @Autowired
    private StepBuilderFactory stepBuilderFactory;

    @Bean
    public Job job() {
        return this.jobBuilderFactory.get("job")
                    .incrementer(new RunIdIncrementer())
                    .start(step1())
                    .build();
    }
```

```java
@Bean
public Step step1() {
        return this.stepBuilderFactory.get("step1")
                        .tasklet((stepContribution, chunkContext) -> {
                                System.out.println("step1 ran today!");
                                return RepeatStatus.FINISHED;
                        }).build();
}
@RestController
public static class JobLaunchingController {

        @Autowired
        private JobLauncher jobLauncher;

        @Autowired
        private ApplicationContext context;

        @PostMapping(path = "/run")
        public ExitStatus runJob(@RequestBody JobLaunchRequest request) throws Exception {
                Job job = this.context.getBean(request.getName(), Job.class);

                return this.jobLauncher.run(job, request.getJobParameters()).
                getExitStatus();
        }
}

public static class JobLaunchRequest {
        private String name;

        private Properties jobParameters;

        public String getName() {
                return name;
        }

        public void setName(String name) {
                this.name = name;
        }

        public Properties getJobParamsProperties() {
                return jobParameters;
        }

        public void setJobParamsProperties(Properties jobParameters) {
                this.jobParameters = jobParameters;
        }
        public JobParameters getJobParameters() {
                Properties properties = new Properties();
                properties.putAll(this.jobParameters);
```

```
                    return new JobParametersBuilder(properties)
                            .toJobParameters();
        }
    }

    public static void main(String[] args) {
        new SpringApplication(RestApplication.class).run(args);
    }
}
```

There is a lot in Listing 6-4, but most of the pieces should be familiar. Starting at the top, we have our normal annotations for a Spring Boot application running Spring Batch (@SpringBootApplication and @EnableBatchProcessing). The next section is the building of our batch job. It's a simple, single step job that emits a "step1 ran!". The goal here isn't to see something complex from Spring Batch, but just to prove that our job did run. This example has two inner classes. The reason for the inner classes is just to make the example self-contained. In a normal application, these would be broken out. The first inner class is the Controller itself. Here we are autowiring in the JobLauncher that @EnableBatchProcessing provides as well as the current ApplicationContext so that we can retrieve the Job bean to be executed within our request.

The @PostMapping allows us to map the URL our HTTP POST needs to go to in order to be called, /run in our case. The body of the post will have two main components, the name of the job to execute and a Map of any parameters to be passed to the job. That structure is modeled in the next inner class, JobLaunchRequest. We'll call our API passing it a JSON payload that Spring will map for us to a JobLaunchRequest instance.

Within the runJob method of our controller, we do two things. The first is we get the Job to be executed. We do that by asking the ApplicationContext for it. Once we have our Job and JobParameters, we can execute the Job by passing both to the JobLauncher. By default, the JobLauncher will execute the job synchronously so we can return the ExitStatus to the user. It's important to note that most batch jobs don't run this fast due to the amount of processing involved. Because of this, running them asynchronously in this case would be a better fit (in which case, we'd just return the JobExecution's id).

The final section of this example is the main method used to bootstrap it. Just like in Listing 6-1, we configure the batch job to not run since we don't want to run it on startup, we want it to launch when the API is called. Listing 6-5 contains an example of how to launch the job in Listing 6-4 via curl.

Listing 6-5. Curl Command and Output for Listing 6-4

```
$ curl -H "Content-Type: application/json" -X POST -d '{"name":"job", "jobParameters":
{"foo":"bar", "baz":"quix"}}' http://localhost:8080/run
{"exitCode":"COMPLETED","exitDescription":"","running":false}
```

Listing 6-5 does a HTTP POST of the included JSON to http://localhost:8080/run with the Content-Type header set to application/json. If you run the application in Listing 6-4 and execute the curl command in Listing 6-5, you'll see the output in Listing 6-6.

Listing 6-6. REST API Output

```
2018-02-08 12:07:56.327  INFO 22104 --- [nio-8080-exec-1] o.s.b.c.l.support.SimpleJob
Launcher        : Job: [SimpleJob: [name=job]] launched with the following parameters:
[{baz=quix, foo=bar}]
2018-02-08 12:07:56.345  INFO 22104 --- [nio-8080-exec-1] o.s.batch.core.job.SimpleStep
Handler         : Executing step: [step1]
step1 ran!
2018-02-08 12:07:56.362  INFO 22104 --- [nio-8080-exec-1] o.s.b.c.l.support.SimpleJob
Launcher        : Job: [SimpleJob: [name=job]] completed with the following parameters:
[{baz=quix, foo=bar}] and the following status: [COMPLETED]
```

It's important to note that we did not include any logic around restarting a job, handling job parameter incrementing, and so on in our API. We'll look at restarting a job later in this chapter. But before we move off of this API, let's look at how to handle incrementing job parameters for subsequent runs.

When using a JobParametersIncrementer, it is the caller of the JobLauncher's responsibility to apply those changes to the parameters. Once parameters have made it to the Job, they are immutable. Spring Batch provides a convenient method on the JobParametersBuilder for incrementing the parameters, getNextJobParameters(Job job). Listing 6-7 shows the controller from our previous application updated with the use of the JobParametersBuilder#getNextJobParameters call.

Listing 6-7. Incrementing Job Parameters Before a Launch

```
...
@Bean
public Job job() {
    return this.jobBuilderFactory.get("job")
                .incrementer(new RunIdIncrementer())
                .start(step1())
                .build();
}
...
@RestController
public static class JobLaunchingController {

    @Autowired
    private JobLauncher jobLauncher;

    @Autowired
    private ApplicationContext context;

    @Autowired
    private JobExplorer jobExplorer;

        @PostMapping(path = "/run")
        public ExitStatus runJob(@RequestBody JobLaunchRequest request) throws Exception
{
            Job job = this.context.getBean(request.getName(), Job.class);
```

```
                JobParameters jobParameters =
                        new JobParametersBuilder(request.getJobParameters(),
                                        this.jobExplorer)
                                .getNextJobParameters(job)
                                .toJobParameters();

                return this.jobLauncher.run(job, jobParameters).getExitStatus();
    }
}
...
```

Listing 6-7 starts off with our job definition. This is identical to the previous listing; however, in our previous example, the RunIdIncrementer was not actually being activated. That's where the updates to our controller come in.

In the controller, we have a new line. We create a new JobParameters instance that adds the run.id to it by calling the JobParametersBuilder#getNextJobParameters(job) method. This method looks at the Job and determines if there is a JobParametersIncrementer on it. If so, it applies it to the JobParameters used in the last JobExecution. It also determines if this execution is a restart or not and handles the JobParameters appropriately. If neither of these scenarios exist, nothing is changed.

With the changes in Listing 6-7, if we run our application and call it again, we get something slightly different in our console. Listing 6-8 illustrates that we get a new parameter added to our output, run.id=1 has been added to the JobParameters. If we run it again, we'll see run.id=2. This is an important change. If we tried the same thing in the first version of this example, we would have gotten an exception about the JobInstance for those parameters has already been completed.

Listing 6-8. Output of Running the Job with RunIdIncrementer

```
2018-02-08 16:21:34.658  INFO 22990 --- [nio-8080-exec-1] o.s.b.c.l.support.SimpleJob
Launcher     : Job: [SimpleJob: [name=job]] launched with the following parameters:
[{baz=quix, foo=bar, run.id=1}]
2018-02-08 16:21:34.669  INFO 22990 --- [nio-8080-exec-1] o.s.batch.core.job.SimpleStep
Handler      : Executing step: [step1]
step1 ran today!
2018-02-08 16:21:34.679  INFO 22990 --- [nio-8080-exec-1] o.s.b.c.l.support.SimpleJob
Launcher     : Job: [SimpleJob: [name=job]] completed with the following parameters:
[{baz=quix, foo=bar, run.id=1}] and the following status: [COMPLETED]
```

Launching jobs on demand either via executing the java -jar command for the uber jar or via a REST API are both useful ways of doing things; however, in most enterprises, batch processing is executed via a schedule. In the next section, we'll take a look at integrating the execution of a Spring Batch job with a third-party library, in this case, Quartz.

Scheduling with Quartz

Many enterprise schedulers are available. They range from the crude but very effective crontab to enterprise automation platforms that can run into the millions of dollars. The scheduler we will use here is an open source scheduler called Quartz (www.quartz-scheduler.org/). This scheduler is commonly used in Java environments of all sizes. In addition to its power and solid community support, it has an established history of Spring integration including Spring Boot support that is helpful in executing jobs.

Given the scope of Quartz, this book won't cover all of it here. However, a brief introduction to how it works and how it integrates with Spring is warranted. Figure 6-1 shows the components of Quartz and their relationships.

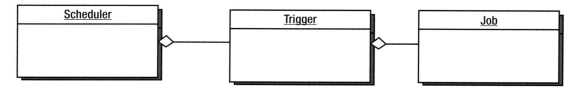

Figure 6-1. *The Quartz scheduler*

As you can see, scheduling via Quartz has three main components: a scheduler, a trigger, and a job. A scheduler, which is obtained from a `SchedulerFactory`, serves as a registry of `JobDetails` (a reference to a Quartz job) and triggers and is responsible for executing a job when its associated trigger fires. A *job* is a unit of work that can be executed. A *trigger* defines when a job is to be run. When a trigger fires, telling Quartz to execute a job, a `JobDetails` object is created to define the individual execution of the job.

Does this sound familiar? It should. The model of defining a `Job` and a `JobDetails` object is very similar to the way Spring Batch defines a `Job` and a `JobInstance`. In order to integrate Quartz with your Spring Batch process, you need to do the following:

- Create a project from Spring Initializr with the correct starters.

- Write a Spring Batch `Job`.

- Write your own Quartz `Job` to launch your job using Spring's `QuartzJobBean`.

- Configure a `JobDetailBean` provided by Spring to create a Quartz `JobDetail`.

- Configure a trigger to define when your job should run.

To show how Quartz can be used to periodically execute a job, we'll start by creating a new project from `https://start.spring.io`. We'll select the Batch, MySQL, JDBC, and Quartz Scheduler dependencies. This will provide us with what we need. Quartz does have the option to store metadata in a database, but that is outside the scope of this book.

With the new project loaded in our IDE, the next step is to create our Spring Batch `Job`. We're going to keep it simple here. Listing 6-9 shows the configuration for the `Job` we'll schedule.

Listing 6-9. Our Scheduled Job

```
...
@Configuration
public class BatchConfiguration {

    @Autowired
    private JobBuilderFactory jobBuilderFactory;

    @Autowired
    private StepBuilderFactory stepBuilderFactory;

    @Bean
    public Job job() {
```

```
            return this.jobBuilderFactory.get("job")
                        .incrementer(new RunIdIncrementer())
                        .start(step1())
                        .build();
    }

    @Bean
    public Step step1() {
        return this.stepBuilderFactory.get("step1")
                    .tasklet((stepContribution, chunkContext) -> {
                            System.out.println("step1 ran!");
                            return RepeatStatus.FINISHED;
                    }).build();
    }
}
```

Nothing in Listing 6-9 should be new. We have a single-step job that will print to the console "step1 ran!". It is important to note that we are using an incrementer here. That is basically a requirement since we'll be running a job multiple times without the ability to provide unique job parameters in another way.

With a job defined, we now need to create a Quartz Job. This will be the piece of code that does the mechanics of launching our Job when the schedule event fires. The code within it should look familiar…it's the same code we used in our REST controller to launch a job. Listing 6-10 shows our BatchScheduledJob.

Listing 6-10. BatchScheduledJob

```
...
public class BatchScheduledJob extends QuartzJobBean {

    @Autowired
    private Job job;

    @Autowired
    private JobExplorer jobExplorer;

    @Autowired
    private JobLauncher jobLauncher;

    @Override
    protected void executeInternal(JobExecutionContext context) {
        JobParameters jobParameters = new JobParametersBuilder(this.jobExplorer)
                    .getNextJobParameters(this.job)
                    .toJobParameters();

        try {
                this.jobLauncher.run(this.job, jobParameters);
        }
        catch (Exception e) {
                e.printStackTrace();
        }
    }
}
```

Listing 6-10 is a class that extends Spring's `QuartzJobBean` class. This class handles most of the boiler plate required for running a Quartz Job. We extend it for our own purposes by overriding the `executeInternal` (`JobExecutionContext context`) method. This is the place where we add the same code we used in our REST controller to launch our job. The only real difference is that we don't need to dynamically choose what job we're going to run so we can just autowire it into the Quartz Job. Our `executeInternal` method will be called once each time the scheduled event fires. All that is left is to configure that schedule.

To configure that schedule, we need to do two things. The first is to configure a bean for our Quartz Job. Quartz provides a `org.quartz.JobBuilder` to help us do that. Using that builder, we provide the `Job` class we created (`BatchScheduledJob` in our case), we tell Quartz to not delete the job definition if it is not associated with a trigger (since we'll use Spring to configure the trigger independently), and tell it to build our `JobDetail`.

Once our `JobDetail` is created, we can create a trigger and schedule it. We'll use Quartz's `org.quartz.SimpleScheduleBuilder` to define a schedule that will launch our job once every 5 seconds and repeat four times (for a total of five executions). A `JobDetail` is the meta data around a Quartz Job to run. A `Schedule` is how often to run a `JobDetail`. A `Trigger` is the association of the two and the final thing we need to create. Using Quartz's `TriggerBuilder`, we create new `Trigger`, passing it the `Job` and `Schedule`. Listing 6-11 shows the full configuration for our Quartz components.

Listing 6-11. Quartz Configuration

```
...
@Configuration
public class QuartzConfiguration {

    @Bean
    public JobDetail quartzJobDetail() {
        return JobBuilder.newJob(BatchScheduledJob.class)
                    .storeDurably()
                    .build();
    }

    @Bean
    public Trigger jobTrigger() {
        SimpleScheduleBuilder scheduleBuilder = SimpleScheduleBuilder.simpleSchedule()
                    .withIntervalInSeconds(5).withRepeatCount(4);

        return TriggerBuilder.newTrigger()
                    .forJob(quartzJobDetail())
                    .withSchedule(scheduleBuilder)
                    .build();
    }
}
```

With this configuration added to our project, we have all that we need to run our job. When we run our Spring Boot application, we'll see the output in Listing 6-12 repeated five times.

Listing 6-12. Quartz Output

```
...
2018-02-16 12:00:13.723  INFO 78906 --- [           main] i.s.b.quartzdemo.QuartzDemo
Application    : Started QuartzDemoApplication in 1.577 seconds (JVM running for 2.05)
2018-02-16 12:00:13.759  INFO 78906 --- [eduler_Worker-1] o.s.b.c.l.support.SimpleJob
Launcher       : Job: [SimpleJob: [name=job]] launched with the following parameters: [{run.id=1}]
2018-02-16 12:00:13.769  INFO 78906 --- [eduler_Worker-1] o.s.batch.core.job.SimpleStep
Handler        : Executing step: [step1]
step1 ran!
2018-02-16 12:00:13.779  INFO 78906 --- [eduler_Worker-1] o.s.b.c.l.support.SimpleJob
Launcher       : Job: [SimpleJob: [name=job]] completed with the following parameters: [{run.
id=1}] and the following status: [COMPLETED]
...
```

Starting a job isn't the only aspect of the job execution lifecycle. While ideally we want our jobs to come to a natural end, there are scenarios where that may not be the case. We may need to stop them. In the next section, we'll look at various mechanisms for stopping a running batch job.

Stopping a Job

A job can stop for a number of reasons, each of which has its own effect on what happens next. It can run to completion naturally (as all the examples have up to this point). You can programmatically stop the execution of a job during processing for some reason. You can stop a job externally (say, someone realizes something is wrong, and they need to stop the job to fix it). And of course, although you may never admit it, errors can occur that cause a job to stop execution. This section looks at how each of these scenarios plays out using Spring Batch and your options for what to do when each occurs. Let's begin with the most basic: a job running to its natural completion.

The Natural End

Up to this point, all of your jobs have run to their natural completion. That is, each Job has run all of its Steps until they returned a COMPLETED status and the Job itself returned an exit code of COMPLETED. What does this mean for a Job?

As you've seen, a Job can't be executed with the same parameter values more than once successfully. This is the *successful* part of that statement. When a Job has been run to the COMPLETED BatchStatus, a new JobInstance can't be created using the same JobParameters again. This is important to note because it dictates how you execute jobs. You've used the JobParametersIncrementer to increment parameters based on their run, which is a good idea, especially in jobs that are run based on a schedule of some kind. For example, if you have a job that is run daily, developing a JobParametersIncrementer implementation that increments a timestamp as a parameter makes sense. That way, each time the job is executed via the schedule, the job is incremented accordingly as we saw in the previous section.

Not all jobs execute to their natural ending every time. There are situations when you want to stop a job based on something that happens during processing (an integrity check at the end of a step fails, for example). In cases like this, you want to stop the job programmatically. The next section goes over this technique.

Programmatic Ending

Batch processing requires a series of checks and balances to be effective. When you're dealing with large amounts of data, you need to be able to validate what is happening as things are processing. It's one thing for a user to update their profile with the wrong address on a web application. That affects one user. However, what if your job is to import a file containing one million records, and the import step completes after importing only 10,000? Something is wrong, and you need to fix it before the job goes any further. This section looks at how to stop a job programmatically. First you look at a more real-world example of using the stop transition introduced in Chapter 4; you join its use with some new attributes in order to restart the job. You also look at how to set a flag to end a job.

Using the Stop Transition

To begin, let's look at constructing a job that is configured to stop using the stop transition and how where to restart is addressed. Let's create a three-step job to see this in action:

1. Import a simple transaction file (transaction.csv). Each transaction consists of an account number, a timestamp, and an amount (positive is a credit, negative is a debit). The file ends with a single summary record containing the number of records in the file.

2. After importing the transactions into a transaction table, apply them to a separate account summary table that consists of the account number and the current account balance.

3. Generate a summary file (summary.csv) that lists the account number and balance for each account.

Looking at these steps from a design perspective, you want to validate that the number of records you import matches the summary record before applying the transactions to each user's account. This integrity check can save you many hours of recovery and reprocessing when dealing with large amounts of data.

To start this job, let's look at the file formats and data model. The file format for this job is simple comma-separated value (CSV) files. This lets you easily configure the appropriate readers and writers with no code. Listing 6-13 shows example record formats for each of the two files you're using (transaction.csv and summary.csv, respectively).

Listing 6-13. Sample Records for Each of the Two Files

```
Transaction file:
3985729387,2010-01-08 12:15:26,523.65
3985729387,2010-01-08 1:28:58,-25.93
2

Summary File:
3985729387,497.72
```

For this example, you also keep the data model simple, consisting of only two tables: TRANSACTION and ACCOUNT_SUMMARY. Figure 6-2 shows the data model.

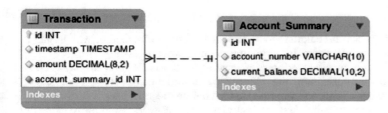

Figure 6-2. *Transaction data model*

To create the Job, we'll begin with a fresh project from Spring Initializr loaded into our IDE with Batch, JDBC, and MySQL dependencies selected. With that in place, we can create a configuration containing all we need for our Steps in the Job.

To start, configure application.yml to use the MySQL as we have in previous examples (Listing 6-3, for example). There are a few custom components we'll need for this Job. Specifically, a custom ItemReader, a custom ItemProcessor, two domain objects, and a data access object (DAO). Let's walk through each of these before assembling them (and other components) into our batch job.

We'll start with the domain objects. As you'd expect, each domain object maps one to one to the tables in our database. They also happen to map well to the files we'll be working with. Listing 6-14 shows the listings for both the Transaction and AccountSummary domain objects.

Listing 6-14. Domain Objects

```
...
public class Transaction {

    private String accountNumber;

    private Date timestamp;

    private double amount;

    public String getAccountNumber() {
        return accountNumber;
    }

    public void setAccountNumber(String accountNumber) {
        this.accountNumber = accountNumber;
    }

    public Date getTimestamp() {
        return timestamp;
    }

    public void setTimestamp(Date timestamp) {
        this.timestamp = timestamp;
    }
```

```
        public double getAmount() {
                return amount;
        }

        public void setAmount(double amount) {
                this.amount = amount;
        }
}

...
public class AccountSummary {

        private int id;

        private String accountNumber;

        private Double currentBalance;

        public int getId() {
                return id;
        }

        public void setId(int id) {
                this.id = id;
        }

        public String getAccountNumber() {
                return accountNumber;
        }

        public void setAccountNumber(String accountNumber) {
                this.accountNumber = accountNumber;
        }

        public Double getCurrentBalance() {
                return currentBalance;
        }
        public void setCurrentBalance(Double currentBalance) {
                this.currentBalance = currentBalance;
        }
}
```

The next component we'll look at is our custom ItemReader. We're reading in a csv, which Spring Batch has robust facilities to do, so why do we need a custom ItemReader? The reason is that the ExitStatus of the step is tied to the state of the reader. If we don't read in the same number of records as the footer record specifies, we should not continue. So we're going to wrap a FlatFileItemReader with our custom ItemReader. This custom ItemReader will count the number of records read in. Once it gets to the footer record, if the number of expected records match, processing will continue. However, if they do not, our custom ItemReader will also provide an AfterStep method that will set the ExitStatus to STOPPED. Listing 6-15 illustrates how this is accomplished.

Listing 6-15. TransactionReader

```
...
public class TransactionReader implements ItemStreamReader<Transaction> {

    private ItemStreamReader<FieldSet> fieldSetReader;
    private int recordCount = 0;
    private int expectedRecordCount = 0;

    public TransactionReader(ItemStreamReader<FieldSet> fieldSetReader) {
        this.fieldSetReader = fieldSetReader;
    }

    public Transaction read() throws Exception {
        return process(fieldSetReader.read());
    }

    private Transaction process(FieldSet fieldSet) {
        Transaction result = null;

        if(fieldSet != null) {
            if(fieldSet.getFieldCount() > 1) {
                result = new Transaction();
                result.setAccountNumber(fieldSet.readString(0));
                result.setTimestamp(fieldSet.readDate(1, "yyyy-MM-DD HH:mm:ss"));
                result.setAmount(fieldSet.readDouble(2));

                recordCount++;
            } else {
                expectedRecordCount = fieldSet.readInt(0);
            }
        }

        return result;
    }
    public void setFieldSetReader(ItemStreamReader<FieldSet> fieldSetReader) {
        this.fieldSetReader = fieldSetReader;
    }

    @AfterStep
    public ExitStatus afterStep(StepExecution execution) {
        if(recordCount == expectedRecordCount) {
            return execution.getExitStatus();
        } else {
            return ExitStatus.STOPPED;
        }
    }

    @Override
    public void open(ExecutionContext executionContext) throws ItemStreamException {
        this.fieldSetReader.open(executionContext);
    }
```

```
    @Override
    public void update(ExecutionContext executionContext) throws ItemStreamException {
        this.fieldSetReader.update(executionContext);
    }

    @Override
    public void close() throws ItemStreamException {
        this.fieldSetReader.close();
    }
}
```

From the top of the TransactionReader listed in Listing 6-15, the ItemReader#read() method delegates to the reader we inject. In our case, it will be a FlatFileItemReader. We let the delegate return a FieldSet because we actually have two record formats. One for the data we're importing and one for the footer that contains the number of records in the file. We pass the FieldSet returned from the delegate ItemReader to a method called process. There we're determining what type of record it is. If there is more than one value in the record, it's a data record. If there is only one field in the record, it's the footer record. Data records are transformed into a Transaction instance and returned. The footer record's value is recorded and null is returned indicating that we're done processing the file.

After the process method, we have the implementation of the StepExecutionListener. afterStep(StepExecution stepExecution). This method will be called once our step is complete giving us the opportunity to return a specific ExitStatus. In our case, this method will look at the number of records read and compare that with the value saved in the footer of the file. If they match, it will return the ExitStatus set by the framework. Otherwise, it will return ExitStatus.STOPPED. This will allow us to stop the Job from continuing if the file is invalid.

The rest of the methods in our TransactionItemReader are the implementation of the ItemStream interface. Spring Batch will automatically look at the ItemReader, ItemProcessor, and ItemWriter to see if it's an ItemStream and register it automatically to have the callbacks executed at the appropriate time. However, we have a delegate that implements ItemStream. This delegate ItemReader won't be explicitly registered with Spring Batch so the framework won't look to see if it implements ItemStream. This leaves us with two options. Either we remember to explicitly register the delegate as an ItemStream on our job (an error-prone approach since it requires us to remember to do the registration) or implement ItemStream in our TransactionItemReader and have it call the appropriate lifecycle methods on the delegate which is what this does.[2]

The next custom component is the TransactionDao. This is an interface with a single method, getTransactionsByAccountNumber(String accountNumber). This method returns a List of Transactions associated with the account number provided. Listing 6-16 illustrates the implementation of this DAO.

Listing 6-16. TransactionDaoSupport

```
...
public class TransactionDaoSupport extends JdbcTemplate implements TransactionDao {

    public TransactionDaoSupport(DataSource dataSource) {
        super(dataSource);
    }
```

---

[2]The ItemStream interface will be covered later in Chapters 7 and 9.

```
@SuppressWarnings("unchecked")
public List<Transaction> getTransactionsByAccountNumber(String accountNumber) {
    return query(
                "select t.id, t.timestamp, t.amount " +
                        "from transaction t inner join account_summary a on " +
                        "a.id = t.account_summary_id " +
                        "where a.account_number = ?",
                new Object[] { accountNumber },
                (rs, rowNum) -> {
                    Transaction trans = new Transaction();
                    trans.setAmount(rs.getDouble("amount"));
                    trans.setTimestamp(rs.getDate("timestamp"));
                    return trans;
                }
    );
    }
}
```

The DAO in Listing 6-16 selects all `Transaction` records associated with the `accountNumber` provided and returns them. We'll use it in an `ItemProcessor` that will apply all the transactions to a given account to determine their current balance. Listing 6-17 shows the `ItemProcessor` that uses the `TransactionDAO` in use.

Listing 6-17. TransactionApplierProcessor

```
...
public class TransactionApplierProcessor implements
            ItemProcessor<AccountSummary, AccountSummary> {

    private TransactionDao transactionDao;

    public TransactionApplierProcessor(TransactionDao transactionDao) {
        this.transactionDao = transactionDao;
    }

    public AccountSummary process(AccountSummary summary) throws Exception {
        List<Transaction> transactions = transactionDao
                    .getTransactionsByAccountNumber(summary.getAccountNumber());

        for (Transaction transaction : transactions) {
            summary.setCurrentBalance(summary.getCurrentBalance()
                            + transaction.getAmount());
        }
        return summary;
    }
}
```

As Listing 6-17 shows, for each `AccountSummary` record that is passed to this `ItemProcessor`, all the transactions will be looked up with the `TransactionDao`, and the current balance of the account will be incremented or decremented according to the `Transaction`.

That's all the custom batch components we need to write for this job. The next step is to configure them all. We'll start by configuring each `Step`, then assembling the `Steps` into our `Job`. The first `Step` is the `importTransactionFileStep` as shown in Listing 6-18.

Listing 6-18. importTransactionFileStep

```
...
    @Bean
    @StepScope
    public TransactionReader transactionReader() {
        return new TransactionReader(fileItemReader(null));
    }

    @Bean
    @StepScope
    public FlatFileItemReader<FieldSet> fileItemReader(
    @Value("#{jobParameters['transactionFile']}") Resource inputFile) {
        return new FlatFileItemReaderBuilder<FieldSet>()
                    .name("fileItemReader")
                    .resource(inputFile)
                    .lineTokenizer(new DelimitedLineTokenizer())
                    .fieldSetMapper(new PassThroughFieldSetMapper())
                    .build();
    }

    @Bean
    public JdbcBatchItemWriter<Transaction> transactionWriter(DataSource dataSource) {
        return new JdbcBatchItemWriterBuilder<Transaction>()
                    .itemSqlParameterSourceProvider(
                        new BeanPropertyItemSqlParameterSourceProvider<>())
                    .sql("INSERT INTO TRANSACTION " +
                        "(ACCOUNT_SUMMARY_ID, TIMESTAMP, AMOUNT) " +
                        "VALUES ((SELECT ID FROM ACCOUNT_SUMMARY " +
                        "    WHERE ACCOUNT_NUMBER = :accountNumber), " +
                        ":timestamp, :amount)")
                    .dataSource(dataSource)
                    .build();
    }

    @Bean
    public Step importTransactionFileStep() {
        return this.stepBuilderFactory.get("importTransactionFileStep")
                    .<Transaction, Transaction>chunk(100)
                    .reader(transactionReader())
                    .writer(transactionWriter(null))
                    .allowStartIfComplete(true)
                    .listener(transactionReader())
                    .build();
    }
...
```

The configuration for this first Step begins with the definition of the TransactionReader. This is the custom ItemReader we reviewed in Listing 6-15. From there, we configure the FlatFileItemReader. We'll cover the details of this more in Chapter 7 when review ItemReaders in detail. The JdbcBatchItemWriter configured next is the way we will write the values to the database. Again, this will be covered in more detail in Chapter 9 so we'll worry about the details then. For now, just realize that it's the way we write to the database. The final bean definition in Listing 6-18 is the definition of the Step itself. Using the StepBuilderFactory, we get a builder and configure it to be a chunk-based Step with the transaction reader and jdbc writer we just configured. We configure this Step to be rerunnable if the Job is restarted. The reason for this is that if the file we import is invalid (meaning the number of records doesn't match the footer record), we'd want to clear out this import and rerun it with a valid file. After configuring the ability to rerun this Step, we register our TransactionReader as a listener before building our Step.

That first Step was to import our file into the database. The second Step is to apply the transactions that were found in the file to the accounts. Listing 6-19 walks us through the configuration of that Step.

Listing 6-19. applyTransactionsStep

```
...
@Bean
@StepScope
public JdbcCursorItemReader<AccountSummary> accountSummaryReader(DataSource dataSource) {
    return new JdbcCursorItemReaderBuilder<AccountSummary>()
                .name("accountSummaryReader")
                .dataSource(dataSource)
                .sql("SELECT ACCOUNT_NUMBER, CURRENT_BALANCE " +
                        "FROM ACCOUNT_SUMMARY A " +
                        "WHERE A.ID IN (" +
                        "    SELECT DISTINCT T.ACCOUNT_SUMMARY_ID " +
                        "    FROM TRANSACTION T) " +
                        "ORDER BY A.ACCOUNT_NUMBER")
                .rowMapper((resultSet, rowNumber) -> {
                    AccountSummary summary = new AccountSummary();

                    summary.setAccountNumber(resultSet.getString("account_number"));
                    summary.setCurrentBalance(resultSet.getDouble("current_balance"));

                    return summary;
                }).build();
    }

@Bean
public TransactionDao transactionDao(DataSource dataSource) {
    return new TransactionDaoSupport(dataSource);
}

@Bean
public TransactionApplierProcessor transactionApplierProcessor() {
    return new TransactionApplierProcessor(transactionDao(null));
}
```

```
@Bean
public JdbcBatchItemWriter<AccountSummary> accountSummaryWriter(DataSource dataSource) {
        return new JdbcBatchItemWriterBuilder<AccountSummary>()
                        .dataSource(dataSource)
                        .itemSqlParameterSourceProvider(
                                        new BeanPropertyItemSqlParameterSourceProvider<>())
                        .sql("UPDATE ACCOUNT_SUMMARY " +
                                        "SET CURRENT_BALANCE = :currentBalance " +
                                        "WHERE ACCOUNT_NUMBER = :accountNumber")
                        .build();
}
@Bean
public Step applyTransactionsStep() {
return this.stepBuilderFactory.get("applyTransactionsStep")
                .<AccountSummary, AccountSummary>chunk(100)
                .reader(accountSummaryReader(null))
                .processor(transactionApplierProcessor())
                .writer(accountSummaryWriter(null))
                .build();
}
...
```

Beginning at the top of Listing 6-19, we define a JdbcCursorItemReader to read the AccountSummary records from the database. The next two bean definitions are for the TransactionDao, which looks up the transactions, and the custom ItemProcessor reviewed in Listing 6-17 that applies the transactions to the accounts. Finally, the updated account summary records are written with a JdbcBatchItemWriter. With those components configured, we can assemble them into our Step. The applyTransactionsStep uses the StepBuilderFactory to obtain a builder and configure a chunk-based step with a chunk size of 100 records, and the ItemReader, ItemProcessor, and ItemWriter configured previously.

The final Step, the generateAccountSummaryStep, actually reuses the ItemReader from the applyTransactionsStep since we're reading the same data (just doing something different with it). This is why the accountSummaryReader is step scoped, so we get a new instance for each step. So all we really need to configure for the generateAccountSummaryStep is the ItemWriter and the Step itself. Listing 6-20 has that code.

Listing 6-20. generateAccountSummaryStep

```
...
    @Bean
    @StepScope
    public FlatFileItemWriter<AccountSummary> accountSummaryFileWriter(
                    @Value("#{jobParameters['summaryFile']}") Resource summaryFile) {

            DelimitedLineAggregator<AccountSummary> lineAggregator =
                            new DelimitedLineAggregator<>();
            BeanWrapperFieldExtractor<AccountSummary> fieldExtractor =
                            new BeanWrapperFieldExtractor<>();
            fieldExtractor.setNames(new String[] {"accountNumber", "currentBalance"});
            fieldExtractor.afterPropertiesSet();
            lineAggregator.setFieldExtractor(fieldExtractor);
```

```
                return new FlatFileItemWriterBuilder<AccountSummary>()
                        .name("accountSummaryFileWriter")
                        .resource(summaryFile)
                        .lineAggregator(lineAggregator)
                        .build();
    }

@Bean
public Step generateAccountSummaryStep() {
    return this.stepBuilderFactory.get("generateAccountSummaryStep")
                .<AccountSummary, AccountSummary>chunk(100)
                .reader(accountSummaryReader(null))
                .writer(accountSummaryFileWriter(null))
                .build();
}
...
```

Listing 6-20 starts off configuring the ItemWriter. The FlatFileItemWriter generates a CSV with the account number and current balance in each record. The Step is then assembled using the StepBuilderFactory to obtain a builder and configure a chunk-based Step with the reader used in the previous Step (accountSummaryReader) and the ItemWriter we just configured.

The last piece of this puzzle is the configuration of the Job itself. This Job needs to have the three Steps in sequence, but handle the STOPPED ExitStatus that could be returned by the first Step. Listing 6-21 has the code to build our stoppable Job.

Listing 6-21. transactionJob

```
...
@Bean
public Job transactionJob() {
    return this.jobBuilderFactory.get("transactionJob")
                .start(importTransactionFileStep())
                .on("STOPPED").stopAndRestart(importTransactionFileStep())
                .from(importTransactionFileStep()).on("*").to(applyTransactionsStep())
                .from(applyTransactionsStep()).next(generateAccountSummaryStep())
                .end()
                .build();
}
...
```

Listing 6-21 begins by using the JobBuilderFactory to get a builder and configures the job to start with our importTransactionFileStep. From there, it says if the ExitStatus is STOPPED, stop the job and restart it back at the same step (in essence, start the job over again if it is programmatically stopped). On all other conditions, go to the applyTransactionsStep. From the applyTransactionsStep, transition to the generateAccountSummaryStep. The call to end() is required because we were building a flow using the transitions APIs. We then call build() to generate our Job.

Now, execute the job twice. The first time, execute the job with a transaction.csv that has an invalid integrity record. In other words, you run the job with an input file of 100 records plus an integrity record at the end. The integrity record is any number other than 100; here you use the number 20. When the job executes, the StepListener validates that the number of records you read in (100) doesn't match the number expected (20) and returns the value ExitStatus.STOPPED, stopping the job. You can see the results of the job in the console as shown in Listing 6-22.

Listing 6-22. `transactionJob` First Run

```
...
2018-03-01 22:02:35.770  INFO 36810 --- [           main] o.s.b.a.b.JobLauncherCommandLine
Runner    : Running default command line with: [transactionFile=/data/transactions.csv,
summaryFile=file://Users/mminella/tmp/summary.xml]
2018-03-01 22:02:35.873  INFO 36810 --- [           main] o.s.b.c.l.support.SimpleJob
Launcher      : Job: [FlowJob: [name=transactionJob]] launched with the following
parameters: [{transactionFile=/data/transactions.csv, summaryFile=file://Users/mminella/tmp/
summary.xml}]
2018-03-01 22:02:35.918  INFO 36810 --- [           main] o.s.batch.core.job.SimpleStep
Handler     : Executing step: [importTransactionFileStep]
2018-03-01 22:03:16.435  INFO 36810 --- [           main] o.s.b.c.l.support.
SimpleJobLauncher       : Job: [FlowJob: [name=transactionJob]] completed with the following
parameters: [{transactionFile=/data/transactions.csv, summaryFile=file://Users/mminella/tmp/
summary.xml}] and the following status: [STOPPED]
...
```

When the job stops, delete the contents of the TRANSACTION table and update your transaction file to have 100 records and an integrity record say 100 as well. This time, when you execute the job, as Listing 6-23 shows, it runs to completion successfully.

Listing 6-23. `transactionJob` Second Run

```
...
2018-03-01 22:04:17.102  INFO 36815 --- [           main] o.s.b.c.l.support.SimpleJob
Launcher       : Job: [FlowJob: [name=transactionJob]] launched with the following
parameters: [{transactionFile=/data/transactions.csv, summaryFile=file://Users/mminella/tmp/
summary.xml}]
2018-03-01 22:04:17.122  INFO 36815 --- [           main] o.s.batch.core.job.SimpleStep
Handler     : Executing step: [importTransactionFileStep]
2018-03-01 22:05:02.977  INFO 36815 --- [           main] o.s.batch.core.job.SimpleStep
Handler     : Executing step: [applyTransactionsStep]
2018-03-01 22:05:53.729  INFO 36815 --- [           main] o.s.batch.core.job.SimpleStep
Handler     : Executing step: [generateAccountSummaryStep]
2018-03-01 22:05:53.822  INFO 36815 --- [           main] o.s.b.c.l.support.
SimpleJobLauncher       : Job: [FlowJob: [name=transactionJob]] completed with the following
parameters: [{transactionFile=/data/transactions.csv, summaryFile=file://Users/mminella/tmp/
summary.xml}] and the following status: [COMPLETED]
...
```

Using the `stop` transition along with configuring the ability to re-execute steps in the job is a useful way to allow for issues to be fixed based on checks in the execution of a job. In the next section, you refactor the listener to use the `StepExecution#setTerminateOnly()` method to communicate to Spring Batch to end the job.

Stopping with `StepExecution`

In the `transactionJob` example, you manually handled stopping the job by using the `ExitStatus` of a `StepExecutionListener` and the configured transitions in the job. Although this approach works, it requires you to specially configure the job's transitions and override the step's `ExitStatus`.

There is a slightly cleaner approach. We replace the afterStep with a beforeStep to get a handle on the StepExecution. Once you have access to that, when we read the footer record, you can call the StepExecution#setTerminateOnly() method. This method sets a flag that tells Spring Batch to end after the step is complete as shown in Listing 6-24.

Listing 6-24. TransactionReader with setTerminateOnly() Call

```
...
public class TransactionReader implements ItemStreamReader<Transaction> {

    private ItemStreamReader<FieldSet> fieldSetReader;
    private int recordCount = 0;
    private int expectedRecordCount = 0;

    private StepExecution stepExecution;

    public TransactionReader(ItemStreamReader<FieldSet> fieldSetReader) {
        this.fieldSetReader = fieldSetReader;
    }

    public Transaction read() throws Exception {
        Transaction record = process(fieldSetReader.read());

        return record;
    }

    private Transaction process(FieldSet fieldSet) {
        Transaction result = null;

        if(fieldSet != null) {
            if(fieldSet.getFieldCount() > 1) {
                result = new Transaction();
                result.setAccountNumber(fieldSet.readString(0));
                result.setTimestamp(fieldSet.readDate(1, "yyyy-MM-DD HH:mm:ss"));
                result.setAmount(fieldSet.readDouble(2));

                recordCount++;
            } else {
                expectedRecordCount = fieldSet.readInt(0);

                if(expectedRecordCount != this.recordCount) {
                    this.stepExecution.setTerminateOnly();
                }
            }
        }

        return result;
    }
```

```
@BeforeStep
public void beforeStep(StepExecution execution) {
        this.stepExecution = execution;
}

@Override
public void open(ExecutionContext executionContext) throws ItemStreamException {
        this.fieldSetReader.open(executionContext);
}

@Override
public void update(ExecutionContext executionContext) throws ItemStreamException {
        this.fieldSetReader.update(executionContext);
}

@Override
public void close() throws ItemStreamException {
        this.fieldSetReader.close();
}
}
```

Although the code is only marginally cleaner (you move the check of the record count from the afterStep(StepExecution execution) to the read() method), the configuration becomes cleaner as well by allowing you to remove the configuration required for the transitions. Listing 6-25 shows the updated Job configuration.

Listing 6-25. Reconfigured transactionJob

```
...
@Bean
public Job transactionJob() {
      return this.jobBuilderFactory.get("transactionJob")
                  .start(importTransactionFileStep())
                  .next(applyTransactionsStep())
                  .next(generateAccountSummaryStep())
                  .build();
}
...
```

You can now execute the job again with the same test (running it the first time with an incorrect number of records in the transaction file and then a second time with the correct number) and see the same results. The only difference is in the output of the job on the console. Instead of the job returning a STOPPED status, Spring Batch throws a JobInterruptedException, as shown in Listing 6-26.

Listing 6-26. Results of the First Execution of Your Updated Job

```
2018-03-01 22:25:19.070  INFO 36931 --- [          main] o.s.b.c.l.support.SimpleJob
Launcher     : Job: [SimpleJob: [name=transactionJob]] launched with the following
parameters: [{transactionFile=/data/transactions.csv, summaryFile=file://Users/mminella/tmp/
summary.csv}]
2018-03-01 22:25:19.118  INFO 36931 --- [          main] o.s.batch.core.job.SimpleStep
Handler      : Executing step: [importTransactionFileStep]
```

```
2018-03-01 22:26:05.265  INFO 36931 --- [            main] o.s.b.c.s.ThreadStepInterruption
Policy   : Step interrupted through StepExecution
2018-03-01 22:26:05.266  INFO 36931 --- [            main] o.s.batch.core.step.AbstractStep
: Encountered interruption executing step importTransactionFileStep in job transactionJob :
Job interrupted status detected.
2018-03-01 22:26:05.274  INFO 36931 --- [            main] o.s.batch.core.job.AbstractJob
: Encountered interruption executing job: Job interrupted by step execution
2018-03-01 22:26:05.277  INFO 36931 --- [            main] o.s.b.c.l.support.
SimpleJobLauncher       : Job: [SimpleJob: [name=transactionJob]] completed with the
following parameters: [{transactionFile=/data/transactions.csv, summaryFile=file://Users/
mminella/tmp/summary.csv}] and the following status: [STOPPED]
```

Stopping a job programmatically is an important tool when you're designing batch jobs. Unfortunately, not all batch jobs are perfect, how does Spring Batch handle errors? Let's take a look at the basics in the next section.

Error Handling

No job is perfect—not even yours. Errors happen. You may receive bad data. You may forget one null check that causes a NullPointerException at the worst of times. How you handle errors using Spring Batch is important. This section discusses the options for what to do when an exception is thrown during your batch job and how to implement them.

Job Failure

It should come as little surprise that the default behavior of Spring Batch is probably the safest: stopping the job and rolling back the current commit. This is one of the driving concepts of chunk-based processing. It allows you to commit the work you've successfully completed and pick up where you left off when you restart.

By default, Spring Batch considers a step and job failed when any exception is thrown. You can see this in action by tweaking TransactionReader as shown in Listing 6-27. In this case, you throw a org. springframework.batch.item.ParseException after reading 510 records, stopping the job in a FAILED status.

Listing 6-27. TransactionReader Set Up to Throw an Exception

```
...
public class TransactionReader implements ItemStreamReader<Transaction> {

    private ItemStreamReader<FieldSet> fieldSetReader;
    private int recordCount = 0;
    private int expectedRecordCount = 0;

    private StepExecution stepExecution;

    public TransactionReader(ItemStreamReader<FieldSet> fieldSetReader) {
        this.fieldSetReader = fieldSetReader;
    }

    public Transaction read() throws Exception {
        if(this.recordCount == 25) {
            throw new ParseException("This isn't what I hoped to happen");
        }
```

```
            Transaction record = process(fieldSetReader.read());

            return record;
        }

    private Transaction process(FieldSet fieldSet) {
        Transaction result = null;

        if(fieldSet != null) {
                if(fieldSet.getFieldCount() > 1) {
                    result = new Transaction();
                    result.setAccountNumber(fieldSet.readString(0));
                    result.setTimestamp(fieldSet.readDate(1, "yyyy-MM-DD HH:mm:ss"));
                    result.setAmount(fieldSet.readDouble(2));

                    recordCount++;
                } else {
                    expectedRecordCount = fieldSet.readInt(0);

                    if(expectedRecordCount != this.recordCount) {
                        this.stepExecution.setTerminateOnly();
                    }
                }
            }

        return result;
    }

    @BeforeStep
    public void beforeStep(StepExecution execution) {
        this.stepExecution = execution;
    }

    @Override
    public void open(ExecutionContext executionContext) throws ItemStreamException {
        this.fieldSetReader.open(executionContext);
    }

    @Override
    public void update(ExecutionContext executionContext) throws ItemStreamException {
        this.fieldSetReader.update(executionContext);
    }

    @Override
    public void close() throws ItemStreamException {
        this.fieldSetReader.close();
    }
}
```

With no other configuration changes from the previous runs, when you execute transactionJob, it throws a ParseException after it reads the record 25 of the transaction file. After this exception is thrown, Spring Batch considers the step and job failed. If we look at the console, we can see the exception thrown and that the job stops processing.

There is a big difference between the examples of stopping via StepExecution and stopping the job with an exception. That difference is the state in which the job is left. In the StepExecution example, the job was stopped after a step is complete in the STOPPED ExitStatus. In the exception case, the step didn't finish. In fact, it was part way through the step when the exception was thrown. Because of this, the step and job are labeled with the ExitStatus FAILED.

When a step is identified as FAILED, Spring Batch doesn't start the step over from the beginning. Instead, Spring Batch is smart enough to remember what chunk you were on when the exception was thrown. When you restart the job, Spring Batch picks up at the chunk it left off on. As an example, let's say the job is processing chunk five of ten, with each chunk consisting of five items. An exception is thrown on the fourth item of the second chunk. Items one to four of the current chunk are rolled back, and when you restart, Spring Batch skips chunks one and two.

Although Spring Batch's default method of handling an exception is to stop the job in the failed status, there are other options at your disposal. Because most of those depend on input/output-specific scenarios, the next few chapters cover them together with I/O.

Controlling Restart

Spring Batch provides many facilities to address stopping and restarting a job, as you've seen. However, it's up to you to determine what can and can't be restarted. If you have a batch process that imports a file in the first step, and that job fails in the second step, you probably do not want to reimport the file. There are scenarios where you may only want to retry a step a given number of times. This section looks at how to configure a job to be restartable and how to control how it's restarted.

Preventing a Job from Being Rerun

All the jobs up to now could be executed again if they failed or were stopped. This is the default behavior of Spring Batch. But what if you have a job that can't be rerun? You give it one try, and if it works, great. If not, you don't run it again. Spring Batch provides the ability to configure jobs to not be restartable using the preventRestart() call on the JobBuilder.

If you look at the transactionJob configuration, by default you can restart the job. However, if you choose to call the preventRestart() method, as shown in Listing 6-28, then when the job fails or is stopped for any reason, you won't be able to re-execute it.

Listing 6-28. transactionJob Configured to Not Be Restartable

```
...
@Bean
public Job transactionJob() {
        return this.jobBuilderFactory.get("transactionJob")
                    .preventRestart()
                    .start(importTransactionFileStep())
                    .next(applyTransactionsStep())
                    .next(generateAccountSummaryStep())
                    .build();
}
...
```

Now if you attempt to run the job after a failure, you're told by Spring Batch that the JobInstance already exists and isn't restartable, as shown in Listing 6-29.

Listing 6-29. Results from Re-executing a Nonrestartable Job

```
2018-03-01 23:08:49.251  INFO 37017 --- [                main] ConditionEvaluationReportLogging
Listener :

Error starting ApplicationContext. To display the conditions report re-run your application
with 'debug' enabled.
2018-03-01 23:08:49.271 ERROR 37017 --- [                main] o.s.boot.SpringApplication        :
Application run failed

java.lang.IllegalStateException: Failed to execute CommandLineRunner
        at org.springframework.boot.SpringApplication.callRunner(SpringApplication.java:793)
        [spring-boot-2.0.0.RELEASE.jar:2.0.0.RELEASE]
        at org.springframework.boot.SpringApplication.callRunners(SpringApplication.java:774)
        [spring-boot-2.0.0.RELEASE.jar:2.0.0.RELEASE]
        at org.springframework.boot.SpringApplication.run(SpringApplication.java:335) [spring-
        boot-2.0.0.RELEASE.jar:2.0.0.RELEASE]
        at org.springframework.boot.SpringApplication.run(SpringApplication.java:1246)
        [spring-boot-2.0.0.RELEASE.jar:2.0.0.RELEASE]
        at org.springframework.boot.SpringApplication.run(SpringApplication.java:1234)
        [spring-boot-2.0.0.RELEASE.jar:2.0.0.RELEASE]
        at io.spring.batch.transaction_stop.TransactionStopApplication.
        main(TransactionStopApplication.java:20) [classes/:na]
Caused by: org.springframework.batch.core.repository.JobRestartException: JobInstance
already exists and is not restartable
        at org.springframework.batch.core.launch.support.SimpleJobLauncher.
        run(SimpleJobLauncher.java:101) ~[spring-batch-core-4.0.0.RELEASE.jar:4.0.0.RELEASE]
        at sun.reflect.NativeMethodAccessorImpl.invoke0(Native Method) ~[na:1.8.0_131]
        at sun.reflect.NativeMethodAccessorImpl.invoke(NativeMethodAccessorImpl.java:62)
        ~[na:1.8.0_131]
        at sun.reflect.DelegatingMethodAccessorImpl.invoke(DelegatingMethodAccessorImpl.java:43)
        ~[na:1.8.0_131]
        at java.lang.reflect.Method.invoke(Method.java:498) ~[na:1.8.0_131]
```

Being able to execute a job once or else may be a bit extreme for some scenarios. Spring Batch also lets you configure the number of times a job can be run, as you see next.

Configuring the Number of Restarts

There can be situations where a job doesn't run successfully for some reason outside of your control. Say, for example, that the job downloads a file from a web site as one of its steps, and the web site is down. If the download fails the first time, it may work if you try again in 10 minutes. However, you probably don't want to try that download indefinitely. Because of this, you may want to configure the job so it can be executed only five times. After the fifth time, it can't be rerun any more.

Spring Batch provides this facility at the step level instead of the job level. Again, looking at the transactionJob example, if you only want to attempt to import an input file twice, you modify the step configuration as in Listing 6-30.

Listing 6-30. Allowing the File Import to Be Attempted Only Twice

```
...
@Bean
public Step importTransactionFileStep() {
        return this.stepBuilderFactory.get("importTransactionFileStep")
                        .startLimit(2)
                        .<Transaction, Transaction>chunk(100)
                        .reader(transactionReader())
                        .writer(transactionWriter(null))
                        .allowStartIfComplete(true)
                        .listener(transactionReader())
                        .build();
}
...
```

In this case, if you attempt to restart this job more than once, because the `.startLimit(int limit)` has been configured to 2, you can't re-execute this job. The initial run takes up one attempt you're allowed allowing you one other try. You receive an `org.springframework.batch.core.StartLimitExceededException`, as shown in Listing 6-31, if you attempt to execute the job again.

Listing 6-31. Results from Re-executing transactionJob More Than Once

```
...
2018-03-01 23:12:17.205 ERROR 37027 --- [          main] o.s.batch.core.job.Abstract
Job          : Encountered fatal error executing job

org.springframework.batch.core.StartLimitExceededException: Maximum start limit exceeded for
step: importTransactionFileStepStartMax: 2
        at org.springframework.batch.core.job.SimpleStepHandler.shouldStart(SimpleStepHandler.
        java:229) ~[spring-batch-core-4.0.0.RELEASE.jar:4.0.0.RELEASE]
...
```

The last configuration aspect that you can use when determining what should happen when your batch job is re-executed, you've seen before: the `allowStartIfComplete()` method.

Rerunning a Complete Step

One of Spring Batch's features (or detriments, depending on how you choose to look at it) is that the framework allows you to execute a job only once successfully with the same identifying parameters. There is no way around this. However, that rule doesn't necessarily apply to steps.

You can override the framework's default configuration and execute a step that has been completed more than once. You did it previously using `transactionJob`. To tell the framework that you want to be able to re-execute a step even if it has been completed, you use the `allowStartIfComplete()` method on the `StepBuilder`. Listing 6-32 shows an example.

Listing 6-32. Configuring a Step to Be Re-executed if Complete

```
...
@Bean
public Step importTransactionFileStep() {
        return this.stepBuilderFactory.get("importTransactionFileStep")
                        .allowStartIfComplete(true)
                        .<Transaction, Transaction>chunk(100)
                        .reader(transactionReader())
                        .writer(transactionWriter(null))
                        .allowStartIfComplete(true)
                        .listener(transactionReader())
                        .build();
}
...
```

In this case, when the step is executed for the second time within a job that failed or was stopped on the previous execution, the step starts over. Because it completed the previous time, there is no middle ground at which to restart, which is why it begins again at the beginning.

■ **Note**   If the job has the BatchStatus of COMPLETE, the JobInstance can't be rerun regardless of whether you configure all the steps to allowStartIfComplete(true);

When you're configuring batch processes, Spring Batch offers many different options for stopping and restarting jobs. Some scenarios can have the full job re-executed. Others can be tried again, but only a given number of times. And some can't be restarted at all. However, it's you, the developer, who must design your batch jobs in a way that is safe for your scenario.

Summary

Starting or stopping a program isn't a topic that typically gets much press. But as you've seen, controlling the execution of a Spring Batch process provides many options. And when you think about the variety of scenarios that must be supported by batch processes, those options make sense.

The next section of this book covers the meat of the framework: ItemReaders, ItemProcessors, and ItemWriters.

CHAPTER 7

■ ■ ■

ItemReaders

The three Rs, Reading, wRiting, and aRithmetic, are considered the basis of the skills children learn in schools. When you think about it, these same concepts apply to software as well. The foundations of any program—whether web applications, batch jobs, or anything else—are the input of data, the processing of it in some way, and the output of data.

This concept is no more obvious than when you use Spring Batch. Each chunk-based step consists of an `ItemReader`, an `ItemProcessor`, and an `ItemWriter`. Reading in any system is not always straightforward, however. There are a number of different formats in which input can be provided; flat files, XML, and databases of all kinds are just some of the potential input sources.

Spring Batch provides standard ways to handle most forms of input without the need to write code as well as the ability to develop your own readers for formats that are not supported, like reading a web service. This chapter will walk through the different features the different `ItemReader` implementations provide within the Spring Batch framework.

The `ItemReader` Interface

Up to this chapter we have vaguely discussed the concept of an `ItemReader` but we have not looked at the interface that Spring Batch uses to define input operations. The `org.springframework.batch.item.ItemReader<T>` interface defines a single method—`read()`—which is used to provide input for a step. Listing 7-1 shows the `ItemReader` interface.

Listing 7-1. `org.springframework.batch.item.ItemReader<T>`

```
package org.springframework.batch.item;

public interface ItemReader<T> {

    T read() throws Exception, UnexpectedInputException, ParseException,
                    NonTransientResourceException;
}
```

The `ItemReader` interface shown in Listing 7-1 is a strategy interface. Spring Batch provides a number of implementations based on the type of input to be processed. Flat files, databases, JMS resources, and other sources of input all have implementations provided by Spring Batch. You can also implement your own `ItemReader` by implementing the `ItemReader` or any one of its subinterfaces.

© Michael T. Minella 2019
M. T. Minella, *The Definitive Guide to Spring Batch*, https://doi.org/10.1007/978-1-4842-3724-3_7

■ **Note** Spring Batch's `org.springframework.batch.item.ItemReader` interface is different from `javax.batch.api.chunk.ItemReader` from JSR-352 (JBatch). The main differences being that the Spring Batch version provides generic support and the JSR version combines the `ItemStream` and `ItemReader` interfaces. This book will only use the one from Spring Batch.

The `read()` method of the `ItemReader` interface returns a single item to be processed by your step as it is called by Spring Batch. This item is what your step will count as it maintains how many items within a chunk have been processed. The item will be passed to any configured `ItemProcessor` before being sent as part of a chunk to the `ItemWriter`.

The best way to understand how to use the `ItemReader` interface is to put it to use. In the next section you will begin to look at the many `ItemReader` implementations provided by Spring Batch by working with the `FlatFileItemReader`.

File Input

When I think of file IO in Java, I can't help but cringe. The API for IO is marginally better than the API for handling dates in this language (although that has gotten much better in recent years), and you all know how good that is. Luckily, the guys at Spring Batch have addressed most of this by providing a number of declarative readers that allow you to declare the format of what you're going to read and they handle the rest. In this section, you'll be looking at the declarative readers that Spring Batch provides and how to configure them for file-based IO.

Flat Files

When I talk about flat files in the case of batch processes, I'm talking about any file that has one or more records. Each record can take up one or more lines. The difference between a flat file and an XML file is that the data within a flat file is nondescriptive. In other words, there is no metainformation within the file itself to define the format or meaning of the data. In contrast, in XML, you use tags to give the data meaning.

Before you get into actually configuring an `ItemReader` for a flat file, let's take a look at the components of reading a file in Spring Batch. Figure 7-1 shows the components of the `FlatFileItemReader`. The `org.springframework.batch.item.file.FlatFileItemReader` consists of two main components: a Spring `Resource` that represents the file to be read and an implementation of the `org.springfamework.batch.item.file.LineMapper` interface. The `LineMapper` serves a similar function as the `RowMapper` does in Spring JDBC. When using a `RowMapper` in Spring JDBC, a `ResultSet` representing a collection of fields is provided for you map to objects.

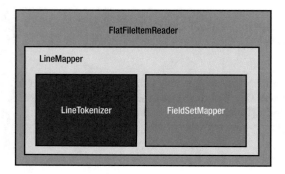

Figure 7-1. `FlatFileItemReader` *pieces*

The FlatFileItemReader allows you to configure a number of attributes about the file you're reading. Table 7-1 shows the options that you are likely to use and explains their meanings.

Table 7-1. *FlatFileItemReader Configuration Options*

Option	Type	Default	Description
comments	String []	null	This array of strings indicates what prefixes will be considered line comments and skipped during file parsing.
currentItemCount	int	0	The current index of the item being read. Used on restarts.
encoding	String	The default Charset for the platform.	The character encoding for the file.
lineMapper	LineMapper	null (required)	This class will take each line of a file as a String and convert it into a domain object (item) to be processed.
linesToSkip	int	0	When running a job, the FlatFileItemReader can be configured to skip lines at the beginning of the file before parsing. This number indicates how many.
maxItemCount	int	Integer.MAX_VALUE	Indicates the maximum number of items to be read from the file.
name	String	null	Used to create a unique key for the values persisted in the ExecutionContext.
record SeparatorPolicy	Record SeparatorPolicy	DefaultRecord SeparatorPolicy	Used to determine the end of each record. By default, an end of line character indicates the end of a record; however, this can be used to determine how to handle things like quoted strings across lines.
Resource	Resource	null (required)	The resource to be read.
saveState	boolean	true	Indicates of the state of the ItemReader should be saved after each chunk for restartability. This should be turned to false if used in a multithreaded environment.
skipped LinesCallback	LineCallback Handler	null	Callback interface called with the line skipped. Every line skipped will be passed to this callback.
strict	boolean	false	An Exception will be thrown if the resource is not found in strict mode.

When reading a file, a `String` is provided to the `LineMapper` implementation, representing a single record from a file. The most common `LineMapper` implementation used is the `DefaultLineMapper`. With the raw `String` from the file, there is a two-step process used by the `DefaultLineMapper` for getting it to the domain object you will later work with. These two steps are handled by the `LineTokenizer` and `FieldSetMapper`:

- A `LineTokenizer` implementation parses the line into a `org.springframework.batch.item.file.FieldSet`. The provided `String` represents the entire line from the file. In order to be able to map the individual fields of each record to your domain object, you need to parse the line into a collection of fields. The `FieldSet` in Spring Batch represents that collection of fields for a single row (similar to a `java.sql.ResultSet` when working with a database).

- The `FieldSetMapper` implementation maps the `FieldSet` to a domain object. With the line divided into individual fields, you can now map each input field to the field of your domain object just like a `RowMapper` would map a `ResultSet` row to the domain object.

Sounds simple doesn't it? It really is. The intricacies come from how to parse the line and when you look at objects that are built out of multiple records from your file. Let's take a look at reading files with fixed-width records first.

Fixed-Width Files

When dealing with legacy mainframe systems, it is common to have to work with fixed-width files due to the way COBOL, big data, and other technologies declare their storage. Because of this, you need to be able to handle fixed-width files as well.

You can use a customer file as your fixed-width file. Consisting of a customer's name and address, Table 7-2 outlines the format of your customer file.

Table 7-2. *Customer File Format*

Field	Length	Description
First Name	11	Your customer's first name.
Middle Initial	1	The customer's middle initial.
Last Name	10	The last name of the customer.
Address Number	4	The street number piece of the customer's address.
Street	20	The name of the street where the customer lives.
City	16	The city the customer is from.
State	2	The two-letter state abbreviation.
Zip Code	5	The customer's postal code.

Defining the format for a fixed-width file is important. A delimited file describes its fields with its delimiters. XML or other structured files are self-describing given the metadata the tags provide. Database data has the metadata from the database describing it. However, fixed-width files are different. They provide zero metadata to describe their format. If you look at Listing 7-2, you can see an example of what the previous description looks like as your input file.

Listing 7-2. customer.txt, the Fixed-Width File

```
Aimee      CHoover     7341Vel Avenue      Mobile       AL35928
Jonas      UGilbert    8852In St.          Saint Paul   MN57321
Regan      MBaxter     4851Nec Av.         Gulfport     MS3319
Sydnee     NRobinson   894 Ornare. Ave     Olathe       KS25606
```

In order for us to read this file, we will need a domain object to represent our record. The Customer object we will use can be found in Listing 7-3.

Listing 7-3. Customer.java

```
...
public class Customer {

    private String firstName;
    private String middleInitial;
    private String lastName;
    private String addressNumber;
    private String street;
    private String city;
    private String state;
    private String zipCode;

    public Customer() {
    }

    public Customer(String firstName, String middleInitial, String lastName, String
    addressNumber, String street, String city, String state, String zipCode) {
            this.firstName = firstName;
            this. middleInitial = middleInitial;
            this.lastName = lastName;
            this.addressNumber = addressNumber;
            this.street = street;
            this.city = city;
            this.state = state;
            this.zipCode = zipCode;
    }

    // Getters and setters removed for brevity
    ...

    @Override
    public String toString() {
            return "Customer{" +
                        "firstName='" + firstName + '\" +
                        ", middleInitial ='" + middleInitial + '\" +
                        ", lastName='" + lastName + '\" +
                        ", addressNumber='" + addressNumber + '\" +
                        ", street='" + street + '\" +
                        ", city='" + city + '\" +
```

```
                        ", state='" + state + '\" +
                        ", zipCode='" + zipCode + '\" +
                        '}';
        }
}
```

To demonstrate how each of these readers work, you will create a single-step job that reads in a file and writes it right back out. For this job, copyJob, you will create a BatchConfiguration configuration class with the following beans:

- *customerReader*: The FlatFileItemReader

- *outputWriter*: The FlatFileItemWriter

- *copyStep*: The Step definition for your job

- *copyJob*: The Job definition

The customerReader is an instance of the FlatFileItemReader. As covered previously, the FlatFileItemReader consists of two pieces, a Resource to read in (in this case, the customerFile) and a way to map each line of the file (a LineMapper implementation).

For the LineMapper implementation, you are going to use Spring Batch's org.springframework.batch. item.file.DefaultLineMapper. This LineMapper implementation is intended for the two-step process of mapping lines to domain objects you talked about previously: parsing the line into a FieldSet and then mapping the fields of the FieldSet to a domain object, the Customer object in your case.

To support the two-step mapping process, the DefaultLineMapper takes two dependencies: a LineTokenizer implementation which will parse the String that is read in from your file into a FieldSet and a FieldSetMapper implementation to map the fields in your FieldSet to the fields in your domain object.

That probably sounds like quite a bit of code. And it was...before Spring Batch 4. However, in Spring Batch 4, a collection of builders was created to simplify the configuration of common use cases. We'll be using the builders for all examples in this chapter. Listing 7-4 shows the customerReader being created via the FlatFileItemReaderBuilder.

Listing 7-4. customerReader in BatchConfiguration

```
...
@Bean
@StepScope
public FlatFileItemReader<Customer> customerItemReader(
@Value("#{jobParameters['customerFile']}")Resource inputFile) {

    return new FlatFileItemReaderBuilder<Customer>()
                .name("customerItemReader")
                .resource(inputFile)
                .fixedLength()
                .columns(new Range[]{new Range(1,11), new Range(12, 12), new Range(13, 22),
                    new Range(23, 26), new Range(27,46), new Range(47,62),
                    new Range(63,64), new Range(65,69)})
                .names(new String[] {"firstName", "middleInitial", "lastName",
                    "addressNumber", "street", "city", "state","zipCode"})
                .targetType(Customer.class)
                .build();
    }
...
```

Listing 7-4 contains all you need to configure the reading of a fixed-width file. However, the builder hides a lot of what is actually going on, so let's walk through what we're configuring here.

Let's start off with the parameter to the method inputFile. This comes from passing the path to our input file as a job parameter to our Boot application. Spring will automatically create a Resource for it and inject it for us.

From there, we create our builder. The first thing we configure with the builder is a name. The ItemStream interface, which we will look at in more detail later in this chapter, requires us to provide a name that serves as the prefix for any keys that are added to our step's ExecutionContext. This allows you to use, for example, two FlatFileItemReader instances in the same step and not have their state persistence step on each other. This value is not required if the saveState configuration of this reader is set to false (in which case, no data will be stored in the ExecutionContext making the reader start at the beginning on a restart). The next thing we do with our builder is to let it know that we're working with a fixed-width file. This call returns a builder specifically for building a FixedLengthTokenizer. This is an implementation of the LineTokenzier interface that the reader will use to parse each line into a FieldSet. The FixedLengthTokenizer requires the configuration of two things: the names of each column in a record and an array of Range objects. Each Range instance represents a start index and an end index for the columns being parsed. The other options you can configure with the FixedLengthTokenizer are a FieldSetFactory which is used to create the FieldSet (a DefaultFieldSetFactory is provided by default) and a strict flag that indicates how to handle a record that has a more tokens than are defined to parse (the flag is true by default which means an Exception is thrown in this condition). In Listing 7-4, we ignore the optional values since the defaults are good enough and configure the ranges and names for our input file.

If you remember, we said that a FlatFileItemReader uses a LineMapper to convert a record of a file to an object. The builder configured in 7-3 is using the DefaultLineMapper which takes two dependencies: the LineTokenizer which we just configured, and the FieldSetMapper (you can specify your own via the builder as well if you want). The builder we're using will create a new BeanWrapperFieldSetMapper when we call the .targetType(Class targetType) method. This FieldSetMapper, provided by Spring Batch, will use the names of the columns to set the values on the class configured. So, for example, the BeanWrapperFieldSetMapper will call Customer#setFirstName, Customer#setMiddleInitial, and so on, based on the names of the columns configured in the LineTokenizer.

■ **Note** The FixedLengthTokenizer doesn't trim any leading or trailing characters (spaces, zeros, etc.) within each field. To do this, you'll have to implement your own LineTokenizer or you can trim in your own FieldSetMapper.

To put your reader to use, you need to configure your Step and Job. You will also need to configure a writer so that you can see that everything works. You will be covering writers in depth in Chapter 9 so you can keep the writer for this example simple. Listing 7-5 shows how to configure a simple writer to output the domain objects to standard out.

Listing 7-5. A Simple Writer

```
...
@Bean
public ItemWriter<Customer> itemWriter() {
      return (items) -> items.forEach(System.out::println);
}
...
```

Looking at the writer in Listing 7-5, we use the fact that `ItemWriter` is a functional interface to return a lambda as its implementation. In this case, for each item in the `List` passed to the `ItemWriter.write(List<T> items)` method, the `.toString()` method will be called via `System.out.println`, displaying the output in the console.

Your job configuration is very simple. As shown in Listing 7-6, a simple step that consists of the reader and writer with a commit count of 10 records is all you need. Your `Job` uses that single `Step`.

Listing 7-6. The `copyFileStep` and `copyFileJob`

```
...
@Bean
public Step copyFileStep() {
    return this.stepBuilderFactory.get("ccpyFileStep")
                    .<Customer, Customer>chunk(10)
                    .reader(customerItemReader(null))
                    .writer(outputWriter(null))
                    .build();
}

@Bean
public Job job() {
    return this.jobBuilderFactory.get("job")
                .start(copyFileStep())
                .build();
}
...
```

The interesting piece of all of this is we've written zero application codes beyond our domain object (`Customer`). The only code we have written is the code to configure our application. Once you build your application, you can execute it with the command shown in Listing 7-7.

Listing 7-7. Executing the `copyJob`

```
java -jar copyJob.jar customerFile=/path/to/customer/file.txt
```

The output of the job is the same content of the input file formatted according to the format string of the writer, as shown in Listing 7-8.

Listing 7-8. Results of the `copyJob`

```
2019-01-28 16:11:44.089  INFO 54762 --- [           main] o.s.b.c.l.support.
SimpleJobLauncher       : Job: [SimpleJob: [name=job]] launched with the following
parameters: [{customerFile=/input/customerFixedWidth.txt}]
2019-01-28 16:11:44.159  INFO 54762 --- [           main] o.s.batch.core.job.
SimpleStepHandler       : Executing step: [copyFileStep]
Customer{firstName='Aimee', middleInitial='C', lastName='Hoover', addressNumber='7341',
street='Vel Avenue', city='Mobile', state='AL', zipCode='35928'}
Customer{firstName='Jonas', middleInitial='U', lastName='Gilbert', addressNumber='8852',
street='In St.', city='Saint Paul', state='MN', zipCode='57321'}
  ...
```

Fixed-width files are a form of input provided for batch processes in many enterprises. As you can see, parsing the file into objects via FlatFileItemReader and FixedLengthTokenizer makes this process easy. In the next section you will look at a file format that provides a small amount of metadata to tell us how the file is to be parsed.

Delimited Files

Delimited files are files that provide a small amount of metadata within the file to tell us what the format of the file is. In this case, a character acts as a divider between each field in your record. This metadata provides us with the ability to not have to know what defines each individual field. Instead, the file dictates to use what each field consists of by dividing each record with a delimiter.

As with fixed-width records, the process is the same to read a delimited record. The record will first be tokenized by the LineTokenizer into a FieldSet. From there, the FieldSet will be mapped into your domain object by the FieldSetMapper. With the process being the same, all you need to do is update the LineTokenizer implementation you use to parse your file based upon a delimiter instead of premapped columns. Let's start by looking at an updated customerFile that is delimited instead of fixed-width. Listing 7-9 shows your new input file.

Listing 7-9. A Delimited customerFile

```
Aimee,C,Hoover,7341,Vel Avenue,Mobile,AL,35928
Jonas,U,Gilbert,8852,In St.,Saint Paul,MN,57321
Regan,M,Baxter,4851,Nec Av.,Gulfport,MS,33193
```

You'll notice right away that there are two changes between the new file and the old one. First, you are using commas to delimit the fields. Second, you have trimmed all of the fields. Typically, when using delimited files, each field is not padded to a fixed-width like they are in fixed-width files. Because of that, the record length can vary, unlike the fixed-width record length.

As mentioned, the only configuration update you need to make to use the new file format is how each record is parsed. For fixed-width records, you used the FixedLengthTokenizer to parse each line. For the new delimited records, you will use the org.springframework.batch.item.file.transform. DelimitedLineTokenizer to parse the records into a FieldSet. Listing 7-10 shows the configuration of the reader updated with the DelimitedLineTokenizer.

Listing 7-10. customerFileReader with the DelimitedLineTokenizer

```
...
@Bean
@StepScope
public FlatFileItemReader<Customer> customerItemReader(@Value("#{jobParameters
['customerFile']}")Resource inputFile) {
    return new FlatFileItemReaderBuilder<Customer>()
            .name("customerItemReader")
            .delimited()
            .names(new String[] {"firstName",
                    "middleInitial",
                    "lastName",
                    "addressNumber",
                    "street",
                    "city",
                    "state",
                    "zipCode"})
```

```
            .targetType(Customer.class)
            .resource(inputFile)
            .build();
}
...
```

The `DelimitedLineTokenizer` allows for two options that you'll find very useful. The first is the ability to configure the delimiter. A comma is the default value; however, any `String` can be used. The second option is the ability to configure what value will be used as a quote character. When this option is used, that value will be used instead of " as the character to indicate quotes. This character will also be able to escape itself. Listing 7-11 shows an example of how a `String` is parsed when you use the # character as the quote character.

Listing 7-11. Parsing a Delimited File with the Quote Character Configured

```
Michael,T,Minella,#123,4th Street#,Chicago,IL,60606
```

Is parsed as

```
Michael
T
Minella
123,4th Street
Chicago
IL
60606
```

Although that's all that is required to process delimited files, it's not the only option you have. The current example maps address numbers and streets to two different fields. However, what if you wanted to map them together into a single field as represented in the domain object in Listing 7-12?

Listing 7-12. Customer with a Single Street Address Field

```
package com.apress.springbatch.chapter7;

public class Customer {
    private String firstName;
    private String middleInitial;
    private String lastName;
    private String address;
    private String city;
    private String state;
    private String zip;

    // Getters & setters go here
...
}
```

With the new object format, you will need to update how the `FieldSet` is mapped to the domain object. To do this, you will create your own implementation of the `org.springframework.batch.item.file.mapping.FieldSetMapper` interface. The `FieldSetMapper` interface, as shown in Listing 7-13, consists of a single method, `mapFieldSet(FieldSet fieldSet)`, that allows you to map the `FieldSet` as it is returned from the `LineTokenizer` to the domain object fields.

Listing 7-13. The FieldSetMapper Interface

```
package org.springframework.batch.item.file.mapping;

import org.springframework.batch.item.file.transform.FieldSet;
import org.springframework.validation.BindException;

public interface FieldSetMapper<T> {

    T mapFieldSet(FieldSet fieldSet) throws BindException;
}
```

To create your own mapper, you will implement the FieldSetMapper interface with the type defined as Customer. From there, as shown in Listing 7-14, you can map each field from the FieldSet to the domain object, concatenating the addressNumber and street fields into a single address field per your requirements.

Listing 7-14. Mapping Fields from the FieldSet to the Customer Object

```
...
public class CustomerFieldSetMapper implements FieldSetMapper<Customer> {

    public Customer mapFieldSet(FieldSet fieldSet) {
        Customer customer = new Customer();

        customer.setAddress(fieldSet.readString("addressNumber") +
                            " " + fieldSet.readString("street"));
        customer.setCity(fieldSet.readString("city"));
        customer.setFirstName(fieldSet.readString("firstName"));
        customer.setLastName(fieldSet.readString("lastName"));
        customer.setMiddleInitial(fieldSet.readString("middleInitial"));
        customer.setState(fieldSet.readString("state"));
        customer.setZipCode(fieldSet.readString("zipCode"));

        return customer;
    }
}
```

The FieldSet methods are very similar to the ResultSet methods of the JDBC realm. Spring provides a method for each of the primitive data types, String (trimmed or untrimmed), BigDecimal, and java.util. Date. Each of these different methods has two different varieties. The first takes an integer as the parameter where the integer represents the index of the field to be retrieved in the record. The other version, shown in Listing 7-15, takes the name of the field. Although this approach requires you to name the fields in the job configuration, it's a more maintainable model in the long run. Listing 7-15 shows the FieldSet interface.

Listing 7-15. FieldSet Interface

```
package org.springframework.batch.item.file.transform;

import java.math.BigDecimal;
import java.sql.ResultSet;
import java.util.Date;
import java.util.Properties;
```

```java
public interface FieldSet {

        String[] getNames();
        boolean hasNames();
        String[] getValues();
        String readString(int index);
        String readString(String name);
        String readRawString(int index);
        String readRawString(String name);
        boolean readBoolean(int index);
        boolean readBoolean(String name);
        boolean readBoolean(int index, String trueValue);
        boolean readBoolean(String name, String trueValue);
        char readChar(int index);
        char readChar(String name);
        byte readByte(int index);
        byte readByte(String name);
        short readShort(int index);
        short readShort(String name);
        int readInt(int index);
        int readInt(String name);
        int readInt(int index, int defaultValue);
        int readInt(String name, int defaultValue);
        long readLong(int index);
        long readLong(String name);
        long readLong(int index, long defaultValue);
        long readLong(String name, long defaultValue);
        float readFloat(int index);
        float readFloat(String name);
        double readDouble(int index);
        double readDouble(String name);
        BigDecimal readBigDecimal(int index);
        BigDecimal readBigDecimal(String name);
        BigDecimal readBigDecimal(int index, BigDecimal defaultValue);
        BigDecimal readBigDecimal(String name, BigDecimal defaultValue);
        Date readDate(int index);
        Date readDate(String name);
        Date readDate(int index, Date defaultValue);
        Date readDate(String name, Date defaultValue);
        Date readDate(int index, String pattern);
        Date readDate(String name, String pattern);
        Date readDate(int index, String pattern, Date defaultValue);
        Date readDate(String name, String pattern, Date defaultValue);
        int getFieldCount();
        Properties getProperties();
}
```

■ **Note** Unlike the JDBC ResultSet, which begins indexing columns at 1, the index used by Spring Batch's FieldSet is zero based.

To put the `CustomerFieldSetMapper` to use, you need to update the configuration to use it. Replace the `BeanWrapperFieldSetMapper` reference with your own bean reference, as shown in Listing 7-16.

Listing 7-16. `customerFileReader` Configured with the `CustomerFieldSetMapper`

```
...
@Bean
@StepScope
public FlatFileItemReader<Customer> customerItemReader(@Value("#{jobParameters
['customerFile']}")Resource inputFile) {
    return new FlatFileItemReaderBuilder<Customer>()
            .name("customerItemReader")
            .delimited()
            .names(new String[] {"firstName",
                    "middleInitial",
                    "lastName",
                    "addressNumber",
                    "street",
                    "city",
                    "state",
                    "zip"})
            .fieldSetMapper(new CustomerFieldSetMapper())
            .resource(inputFile)
            .build();
}
...
```

Note that with your new `CustomerFieldSetMapper`, you don't need to configure the reference to the `Customer` bean. Since you handle the instantiation yourselves, this is no longer needed.

Parsing files with the standard Spring Batch parsers, as you have shown, requires nothing more than a few lines of Java. However, not all files consist of Unicode characters laid out in a format that is easy for Java to understand. When dealing with legacy systems, it's common to come across data storage techniques that require custom parsing. In the next section, you will look at how to implement your own `LineTokenizer` to be able to handle custom file formats.

Custom Record Parsing

In the previous section you looked at how to address the ability to tweak the mapping of fields in your file to the fields of your domain object by creating a custom `FieldSetMapper` implementation. However, that is not the only option. Instead, you can create your own `LineTokenizer` implementation. This will allow you to parse each record however you need.

Like the `FieldSetMapper` interface, the `org.springframework.batch.item.file.transform.LineTokenizer` interface has a single method: `tokenize`. Listing 7-17 shows the `LineTokenizer` interface.

Listing 7-17. `LineTokenizer` Interface

```
package org.springframework.batch.item.file.transform;

public interface LineTokenizer {

    FieldSet tokenize(String line);
}
```

For this approach you will use the same delimited input file you used previously; however, since the domain object has the address number and the street combined into a single field, you will combine those two tokens into a single field in the FieldSet. Listing 7-18 shows the CustomerFileLineTokenizer.

Listing 7-18. CustomerFileLineTokenizer

```
...
public class CustomerFileLineTokenizer implements LineTokenizer {

    private String delimiter = ",";
    private String[] names = new String[] {"firstName",
                    "middleInitial",
                    "lastName",
                    "address",
                    "city",
                    "state",
                    "zipCode"};

    private FieldSetFactory fieldSetFactory = new DefaultFieldSetFactory();

    public FieldSet tokenize(String record) {

        String[] fields = record.split(delimiter);

        List<String> parsedFields = new ArrayList<>();

        for (int i = 0; i < fields.length; i++) {
            if (i == 4) {
                parsedFields.set(i - 1,
                            parsedFields.get(i - 1) + " " + fields[i]);
            } else {
                parsedFields.add(fields[i]);
            }
        }

        return fieldSetFactory.create(parsedFields.toArray(new String [0]),
                            names);
    }
}
```

The tokenize(String line) method of the CustomerFileLineTokenizer takes each record and splits it based upon the delimiter that was configured with Spring. You loop through the fields, combining the third and fourth fields together so that they are a single field. You then create a FieldSet using the DefaultFieldSetFactory, passing it the one required parameter (an array of values to be your fields) and one optional parameter (an array of names for the fields). This LineTokenizer names your fields so that you can use the BeanWrapperFieldSetMapper to do your FieldSet to domain object mapping without any additional code.

Configuring the CustomerFileLineTokenizer is similar to how we configured our custom FieldSetMapper by removing the other configuration and replacing it with a single method call. Listing 7-19 shows the updated configuration.

Listing 7-19. Configuring the `CustomerFileLineTokenizer`

```
...
@Bean
@StepScope
public FlatFileItemReader<Customer> customerItemReader(@Value("#{jobParameters['customer
File']}")Resource inputFile) {
    return new FlatFileItemReaderBuilder<Customer>()
            .name("customerItemReader")
            .lineTokenizer(new CustomerFileLineTokenizer())
            .fieldSetMapper(new CustomerFieldSetMapper())
            .resource(inputFile)
            .build();
}
...
```

The sky's the limit with what you can do with your own `LineTokenizer` and `FieldSetMapper`. Other uses for custom `LineTokenizer`s could include

- Parsing unusual file formats

- Parsing third party file formats like Microsoft's Excel Worksheets

- Handling special type conversion requirements

However, not all files are as simple as the customer one you have been working with. What if your file contains multiple record formats? The next section will discuss how Spring Batch can choose the appropriate `LineTokenizer` to parse each record it comes across.

Multiple Record Formats

Up to this point you have been looking at a customer file that contains a collection of customer records. Each record in the file has the exact same format. However, what if you received a file that had customer information as well as transaction information? Yes, you could implement a single custom `LineTokenizer`. However, there are two issues with this approach:

1. *Complexity*: If you have a file that has three, four, five, or more line formats—each with a large number of fields—this single class can get out of hand quickly.

2. *Separation of concerns*: The `LineTokenizer` is intended to parse a record. That's it. It should not need to determine what the record type is prior to the parsing.

With this in mind, Spring Batch provides another `LineMapper` implementation: the `org.springframework.batch.item.file.mapping.PatternMatchingCompositeLineMapper`. The previous examples used the `DefaultLineMapper`, which provided the ability to use a single `LineTokenizer` and a single `FileSetMapper`. With the `PatternMatchingCompositeLineMapper`, you will be able to define a `Map` of `LineTokenizer`s and a corresponding `Map` of `FieldSetMapper`s. The key for each `Map` will be a pattern that the `LineMapper` will use to identify which `LineTokenizer` to use to parse each record.

Let's start this example by looking at the updated input file. In this case, you still have the same customer records. However, interspersed between each customer record is a random number of transaction records. To help identify each record, you have added a prefix to each record. Listing 7-20 shows the updated input file.

Listing 7-20. The Updated customerInputFile

```
CUST,Warren,Q,Darrow,8272 4th Street,New York,IL,76091
TRANS,1165965,2011-01-22 00:13:29,51.43
CUST,Ann,V,Gates,9247 Infinite Loop Drive,Hollywood,NE,37612
CUST,Erica,I,Jobs,8875 Farnam Street,Aurora,IL,36314
TRANS,8116369,2011-01-21 20:40:52,-14.83
TRANS,8116369,2011-01-21 15:50:17,-45.45
TRANS,8116369,2011-01-21 16:52:46,-74.6
TRANS,8116369,2011-01-22 13:51:05,48.55
TRANS,8116369,2011-01-21 16:51:59,98.53
```

In the file shown in Listing 7-20, you have two comma-delimited formats. The first consists of the standard customer format you have been working with up to now with the concatenated address number and street. These records are indicated with the prefix CUST. The other records are transaction records; each of these records, prefixed with the TRANS, prefix, are also comma delimited, with the following three fields:

1. *Account number*: The customer's account number.

2. *Date*: The date the transaction occurred. The transactions may or may not be in date order.

3. *Amount*: The amount in dollars for the transaction. Negative values symbolize debits and positive amounts symbolize credits.

Listing 7-21 shows the code for the Transaction domain object.

Listing 7-21. Transaction Domain Object Code

```
...
public class Transaction {

    private String accountNumber;
    private Date transactionDate;
    private Double amount;

    private DateFormat formatter = new SimpleDateFormat("MM/dd/yyyy");

    // Getters and setters are omitted
      @Override
      public String toString() {
            return "Transaction{" +
                        "accountNumber='" + accountNumber + '\"' +
                        ", transactionDate=" + transactionDate +
                        ", amount=" + amount +
                        '}';
    }

}
```

With the record formats identified, you can look at the reader. Listing 7-22 shows the configuration for the updated customerFileReader. As mentioned, using the PatternMatchingCompositeLineMapper, you map two instances of the DelimitedLineTokenizer, each with the correct record format configured. You'll notice that you have an additional field named prefix for each of the LineTokenizers. This is to address the string at the beginning of each record (CUST and TRANS). Spring Batch will parse the prefix and name it prefix in your FieldSet; however, since you don't have a prefix field in either of your domain objects, it will be ignored in the mapping.

Listing 7-22. Configuring the `customerFileReader` with Multiple Record Formats

```
...
@Bean
@StepScope
public FlatFileItemReader customerItemReader(
        @Value("#{jobParameters['customerFile']}")Resource inputFile) {

    return new FlatFileItemReaderBuilder<Customer>()
            .name("customerItemReader")
            .lineMapper(lineTokenizer())
            .resource(inputFile)
            .build();
}

@Bean
public PatternMatchingCompositeLineMapper lineTokenizer() {
    Map<String, LineTokenizer> lineTokenizers = new HashMap<>(2);

    lineTokenizers.put("CUST*", customerLineTokenizer());
    lineTokenizers.put("TRANS*", transactionLineTokenizer());

    Map<String, FieldSetMapper> fieldSetMappers = new HashMap<>(2);

    BeanWrapperFieldSetMapper<Customer> customerFieldSetMapper =
      new BeanWrapperFieldSetMapper<>();
    customerFieldSetMapper.setTargetType(Customer.class);

    fieldSetMappers.put("CUST*", customerFieldSetMapper);
    fieldSetMappers.put("TRANS*", new TransactionFieldSetMapper());

    PatternMatchingCompositeLineMapper lineMappers =
      new PatternMatchingCompositeLineMapper();

    lineMappers.setTokenizers(lineTokenizers);
    lineMappers.setFieldSetMappers(fieldSetMappers);

    return lineMappers;
}

@Bean
public DelimitedLineTokenizer transactionLineTokenizer() {
    DelimitedLineTokenizer lineTokenizer = new DelimitedLineTokenizer();

    lineTokenizer.setNames("prefix", "accountNumber", "transactionDate", "amount");

    return lineTokenizer;
}
```

```
@Bean
public DelimitedLineTokenizer customerLineTokenizer() {
    DelimitedLineTokenizer lineTokenizer = new DelimitedLineTokenizer();

    lineTokenizer.setNames("firstName",
        "middleInitial",
        "lastName",
        "address",
        "city",
        "state",
        "zip");

    lineTokenizer.setIncludedFields(1, 2, 3, 4, 5, 6, 7);

    return lineTokenizer;
}
...
```

The configuration of the customerFileReader is beginning to get a bit verbose. Let's walk through what will actually happen when this reader is executed. If you look at Figure 7-2, you can follow the flow of how the customerFileReader will process each line.

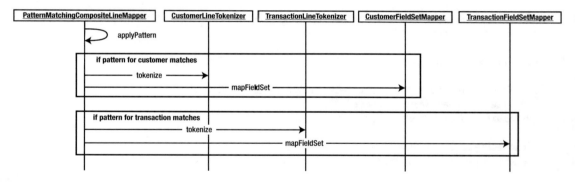

Figure 7-2. *Flow of processing for multiple record formats*

As Figure 7-2 shows, the PatternMatchingCompositeLineMapper will look at each record of the file and apply your pattern to it. If the record begins with CUST,* (where * is zero or more characters), it will pass the record to the customerLineTokenizer for parsing. Once the record is parsed into a FieldSet, it will be passed to a FieldSetMapper. In this case, we are using the BeanWrapperFieldSetMapper from the framework. Because of this and our Customer domain object does not have a prefix field, we want our tokenizer to skip the prefix field. To do that, we do two things. First, we leave the prefix name out of the list of field names we configure in the DelimitedLineTokenizer. Second, we provide a list of indices (0 based) of the fields that we want to include. In our case, we want to include all of the fields except for the prefix.

If the record begins with TRANS,*, it will be passed to the transactionLineTokenizer for parsing with the resulting FieldSet being passed to the custom transactionFieldSetMapper.

But why do you need a custom FieldSetMapper? It's necessary for custom-type conversion. By default, the BeanWrapperFieldSetMapper doesn't do any special type conversion. The Transaction domain object consists of an accountNumber field, which is a String; however, the other two fields, transactionDate and amount, are a java.util.Date and a Double, respectively. Because of this, you will need to create a custom FieldSetMapper to do the required type conversions. Listing 7-23 shows the TransactionFieldSetMapper.

Listing 7-23. TransactionFieldSetMapper

```
package com.apress.springbatch.chapter7;

import org.springframework.batch.item.file.mapping.FieldSetMapper;
import org.springframework.batch.item.file.transform.FieldSet;
import org.springframework.validation.BindException;

public class TransactionFieldSetMapper implements FieldSetMapper<Transaction> {

    public Transaction mapFieldSet(FieldSet fieldSet) {
        Transaction trans = new Transaction();

        trans.setAccountNumber(fieldSet.readString("accountNumber"));
        trans.setAmount(fieldSet.readDouble("amount"));
        trans.setTransactionDate(fieldSet.readDate("transactionDate",
                                        "yyyy-MM-dd HH:mm:ss"));

        return trans;
    }
}
```

As you can see, the FieldSet interface, like the ResultSet interface of the JDBC world, provides custom methods for each data type. In the case of the Transaction domain object, you use the readDouble method to have the String in your file converted into a java.lang.Double, and you use the readDate method to parse the String contained in your file into a java.util.Date. For the date conversion, you specify not only the field's name but also the format of the date to be parsed.

When you execute the job, you're able to read in the two different record formats and parse them into their respective domain objects. A sample of the results of this job is shown in Listing 7-24.

Listing 7-24. Results of Running the copyJob Job with Multiple Record Formats

```
2019-01-28 22:41:09.812  INFO 60498 --- [            main] o.s.batch.core.job.
SimpleStepHandler      : Executing step: [copyFileStep]
Customer{firstName='Warren', middleInitial='Q', lastName='Darrow', address='8272 4th
Street', city='New York', state='IL', zipCode='76091'}
Transaction{accountNumber='1165965', transactionDate=Sat Jan 22 00:13:29 CST 2011,
amount=51.43}
Customer{firstName='Ann', middleInitial='V', lastName='Gates', address='9247 Infinite Loop
Drive', city='Hollywood', state='NE', zipCode='37612'}
Customer{firstName='Erica', middleInitial='I', lastName='Jobs', address='8875 Farnam
Street', city='Aurora', state='IL', zipCode='36314'}
Transaction{accountNumber='8116369', transactionDate=Fri Jan 21 20:40:52 CST 2011,
amount=-14.83}
Transaction{accountNumber='8116369', transactionDate=Fri Jan 21 15:50:17 CST 2011,
amount=-45.45}
Transaction{accountNumber='8116369', transactionDate=Fri Jan 21 16:52:46 CST 2011,
amount=-74.6}
Transaction{accountNumber='8116369', transactionDate=Sat Jan 22 13:51:05 CST 2011,
amount=48.55}
Transaction{accountNumber='8116369', transactionDate=Fri Jan 21 16:51:59 CST 2011,
amount=98.53}
```

The ability to process multiple records from a single file is a common requirement in batch processing. However, this example assumes that there was no real relationship between the different records. What if there is? The next section will look at how to read multiline records into a single item.

Multiline Records

In the last example, you looked at the processing of two different record formats into two different, unrelated items. However, if you take a closer look at the file format you were using, you can see that the records you were reading were actually related. While not related by a field in the file, the transaction records are the transaction records for the customer record above it. Instead of processing each record independently, doesn't it make more sense to have a Customer object that has a collection of Transaction objects on it?

To make this work, you will need to perform a small bit of trickery. The examples provided with Spring Batch use a footer record to identify the true end of a record. Although convenient, many files seen in batch do not have that trailer record. With your file format, you run into the issue of not knowing when a record is complete without reading the next row. To get around this, you can implement your own ItemReader that adds a bit of logic around the customerFileReader you configured in the previous section. Figure 7-3 shows the flow of logic you will use within your custom ItemReader.

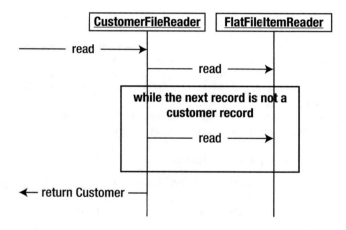

Figure 7-3. CustomerFileReader flow

As Figure 7-3 shows, your read method will begin by determining if a Customer object has already been read. If it hasn't, it will attempt to read one from the FlatFileItemReader. Assuming you read a record (you won't have read one once you reach the end of the file), you will initialize the transaction List on the Customer object. While the next record you read is a Transaction, you will add it to the Customer object.

Before we get to the listing for our custom ItemReader implementation, our domain object needs to change slightly. Instead of two independent domain objects, Customer and Transaction, our new configuration will use a Customer object that contains a List<Transaction> objects in it. Listing 7-25 illustrates the updated Customer object.

Listing 7-25. Updated Customer Object

```
...
public class Customer {

        private String firstName;
        private String middleInitial;
        private String lastName;
        private String address;
        private String city;
        private String state;
        private String zipCode;

        List<Transaction> transactions;

        public Customer() {
        }

        // Getters and setters removed for brevity
        ...
}
```

With our updated domain object, we can take a look at the implementation of the CustomerFileReader in Listing 7-26.

Listing 7-26. CustomerFileReader

```
public class CustomerFileReader implements ItemStreamReader<Customer> {

    private Object curItem = null;

    private ItemStreamReader<Object> delegate;

    public CustomerFileReader(ItemStreamReader<Object> delegate) {
        this.delegate = delegate;
    }

    public Customer read() throws Exception {
        if(curItem == null) {
            curItem = delegate.read();
        }

        Customer item = (Customer) curItem;
        curItem = null;

        if(item != null) {
            item.setTransactions(new ArrayList<>());

            while(peek() instanceof Transaction) {
                item.getTransactions().add((Transaction) curItem);
                curItem = null;
            }
        }

        return item;
    }
```

```
    private Object peek() throws Exception {
        if (curItem == null) {
            curItem = delegate.read();
        }
        return curItem;
    }

    public void close() throws ItemStreamException {
        delegate.close();
    }

    public void open(ExecutionContext arg0) throws ItemStreamException {
        delegate.open(arg0);
    }

    public void update(ExecutionContext arg0) throws ItemStreamException {
        delegate.update(arg0);
    }
}
```

The CustomerFileReader has two key methods that you should look at. The first is the read() method. This method is responsible for implementing the logic involved in reading and assembling a single Customer item including its child transaction records. It does so by reading in a customer record from the file you are reading. It then reads the related transaction records until the next record is the next customer record. Once the next customer record is found, the current customer is considered complete and returned by your ItemReader. This type of logic is called control break logic.

The other method of consequence is the peak method. This method is used to read ahead while still working on the current Customer. It caches the current record. If the record has been read but not processed, it will return the same record again. If the record has been processed (indicated to this method by setting curItem to null), it will read in the next record.[1]

You should notice that your custom ItemReader does not implement the ItemReader interface. Instead, it implements on of its subinterfaces, the ItemStreamReader interface. The reason for this is that when using one of the Spring Batch ItemReader implementations, they handle the opening and closing of the resource being read as well as maintaining the ExecutionContext as records are being read. However, if you implement your own, you need to manage that yourself. Since you are just wrapping a Spring Batch ItemReader (the FlatFileItemReader), you can use it to maintain those resources.

To configure the CustomerFileReader, the only dependency you have is the delegate. The delegate in this case is the reader that will do the actual reading and parsing work for you. Listing 7-27 shows the configuration for the CustomerFileReader.

Listing 7-27. CustomerFileReader Configuration

```
...
@Bean
@StepScope
public FlatFileItemReader customerItemReader(@Value("#{jobParameters['customerFile']}")
Resource inputFile) {
```

---

[1]It is important to note that there is an ItemReader subinterface called the org.springframework.batch.item. PeekableItemReader<T>. Since the CustomerFileReader does not firmly meet the contract defined by that interface, here we do not implement it.

```
        return new FlatFileItemReaderBuilder()
                .name("customerItemReader")
                .lineMapper(lineTokenizer())
                .resource(inputFile)
                .build();
}

@Bean
public CustomerFileReader customerFileReader() {
    return new CustomerFileReader(customerItemReader(null));
}

@Bean
public PatternMatchingCompositeLineMapper lineTokenizer() {
    Map<String, LineTokenizer> lineTokenizers = new HashMap<>(2);

    lineTokenizers.put("CUST*", customerLineTokenizer());
    lineTokenizers.put("TRANS*", transactionLineTokenizer());

    Map<String, FieldSetMapper> fieldSetMappers = new HashMap<>(2);

    BeanWrapperFieldSetMapper<Customer> customerFieldSetMapper = new
    BeanWrapperFieldSetMapper<>();
    customerFieldSetMapper.setTargetType(Customer.class);

    fieldSetMappers.put("CUST*", customerFieldSetMapper);
    fieldSetMappers.put("TRANS*", new TransactionFieldSetMapper());

    PatternMatchingCompositeLineMapper lineMappers = new PatternMatchingCompositeLineMapper();

    lineMappers.setTokenizers(lineTokenizers);
    lineMappers.setFieldSetMappers(fieldSetMappers);

    return lineMappers;
}

@Bean
public DelimitedLineTokenizer transactionLineTokenizer() {
    DelimitedLineTokenizer lineTokenizer = new DelimitedLineTokenizer();

    lineTokenizer.setNames("prefix", "accountNumber", "transactionDate", "amount");

    return lineTokenizer;
}

@Bean
public DelimitedLineTokenizer customerLineTokenizer() {
    DelimitedLineTokenizer lineTokenizer = new DelimitedLineTokenizer();

    lineTokenizer.setNames("prefix", "firstName", "middleInitial", "lastName", "address",
    "city", "state", "zip");
```

```
        return lineTokenizer;
}
...
```

The configuration in Listing 7-27 should look familiar. It's essentially the exact same as the configuration you used for multiple record formats (see Listing 7-19). The only addition, as highlighted in bold, is the configuration of your new CustomerFileReader with its reference to the old ItemReader and renaming the old ItemReader.

With the new CustomerFileReader wrapping the original ItemReader that did the work, we will need to update our step to reference the CustomerFileReader as the ItemReader to do the work with. Listing 7-28 shows the updated configuration for our step.

Listing 7-28. copyFileStep

```
...
@Bean
public Step copyFileStep() {
        return this.stepBuilderFactory.get("copyFileStep")
                        .<Customer, Customer>chunk(10)
                        .reader(customerFileReader())
                        .writer(itemWriter())
                        .build();
}
...
```

For each Customer object, we want to print how many transactions the user has. This will provide enough detail for you to verify that your reading worked correctly. Given the simple ItemWriter we have been using, you only need to override the Customer's toString() method to format the output. Listing 7-29 shows the updated method.

Listing 7-29. Customer's toString() Method

```
...
    @Override
    public String toString() {
        StringBuilder output = new StringBuilder();

        output.append(firstName);
        output.append(" ");
        output.append(middleInitial);
        output.append(". ");
        output.append(lastName);

        if(transactions != null&& transactions.size() > 0) {
            output.append(" has ");
            output.append(transactions.size());
            output.append(" transactions.");
        } else {
            output.append(" has no transactions.");
        }
```

```
        return output.toString();
    }
...
```

With a run of the job, you can see each of your customers and the number of transaction records you read in. It's important to note that when reading records in this way, the customer record and all the subsequent transaction records are considered a single item. The reason for this is that Spring Batch considers an item to be any object that is returned by the `ItemReader`. In this case, the `Customer` object is the object returned by the `ItemReader` so it is the item used for things like commit counts, etc. Each `Customer` object will be processed once by any configured `ItemProcessor` you add and once by any `configured ItemWriter`. The output from the job configured with the new `ItemReaders` can be seen in Listing 7-30.

Listing 7-30. Output from Multiline Job

```
2019-01-28 23:32:17.635  INFO 61271 --- [          main] o.s.batch.core.job.
SimpleStepHandler      : Executing step: [copyFileStep]
Warren Q. Darrow has 1 transactions.
Ann V. Gates has no transactions.
Erica I. Jobs has 5 transactions.
```

Multiline records are a common element in batch processing. Although they are a bit more complex than basic record processing, as you can see from this example, there is still only a minimal amount of actual code that needs to be written to handle these robust situations.

The last piece of the flat file puzzle is to look at input situations where you read in from multiple files. This is a common requirement in the batch world and it's covered in the next section.

Multiple Sources

The examples up to this point have been based around a customer file with transactions for each customer. Many companies have multiple departments or locations that sell things. Take, for example, a restaurant chain with restaurants nationwide. Each location may contribute a file with the same format to be processed. If you were to process each one with a separate writer like you have been up to now, there would be a number of issues from performance to maintainability. So how does Spring Batch provide for the ability to read in multiple files with the same format?

Using a similar pattern to the one you just used in the multiline record example, Spring Batch provides an `ItemReader` called the `MultiResourceItemReader`. This reader wraps another `ItemReader` like the `CustomerFileItemReader` did; however, instead of defining the resource to be read as part of the child `ItemReader`, a pattern that defines all of the files to be read is defined as a dependency of the `MultiResourceItemReader`. Let's take a look.

You can use the same file format as you did in your multi-record example (as shown in Listing 7-20), which will allow you to use the same `ItemReader` configuration you created in the multiline example as well. However, if you have five of these files with the filenames `customerFile1.csv`, `customerFile2.csv`, `customerFile3.csv`, `customerFile4.csv`, and `customerFile5.csv`, you need to make two small updates. The first is to the configuration. You need to tweak your configuration to use the `MultiResourceItemReader` with the correct resource pattern. Listing 7-31 shows the updated configuration.

Listing 7-31. Configuration to Process Multiple Customer Files

```
...
@Bean
@StepScope
public MultiResourceItemReader multiCustomerReader(@Value("#{jobParameters['customerFile']}")
Resource[] inputFiles) {
    return new MultiResourceItemReaderBuilder<>()
            .name("multiCustomerReader")
            .resources(inputFiles)
            .delegate(customerFileReader())
            .build();
}

@Bean
public CustomerFileReader customerFileReader() {
    return new CustomerFileReader(customerItemReader());
}

@Bean
public FlatFileItemReader customerItemReader() {
    return new FlatFileItemReaderBuilder()
            .name("customerItemReader")
            .lineMapper(lineTokenizer())
            .build();
}

@Bean
public PatternMatchingCompositeLineMapper lineTokenizer() {
    Map<String, LineTokenizer> lineTokenizers = new HashMap<>(2);

    lineTokenizers.put("CUST*", customerLineTokenizer());
    lineTokenizers.put("TRANS*", transactionLineTokenizer());

    Map<String, FieldSetMapper> fieldSetMappers = new HashMap<>(2);

    BeanWrapperFieldSetMapper<Customer> customerFieldSetMapper = new
    BeanWrapperFieldSetMapper<>();
    customerFieldSetMapper.setTargetType(Customer.class);

    fieldSetMappers.put("CUST*", customerFieldSetMapper);
    fieldSetMappers.put("TRANS*", new TransactionFieldSetMapper());

    PatternMatchingCompositeLineMapper lineMappers = new PatternMatchingCompositeLineMapper();

    lineMappers.setTokenizers(lineTokenizers);
    lineMappers.setFieldSetMappers(fieldSetMappers);

    return lineMappers;
}
```

```
@Bean
public DelimitedLineTokenizer transactionLineTokenizer() {
    DelimitedLineTokenizer lineTokenizer = new DelimitedLineTokenizer();

    lineTokenizer.setNames("prefix", "accountNumber", "transactionDate", "amount");

    return lineTokenizer;
}

@Bean
public DelimitedLineTokenizer customerLineTokenizer() {
    DelimitedLineTokenizer lineTokenizer = new DelimitedLineTokenizer();

    lineTokenizer.setNames("prefix", "firstName", "middleInitial", "lastName", "address",
    "city", "state", "zip");

    return lineTokenizer;
}
...
```

The `MultiResourceItemReader` takes three main pieces. The first is the name since it is stateful like the other `ItemReader` implementations we have looked at up to now. The second is an array of `Resource` objects. These are the files to be read in. We can use SpEL to have Spring resolve the array for us as shown in the example. The final piece is the delegate that will do the actual work for us. In this example, our delegate is the custom `CustomerFileReader` we used in the previous section.

In order for the `MultiResourceItemReader` to do its job, we need to change the configuration for our `FlatFileItemReader` as well. In the previous listings, we configured the resource to be read at that point. However, in this case, the `MultiResourceItemReader` is going to loop over an array of `Resource` objects, injecting a new one as each one is exhausted. Because of that, we need to remove the configuration of the `Resource` on our `customerItemReader` bean as shown in bold in Listing 7-31.

The other change you need to make is to the `CustomerFileReader` code. Previously, you were able to use the `ItemStreamReader` interface as what you implemented and the delegate's type. However, that won't be specific enough this time around. Instead, you are going to need to use one of the `ItemStreamResource`'s sub interfaces. The `ResourceAwareItemReaderItemStream` interface is for any `ItemReader` that reads its input from `Resources`. The reason you will want to make the two changes is that you will need to be able to inject multiple `Resources` into the `ItemReader`.

By implementing `org.springframework.batch.item.file.ResourceAwareItemReaderItemStream`, you will be required to add one additional method: `setResource(Resource resource)`. Like the open, close and update methods of the `ItemStream` interface, you will just be calling the `setResource` method on the delegate in your implementation. The other change you need to make is to have your delegate be of the type `ResourceAwareItemReaderItemStream`. Since you are using the `FlatFileItemReader` as your delegate, you won't need to use a different `ItemReader` as the delegate. The updated code is listed in Listing 7-32.

Listing 7-32. `CustomerFileReader`

```
public class CustomerFileReader implements ResourceAwareItemReaderItemStream<Customer> {

    private Object curItem = null;

    private ResourceAwareItemReaderItemStream<Object> delegate;

    public CustomerFileReader(ResourceAwareItemReaderItemStream<Object> delegate) {
        this.delegate = delegate;
```

```
        }

    public Customer read() throws Exception {
        if(curItem == null) {
            curItem = delegate.read();
        }

        Customer item = (Customer) curItem;
        curItem = null;

        if(item != null) {
            item.setTransactions(new ArrayList<>());

            while(peek() instanceof Transaction) {
                item.getTransactions().add((Transaction) curItem);
                curItem = null;
            }
        }

        return item;
    }

    private Object peek() throws Exception {
        if (curItem == null) {
            curItem = delegate.read();
        }
        return curItem;
    }

    public void close() throws ItemStreamException {
        delegate.close();
    }

    public void open(ExecutionContext arg0) throws ItemStreamException {
        delegate.open(arg0);
    }

    public void update(ExecutionContext arg0) throws ItemStreamException {
        delegate.update(arg0);
    }

    @Override
    public void setResource(Resource resource) {
        this.delegate.setResource(resource);
    }
}
```

The sole difference from a processing standpoint between what is shown in Listing 7-32 and what you originally wrote in Listing 7-26 is the ability to inject a Resource. This allows Spring Batch to create each of the files as needed and inject them into the ItemReader instead of the ItemReader itself being responsible for file management.

When you run this example with the command `java -jar copyJob.jar customerFile=/input/ customerMulitFormat*`, Spring Batch will iterate through all of the resources that match your provided pattern and execute your reader for each file. The output for this job is nothing more than a larger version of the output from the multiline record example.

Listing 7-33. Output from Multiline Job

```
Warren Q. Darrow has 1 transactions.
Ann V. Gates has no transactions.
Erica I. Jobs has 5 transactions.
Joseph Z. Williams has 2 transactions.
Estelle Y. Laflamme has 3 transactions.
Robert X. Wilson has 1 transactions.
Clement A. Blair has 1 transactions.
Chana B. Meyer has 1 transactions.
Kay C. Quinonez has 1 transactions.
Kristen D. Seibert has 1 transactions.
Lee E. Troupe has 1 transactions.
Edgar F. Christian has 1 transactions.
```

It is important to note that when dealing with multiple files like this, Spring Batch provides no added safety around things like restart. So in this example, if your job started with files `customerFile1.csv`, `customerFile2.csv`, and `customerFile3.csv` and it were to fail after processing `customerFile2.csv`, and you added a `customerFile4.csv` before it was restated, `customerFile4.csv` would be processed as part of this run even though it didn't exist when the job was first executed. To safeguard against this, it's a common practice to have a directory for each batch run. All files that are to be processed for the run go into the appropriate directory and are processed. Any new files go into a new directory so that they have no impact on the currently running execution.

I have covered many scenarios involving flat files—from fixed-width records, delimited records, multiline records, and even input from multiple files. However, flat files are not the only type of files that you are likely to see. XML, while not the most sexy input format, still represents a large amount of the file-based input you will see in an enterprise. Let's see what Spring Batch can do for you when you're faced with XML files.

XML

When I began talking about file-based processing at the beginning of this chapter, I talked about how different file formats have differing amounts of metadata that describe the format of the file. Fixed-width records have the least amount of metadata, requiring the most information about the record format to be known in advance. XML is at the other end of the spectrum. XML uses tags to describe the data in the file, providing a full description of the data it contains.

Two XML parsers are commonly used: DOM and SAX. The DOM parser loads the entire file into memory in a tree structure for navigation of the nodes. This approach is not useful for batch processing due to the performance implications. This leaves you with the SAX parser. SAX is an event-based parser that fires events when certain elements are found.

In Spring Batch, you use a StAX parser. Although this is an event-based parser similar to SAX, it has the advantage of allowing for the ability to parse sections of your document independently. This relates directly with the item-oriented reading you do. A SAX parser would parse the entire file in a single run; the StAX parser allows you to read each section of a file that represents an item to be processed at a time.

Before you look at how to parse XML with Spring Batch, let's look at a sample input file. To see how the XML parsing works with Spring Batch, you will be working with the same input: your customer file. However, instead of the data in the format of a flat file, you will structure it via XML. Listing 7-34 shows a sample of the input.

Listing 7-34. Customer XML File Sample

```
<customers>
    <customer>
        <firstName>Laura</firstName>
        <middleInitial>O</middleInitial>
        <lastName>Minella</lastName>
        <address>2039 Wall Street</address>
        <city>Omaha</city>
        <state>IL</state>
        <zipCode>35446</zipCode>
        <transactions>
            <transaction>
                <accountNumber>829433</accountNumber>
                <transactionDate>2010-10-14 05:49:58</transactionDate>
                <amount>26.08</amount>
            </transaction>
        </transactions>
    </customer>
    <customer>
        <firstName>Michael</firstName>
        <middleInitial>T</middleInitial>
        <lastName>Buffett</lastName>
        <address>8192 Wall Street</address>
        <city>Omaha</city>
        <state>NE</state>
        <zipCode>25372</zipCode>
        <transactions>
            <transaction>
                <accountNumber>8179238</accountNumber>
                <transactionDate>2010-10-27 05:56:59</transactionDate>
                <amount>-91.76</amount>
            </transaction>
            <transaction>
                <accountNumber>8179238</accountNumber>
                <transactionDate>2010-10-06 21:51:05</transactionDate>
                <amount>-25.99</amount>
            </transaction>
        </transactions>
    </customer>
</customers>
```

The customer file is structured as a collection of customer sections. Each of these contains a collection of transaction sections. Spring Batch parses lines in flat files into FieldSets. When working with XML, Spring Batch parses XML fragments that you define into your domain objects. What is a fragment? As Figure 7-4 shows, an XML fragment is a block of XML from open to close tag. Each time the specified fragment exists in your file, it will be considered a single record and converted into an item to be processed.

```
<customers>
  <customer>
    <firstName>Laura</firstName>
    <middleInitial>O</middleInitial>
    <lastName>Minella</lastName>
    <address>2039 Wall Street</address>
    <city>Omaha</city>
    <state>IL</state>
    <zip>35446</zip>
    <transaction>
      <account>829433</account>
      <transactionDate>2010-10-14 05:49:58</transactionDate>
      <amount>26.08</amount>
    </transaction>
  </customer>
  <customer>
    <firstName>Michael</firstName>
    <middleInitial>T</middleInitial>
    <lastName>Buffett</lastName>
    <address>8192 Wall Street</address>
    <city>Omaha</city>
    <state>NE</state>
    <zip>25372</zip>
    <transaction>
      <account>8179238</account>
      <transactionDate>2010-10-27 05:56:59</transactionDate>
      <amount>-91.76</amount>
    </transaction>
    <transaction>
      <account>8179238</account>
      <transactionDate>2010-10-06 21:51:05</transactionDate>
      <amount>-25.99</amount>
    </transaction>
  </customer>
</customers>
```

```
  <customer>
    <firstName>Laura</firstName>
    <middleInitial>O</middleInitial>
    <lastName>Minella</lastName>
    <address>2039 Wall Street</address>
    <city>Omaha</city>
    <state>IL</state>
    <zip>35446</zip>
    <transaction>
      <account>829433</account>
      <transactionDate>2010-10-14 05:49:58</transactionDate>
      <amount>26.08</amount>
    </transaction>
  </customer>
```
Fragment

Figure 7-4. XML fragments as Spring Batch sees them

In the customer input file, you have the same data at the customer level. You also have a collection of transaction elements within each customer, representing the list of transactions you put together in the multiline example previously.

To parse your XML input file, you will use the `org.springframework.batch.item.xml.StaxEventItemReader` that Spring Batch provides. To use it, you define a fragment root element name, which identifies the root element of each fragment considered an item in your XML. In your case, this will be the `customer` tag. It also takes a resource, which will be the same your `customerFile` bean as it has been previously. Finally, it takes an `org.springframework.oxm.Unmarshaller` implementation. This will be used to convert the XML to your domain object. Listing 7-35 shows the configuration of your `customerFileReader` using the `StaxEventItemReader` implementation.

Listing 7-35. `customerFileReader` Configured with the `StaxEventItemReader`

```
...
@Bean
@StepScope
public StaxEventItemReader<Customer> customerFileReader(
            @Value("#{jobParameters['customerFile']}") Resource inputFile) {

    return new StaxEventItemReaderBuilder<Customer>()
                .name("customerFileReader")
                .resource(inputFile)
                .addFragmentRootElements("customer")
                .unmarshaller(customerMarshaller())
                .build();
}
...
```

Spring Batch is not picky about the XML binding technology you choose to use. Spring provides Unmarshaller implementations that use Castor, JAXB, JiBX, XMLBeans, and XStream in their oxm package. For this example, you will use JAXB .

For your customerMarshaller configuration, you will use the org.springframework.oxm. jaxb.Jaxb2Marshaller implementation provided by Spring. To use it, we will need to add a few new dependencies to our project. Listing 7-36 lists out the new dependencies we will need to add JAXB to our classpath.

Listing 7-36. JAXB Dependencies

```
...
<dependency>
      <groupId>org.springframework</groupId>
      <artifactId>spring-oxm</artifactId>
</dependency>
<dependency>
      <groupId>javax.xml.bind</groupId>
      <artifactId>jaxb-api</artifactId>
      <version>2.2.11</version>
</dependency>
<dependency>
      <groupId>com.sun.xml.bind</groupId>
      <artifactId>jaxb-core</artifactId>
      <version>2.2.11</version>
</dependency>
<dependency>
      <groupId>com.sun.xml.bind</groupId>
      <artifactId>jaxb-impl</artifactId>
      <version>2.2.11</version>
</dependency>
<dependency>
      <groupId>javax.activation</groupId>
      <artifactId>activation</artifactId>
      <version>1.1.1</version>
</dependency>
...
```

With the dependencies for both JAXB and the Spring components to use it (via the Spring OXM module), we will need to configure our application to parse the XML. First, we need to add the JAXB annotations to our domain objects. We have both the Customer and Transaction objects. In order for JAXB to understand how they map to the tags in the XML, we need to annotate the classes and tell it. For our Transaction class, we just need to add @XmlType(name = "transaction") to our class. However, for the Customer class, we need to tell JAXB not only the element to expect (via the @XmlRootElement annotation), but also we need to explain to the parser how we've constructed the transaction collection (that the collection is wrapped via the <transactions> element and each element within the collection consists of a <transaction> block). Listing 7-37 illustrates our updated Customer class with the annotations applied.

Listing 7-37. JAXB Annotations for the Customer Class

```
...
@XmlRootElement
public class Customer {

     private String firstName;
     private String middleInitial;
     private String lastName;
     private String address;
     private String city;
     private String state;
     private String zipCode;

     private List<Transaction> transactions;

     public Customer() {
     }

     // Other getters and setters were removed for brevity.
// No change to them is required

     @XmlElementWrapper(name = "transactions")
     @XmlElement(name = "transaction")
     public void setTransactions(List<Transaction> transactions) {
             this.transactions = transactions;
     }

     @Override
     public String toString() {
             StringBuilder output = new StringBuilder();

             output.append(firstName);
             output.append(" ");
             output.append(middleInitial);
             output.append(". ");
             output.append(lastName);

             if(transactions != null&& transactions.size() > 0) {
                     output.append(" has ");
                     output.append(transactions.size());
                     output.append(" transactions.");
             } else {
                     output.append(" has no transactions.");
             }

             return output.toString();
     }
}
```

Once we have configured our domain objects with the appropriate mappings, we can configure the actual Unmarshaller that the StaxEventItemReader will use to parse each block. Since our annotations on the domain objects handle the majority of the configuration, we really just need to tell the Jaxb2Marshaller[2] what classes to be aware of.

Listing 7-38 shows the code needed to configure our Unmarshaller.

Listing 7-38. Jaxb2Marshaller Configuration

```
...
@Bean
public Jaxb2Marshaller customerMarshaller() {
    Jaxb2Marshaller jaxb2Marshaller = new Jaxb2Marshaller();

    jaxb2Marshaller.setClassesToBeBound(Customer.class,
                Transaction.class);

    return jaxb2Marshaller;
}
...
```

The last piece of the puzzle is to configure our step to use this new ItemReader as its source of input. Listing 7-39 shows the updated code for our step.

Listing 7-39. copyFileStep

```
...
@Bean
public Step copyFileStep() {
    return this.stepBuilderFactory.get("copyFileStep")
                .<Customer, Customer>chunk(10)
                .reader(customerFileReader(null))
                .writer(itemWriter())
                .build();
}
...
```

That's all you need to parse XML into items in Spring Batch! By running this job, you will get the same output as you did from the multiline record job.

While XML is still found in many enterprises, it is not the serialization format preferred by most these days. JSON has taken over in many ways as the preferred storage format for data. Spring Batch has you covered if you need to read JSON. This next section will take a look at the JsonItemReader and its capabilities for reading JSON documents in a fashion similar to what we just looked at with XML.

JSON

As XML has fallen out of favor by many due to its verbosity, JSON has skyrocketed up as a common replacement. JSON as a data format is less verbose, yet just as flexible as XML is. The rise in complex JavaScript-based web front ends ushered in the need for a common mechanism for communicating data

---

[2]The Jaxb2Marshaller also implements the Unmarshaller interface which is why we are using it here.

between back ends and front ends. Once back ends began using JSON for that piece of communication, it quickly spread into a number of other applications. Because of this, you may see JSON that you need to read in your batch processing. Fortunately, Spring Batch has an `ItemReader` that just fits the bill.

The `JsonItemReader` works with the same concept as the `StaxEventItemReader` does in that it reads chunks of JSON and parses them into objects. The JSON document is expected to be a complete JSON document containing a single array of objects. The parsing that the `JsonItemReader` does is delegated to an implementation of the `JsonObjectReader` interface. This interface is what does the actual parsing from JSON to an object in a similar manner to how the `Unmarshaller` parses XML into an object in the `StaxEventItemReader`. Out of the box, Spring Batch provides two implementations of the `JsonObjectReader` interface, one using Jackson as the parsing engine and one using Gson. For our example, we will use the Jackson implementation.

Before we look at the code, let's take a look at the input file we will be reading in. It is actually the same data as we read in from the customer.xml file in the previous section; however, it is represented via JSON instead. Listing 7-40 shows the file.

Listing 7-40. customer.json

```json
[
  {
    "firstName": "Laura",
    "middleInitial": "O",
    "lastName": "Minella",
    "address": "2039 Wall Street",
    "city": "Omaha",
    "state": "IL",
    "zipCode": "35446",
    "transactions": [
      {
        "accountNumber": 829433,
        "transactionDate": "2010-10-14 05:49:58",
        "amount": 26.08
      }
    ]
  },
  {
    "firstName": "Michael",
    "middleInitial": "T",
    "lastName": "Buffett",
    "address": "8192 Wall Street",
    "city": "Omaha",
    "state": "NE",
    "zipCode": "25372",
    "transactions": [
      {
        "accountNumber": 8179238,
        "transactionDate": "2010-10-27 05:56:59",
        "amount": -91.76
      },
      {
        "accountNumber": 8179238,
        "transactionDate": "2010-10-06 21:51:05",
```

```
      "amount": -25.99
    }
  ]
 }
]
```

To configure our JsonItemReader, we will use the builder that Spring Batch provides to do so. For this file, we will need to configure three dependencies: a name for restartability, the JsonObjectReader we will be using, and the Resource to be read in. Other configuration options on this ItemReader include a flag to indicate if the input must exist (strict, true by default), a flag indicating if state should be saved (saveState, true by default), the maximum number of items to be read (maxItemCount, Integer.MAX_VALUE by default) and the currentItemCount (used on restarts). Listing 7-41 illustrates configuring the JsonItemReader to read in our file.

Listing 7-41. JsonItemReader Configuration

```
...
@Bean
@StepScope
public JsonItemReader<Customer> customerFileReader(
            @Value("#{jobParameters['customerFile']}") Resource inputFile) {

    ObjectMapper objectMapper = new ObjectMapper();
    objectMapper.setDateFormat(new SimpleDateFormat("yyyy-MM-dd hh:mm:ss"));

    JacksonJsonObjectReader<Customer> jsonObjectReader =
            new JacksonJsonObjectReader<>(Customer.class);
    jsonObjectReader.setMapper(objectMapper);

    return new JsonItemReaderBuilder<Customer>()
                  .name("customerFileReader")
                  .jsonObjectReader(jsonObjectReader)
                  .resource(inputFile)
                  .build();
}
...
```

Walking through Listing 7-41, we begin by creating an ObjectMapper instance. This is the main class that Jackson uses to read and write JSON. In many cases, you won't need to do this step; however, in our case we need to specify the format of the dates in our input file. This means that we need to customize the ObjectMapper instance we will be using. Once we create the ObjectMapper instance and configure the format for the transactionDate fields in the input file, we can create our JacksonJsonObjectReader. This class has two dependencies: The first is the class it will be returning (Customer in our case). The second is the customized ObjectMapper instance we just created. Finally, we can configure our JsonItemReader instance. We create a new instance of the JsonItemReaderBuilder, configure our name, the JsonObjectReader instance, the Resource to be read, and call build() to construct our instance.

This is the only change we need to make from the XML-based example.[3] If we run the job using the command java -jar copyJob.jar customerFile=/path/to/customer/customer.json, we see the same output as we saw when running the XML sample as illustrated in Listing 7-42.

---

[3]Jackson is already included on the classpath of a Spring Batch application so there are no dependency updates needed. If you choose to use Gson, you will need to import that as well.

Listing 7-42. `JsonItemReader` Job Output

```
2019-01-30 23:50:27.012  INFO 10451 --- [          main] o.s.b.a.b.JobLauncherCommandLineRu
nner    : Running default command line with: [customerFile=/input/customer.json]
2019-01-30 23:50:27.153  INFO 10451 --- [          main] o.s.b.c.l.support.
SimpleJobLauncher      : Job: [SimpleJob: [name=job]] launched with the following
parameters: [{customerFile=/input/customer.json}]
2019-01-30 23:50:27.222  INFO 10451 --- [          main] o.s.batch.core.job.
SimpleStepHandler      : Executing step: [copyFileStep]
Laura O. Minella has 1 transactions.
Michael T. Buffett has 2 transactions.
2019-01-30 23:50:27.355  INFO 10451 --- [          main] o.s.b.c.l.support.
SimpleJobLauncher      : Job: [SimpleJob: [name=job]] completed with the following
parameters: [{customerFile=/input/customer.json}] and the following status: [COMPLETED]
```

Over the course of this section, you have covered a wide array of file-based input formats. Fixed-length files, delimited files, and various record configurations as well as XML and JSON are all available to be handled via Spring Batch with no or very limited coding, as you have seen. However, not all input will come from a file. Relational databases will provide a large amount of the input for your batch processes. The next section will cover the facilities that Spring Batch provides for database input.

Database Input

Databases serve as a great source of input for batch processes for a number of reasons. They provide transactionality built in, they are typically more performant, and they scale better than files. They also provide better recovery features out of the box than most other input formats. When you consider all of these and the fact that most enterprise data is stored in relational databases to begin with, your batch processes will need to be able to handle input from databases. In this section, you will look at some of the facilities that Spring Batch provide out of the box to handle reading input data from a database including JDBC, Hibernate, and JPA.

JDBC

In the Java world, database connectivity begins with JDBC. We all go through the pain of writing the JDBC connection code when we learn it, then quickly forget those lines when we realize that most frameworks handle things like connections for us. One of the Spring framework's strengths is encapsulating the pain points of things like JDBC in ways that allow developers to concentrate only on the business-specific details.

In this tradition, the developers of the Spring Batch framework have extended the Spring framework's JDBC functionality with the features that are needed in the batch world. But what are those features and how has Spring Batch addressed them?

When working with batch processes, the need to process large amounts of data is common. If you have a query that returns millions of records, you probably don't want all of that data loaded into memory at once. However, if you use Spring's `JdbcTemplate`, that is exactly what you would get. The `JdbcTemplate` loops through the entire `ResultSet`, mapping every row to the required domain object in memory.

Instead, Spring Batch provides two different methods for loading records one at a time as they are processed: a cursor and paging. A cursor is implemented via a standard `java.sql.ResultSet`. When a `ResultSet` is opened, every time the `next()` method is called a batch of records from the database is returned. This allows records to be streamed from the database on demand, which is the behavior that you need for a cursor.

Paging, on the other hand, takes a bit more work. The concept of paging is that you retrieve records from the database in chunks called pages. Each page is created by its own, independent SQL query. As you read through each page, a new page is read from the database via a new query. Figure 7-5 shows the difference between the two approaches.

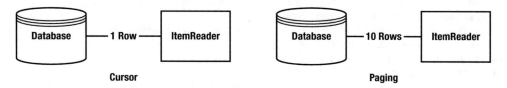

Figure 7-5. *Cursor vs. paging*

As you can see in Figure 7-5, the first read in the cursor returns a single record and advances the record you point at to the next record, streaming a single record at a time, whereas in the pagination approach, you receive 10 records from the database at a time. You will look at both approaches (cursor implementations as well as paging) for each of the database technologies you will look at. Let's start with straight JDBC.

JDBC Cursor Processing

For this example, you'll be using a CUSTOMER table. Using the same fields you have been working with up to now, you will create a database table to hold the data. Figure 7-6 shows the database model for the new CUSTOMER table.

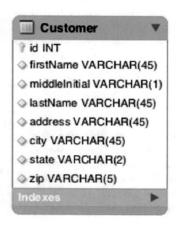

Figure 7-6. *Customer data model*

To implement a JDBC reader (either cursor-based or page-based), you will need to do two things: configure the reader to execute the query that is required and create a RowMapper implementation just like the Spring JdbcTemplate requires to map your ResultSet to your domain object. Before we get into the new components we need to use, let's revisit our domain object (Customer) and address the changes we need to make for it to be compatible with the database table we are using. Listing 7-43 illustrates the Customer domain object updated for this section.

Listing 7-43. `Customer`

```
...
public class Customer {

    private Long id;

    private String firstName;
    private String middleInitial;
    private String lastName;
    private String address;
    private String city;
    private String state;
    private String zipCode;

    public Customer() {}

    // Getters and setters removed

    @Override
    public String toString() {
        return "Customer{" +
                    "id=" + id +
                    ", firstName='" + firstName + '\" +
                    ", middleInitial='" + middleInitial + '\" +
                    ", lastName='" + lastName + '\" +
                    ", address='" + address + '\" +
                    ", city='" + city + '\" +
                    ", state='" + state + '\" +
                    ", zipCode='" + zipCode + '\" +
                    '}';
    }
}
```

The `Customer` object has had all the transaction related code removed as well as any JAXB annotations. We have also added an `id` field for the primary key in our database table. With the domain object properly defined, we can move onto the `RowMapper` implementation.

A standard piece of the core Spring Framework's JDBC support, a `RowMapper` is exactly what it sounds like. It takes a row from a `ResultSet` and maps the fields to a domain object. In your case, you will be mapping the fields of the CUSTOMER table to the `Customer` domain object. Listing 7-44 shows the `CustomerRowMapper` you'll use for your JDBC implementations.

Listing 7-44. `CustomerRowMapper`

```
...
public class CustomerRowMapper implements RowMapper<Customer> {

    @Override
    public Customer mapRow(ResultSet resultSet, int rowNumber) throws
                    SQLException {
        Customer customer = new Customer();
```

```
            customer.setId(resultSet.getLong("id"));
            customer.setAddress(resultSet.getString("address"));
            customer.setCity(resultSet.getString("city"));
            customer.setFirstName(resultSet.getString("firstName"));
            customer.setLastName(resultSet.getString("lastName"));
            customer.setMiddleInitial(resultSet.getString("middleInitial"));
            customer.setState(resultSet.getString("state"));
            customer.setZipCode(resultSet.getString("zipCode"));

            return customer;
        }
}
```

With the ability to map your query results to a domain object, you need to be able to execute a query by opening a cursor to return results on demand. To do that, you will use Spring Batch's org.springframework. batch.item.database.JdbcCursorItemReader. This ItemReader opens a cursor (by creating a ResultSet) and have a row mapped to a domain object each time the read method is called by Spring Batch. To configure the JdbcCursorItemReader, you provide a minimum of three dependencies: a DataSource, the query you want to run, and your RowMapper implementation. Listing 7-45 shows the configuration for your customerItemReader.

Listing 7-45. JDBC Cursor-Based customerItemReader

```
...
@Bean
public JdbcCursorItemReader<Customer> customerItemReader(DataSource dataSource) {
    return new JdbcCursorItemReaderBuilder<Customer>()
            .name("customerItemReader")
            .dataSource(dataSource)
            .sql("select * from customer")
            .rowMapper(new CustomerRowMapper())
            .build();
}
...
```

I should point out that while the rest of the configurations for the job do not need to be changed (the same ItemWriter will work fine), you will need to update the reference to the customerFileReader in the copyFileStep to reference your new customerItemReader instead.

With the configuration you have now, each time Spring Batch calls the read() method on the JdbcCursorItemReader, the database will return a single row to be mapped to your domain object and processed.

■ **Note**   Not all databases stream data in a ResultSet by default. Some will attempt to load all of the rows into memory at once, which can be problematic on larger datasets. Special configuration may be needed. Refer to the documentation of your database for details.

To run your job, you use the command: java -jar copyJob. This command will execute your job generating the same type of output you have in your previous examples.

Although this example is nice, it lacks one key ingredient. The SQL is hardcoded. I can think of very few instances where SQL requires no parameters. Using the JdbcCursorItemReader, you use the same functionality to set parameters in your SQL as you would using the JdbcTemplate and a PreparedStatement. To do this, you need to use an org.springframework.jdbc.core.PreparedStatementSetter implementation. A PreparedStatementSetter is similar to a RowMapper; however, instead of mapping a ResultSet row to a domain object, you are mapping parameters to your SQL statement. You can write your own, however Spring is nice enough to provide a few useful implementations for you. The one we will use is the ArgumentPreparedStatementSetter found in Spring Framework. This instance takes an array of objects. If the objects are not of type SqlParameterValue, then the objects are set as the values (in order of the array) on the PreaparedStatement (where the ? is in the SQL statement). If the values are SqlParameterValue instances, that type provides more metadata on what to do with the value (what index to set it at, what type it is, etc.) and Spring will obey its wishes. Listing 7-46 shows the updated configuration with both the reader builder and the ArgumentPreparedStatementSetter configurations.

Listing 7-46. Processing Only Customers by a Given City

```
...
@Bean
public JdbcCursorItemReader<Customer> customerItemReader(DataSource dataSource) {
    return new JdbcCursorItemReaderBuilder<Customer>()
            .name("customerItemReader")
            .dataSource(dataSource)
            .sql("select * from customer where city = ?")
            .rowMapper(new CustomerRowMapper())
            .preparedStatementSetter(citySetter(null))
            .build();
}

@Bean
@StepScope
public ArgumentPreparedStatementSetter citySetter(
        @Value("#{jobParameters['city']}") String city) {

    return new ArgumentPreparedStatementSetter(new Object [] {city});
}
...
```

This job is executed using a command that includes the job parameter city=Chicago. The full command being java -jar copyJob.jar city=Chicago. The results of the job are only the customers with an address in Chicago are displayed as shown in Listing 7-47.

Listing 7-47. Customers in Chicago

```
...
2019-01-31 22:31:41.939  INFO 33800 --- [           main] o.s.b.c.l.support.
SimpleJobLauncher     : Job: [SimpleJob: [name=job]] launched with the following
parameters: [{city=Chicago}]
2019-01-31 22:31:41.995  INFO 33800 --- [           main] o.s.batch.core.job.
SimpleStepHandler     : Executing step: [copyFileStep]
Customer{id=297, firstName='Hermione', middleInitial='K', lastName='Kirby',
address='599-9125 Et St.', city='Chicago', state='IL', zipCode='95546'}
```

```
Customer{id=831, firstName='Oren', middleInitial='Y', lastName='Benson', address='P.O. Box
201, 1204 Sed St.', city='Chicago', state='IL', zipCode='91416'}
2019-01-31 22:31:42.063  INFO 33800 --- [            main] o.s.b.c.l.support.
SimpleJobLauncher      : Job: [SimpleJob: [name=job]] completed with the following
parameters: [{city=Chicago}] and the following status: [COMPLETED]
...
```

With the ability to not only stream items from the database but also inject parameters into your queries, this approach is useful in the real world. There are good and bad things about this approach. It can be a good thing to stream records in certain cases; however, when processing a million rows, the individual network overhead for each request can add up. Add to that, a ResultSet is not thread safe which means this approach cannot be used in a multithreaded environment. All of this leads you to the other option, paging.

JDBC Paged Processing

When working with a paginated approach, Spring Batch returns the result set in chunks called pages. Each page is a predefined number of records to be returned by the database. It is important to note that when working with pages, the items your job will process will still be processed individually. There is no difference in the processing of the records. What differs is the way they are retrieved from the database. Instead of retrieving records one at a time via a single SQL query, paging will execute a new query for each page, loading only those rows into memory at a time. In this section, you'll update your configuration to return a page of 10 records in a page.

In order for paging to work, you need to be able to query based on a page size and page number (the number of records to return and which page you are currently processing). For example, if your total number of records is 10,000 and your page size is 100 records, you need to be able to specify that you are requesting the 20th page of 100 records (or records 2,000 through 2100). To do this, you provide an implementation of the org.springframework.batch.item.database.PagingQueryProvider interface to the JdbcPagingItemReader. The PagingQueryProvider interface provides all of the functionality required to navigate a paged ResultSet.

Unfortunately, each database offers its own paging implementation. Because of this, you have the following two options:

1. Configure a database-specific implementation of the PagingQueryProvider. As of this writing, Spring Batch provides implementations for DB2, Derby, H2, HSql, MySQL, Oracle, Postgres, SqlServer, and Sybase.

2. Configure your reader to use the org.springframework.batch.item.database. support.SqlPagingQueryProviderFactoryBean. This factory detects what database implementation to use.

Since using the SqlPagingQueryProviderFactoryBean will usually provide us with what we want by autodetecting what database platform you are using and returning the appropriate PagingQueryProvider, we will use it for our examples.

To configure the JdbcPagingItemReader, you have four dependencies: a DataSource, the PagingQueryProvider implementation, your RowMapper implementation, and the size of your page. You also have the opportunity to configure your SQL statement's parameters to be injected by Spring. Listing 7-48 shows the configuration for the JdbcPagingItemReader.

Listing 7-48. `JdbcPagingItemReader` Configuration

```
...
@Bean
@StepScope
public JdbcPagingItemReader<Customer> customerItemReader(DataSource dataSource,
        PagingQueryProvider queryProvider,
        @Value("#{jobParameters['city']}") String city) {

    Map<String, Object> parameterValues = new HashMap<>(1);
    parameterValues.put("city", city);

    return new JdbcPagingItemReaderBuilder<Customer>()
            .name("customerItemReader")
            .dataSource(dataSource)
            .queryProvider(queryProvider)
            .parameterValues(parameterValues)
            .pageSize(10)
            .rowMapper(new CustomerRowMapper())
            .build();
}

@Bean
public SqlPagingQueryProviderFactoryBean pagingQueryProvider(DataSource dataSource) {
    SqlPagingQueryProviderFactoryBean factoryBean = new SqlPagingQueryProviderFactoryBean();

    factoryBean.setDataSource(dataSource);
    factoryBean.setSelectClause("select *");
    factoryBean.setFromClause("from Customer");
    factoryBean.setWhereClause("where city = :city");
    factoryBean.setSortKey("lastName");

    return factoryBean;
}
```

As you can see, to configure your `JdbcPagingItemReader`, you provide it a `DataSource`, `PagingQueryProvider`, the parameters to be injected into your SQL, the size of each page, and the `RowMapper` implementation that will be used to map your results.

Within the `PagingQueryProvider`'s configuration, you provide five pieces of information. The first three are the different pieces of your SQL statement: the select clause, the from clause, and the where clause of your statement. The next property you set is the sort key. It is important to sort your results when paging since instead of a single query being executed and the results being streamed, a paged approach will typically execute a query for each page. In order for the record order to be guaranteed across query executions, an order by is required and is applied to the generated SQL statement for any fields that are listed in the `sortKey`. It is important that this sort key is also required to be unique within the `ResultSet`. The reason for this is that the sort key is the field that Spring Batch uses during its creation of the SQL query to execute. Finally, you have a `DataSource` reference. You may wonder why you need to configure the `DataSource` in both the `SqlPagingQueryProviderFactoryBean` and the `JdbcPagingItemReader`. The `SqlPagingQueryProviderFactoryBean` uses the `DataSource` to determine what type of database it's working with. You can configure the database type explicitly if you want as well via the `setDatabaseType(String databaseType)` method. From there, it provides the appropriate implementation of the `PagingQueryProvider` to be used for your reader.

The use of parameters in a paging context is different than it is in the previous cursor example. Instead of creating a single SQL statement with question marks as parameter placeholders, you build your SQL statement in pieces. Within the whereClause string, you have the option of using either the standard question mark placeholders or you can use the named parameters as I did in the customerItemReader in Listing 7-48. From there, you can inject the values to be set as a Map in your configuration. In this case, the city entry in the parameterValues maps to the named parameter city in your whereClause String. If you wanted to use question marks instead of names, you would use the number of the question mark as the key for each parameter. With all of the pieces in place, Spring Batch will construct the appropriate query for each page each time it is required.

As you can see, straight JDBC interaction with a database for reading the items to be processed is actually quite simple. With not much more than a few lines of Java, you can have a performant ItemReader in place that allows you to input data to your job. However, JDBC isn't the only way to access database records. Object Relational Mapping (ORM) technologies like Hibernate and MyBatis have become popular choices for data access given their well-executed solution for mapping relational database tables to objects. You will take a look at how to use Hibernate for data access next.

Hibernate

Hibernate is the leading ORM technology in Java today. Written by Gaven King back in 2001, Hibernate provides the ability to map the object-oriented model you use in your applications to a relational database. Hibernate uses XML files or annotations to configure mappings of objects to database tables; it also provides a framework for querying the database by object. This provides the ability to write queries based on the object structure with little or no knowledge of the underlying database structure. In this section, you will look at how to use Hibernate as your method of reading items from a database.

Using Hibernate in batch processing is not as straightforward as it is for web applications. For web applications, the typical scenario is to use the session in view pattern. In this pattern, the session is opened as a request comes into the server, all processing is done using the same session, and then the session is closed as the view is returned to the client. Although this works well for web applications that typically have small independent interactions, batch processing is different.

For batch processing, if you use Hibernate naively, you would use the normal stateful session implementation, read from it as you process your items, and write to it as you complete your processing closing the session once the step is complete. However, as mentioned, the standard session within Hibernate is stateful. If you are reading a million items, processing them, then writing those same million items, the Hibernate session will cache the items as they are read and an OutOfMemoryException will occur.

Another issue with using Hibernate as a persistence framework for batch processing is that Hibernate incurs larger overhead than straight JDBC does. When processing millions of records, every millisecond can make a big difference.[4]

Spring Batch's Hibernate based ItemReaders are developed to do the right thing. They do things like flushing the Session on commit as well as other features that are related more for batch processing than web-based use of Hibernate. In environments where Hibernate objects are mapped previously for another system, it can be an efficient way to get things up and running. Hibernate also does solve the fundamental issue of mapping objects to database tables in a very robust way. It's up to you and your requirements to determine if Hibernate or any ORM tool is right for your job.

---

[4]A one millisecond increase per item over the course of a million items can add over 15 minutes of processing time to a single step.

Cursor Processing with Hibernate

To use Hibernate with a cursor, you will need to configure the `SessionFactory`, your `Customer` mapping, the `HibernateCursorItemReader`, and add the Hibernate dependencies to your pom.xml file. Let's start with updating your pom.xml file.

Using Hibernate in your job will require the addition of a new dependency, the spring-boot-starter-jpa dependency. Now this example won't use JPA for data access directly (we will use the JPA annotations for data mapping), but since this starter is backed by Hibernate, we'll get all the Hibernate specific dependencies as well as all the extra stuff Spring Data JPA does for us (register custom type converters, etc.). We'll look at how to use JPA in the next section. Listing 7-49 shows the addition of the starter to our pom.xml.

Listing 7-49. Hibernate Dependencies in POM

```
...
<dependency>
      <groupId>org.springframework.boot</groupId>
      <artifactId>spring-boot-starter-data-jpa</artifactId>
</dependency>
 ...
```

With the Hibernate framework added to your project, you can map your `Customer` object to the Customer table in the database. To keep things simple, you will use Hibernate's annotations to configure the mapping. Listing 7-50 shows the updated `Customer` object mapped to the Customer table.

Listing 7-50. Customer Object Mapped to the Customer Table via Hibernate Annotations

```
...
@Entity
@Table(name = "customer")
public class Customer {

    @Id
    private Long id;

    @Column(name = "firstName")
    private String firstName;
    @Column(name = "middleInitial")
    private String middleInitial;
    @Column(name = "lastName")
    private String lastName;
    private String address;
    private String city;
    private String state;
    private String zipCode;

    public Customer() {
    }

    @Override
    public String toString() {
        return "Customer{" +
                    "id=" + id +
```

```
                            ", firstName='" + firstName + '\" +
                            ", middleInitial='" + middleInitial + '\" +
                            ", lastName='" + lastName + '\" +
                            ", address='" + address + '\" +
                            ", city='" + city + '\" +
                            ", state='" + state + '\" +
                            ", zipCode='" + zipCode + '\" +
                            '}';
        }
}
```

The Customer class's mapping consists of identifying the object as an Entity using the JPA annotation @Entity, specifying the table the entity maps to using the @Table annotation, and finally identifying the ID for the table with the @Id tag. All other attributes on the Customer will be mapped automatically by Hibernate since you have named the columns in the database the same as the attributes in your object. However, we do need to specify two properties within Spring Boot for this mapping to be applied. Since we are using camel case in this example and not the conventional underscore notation, we need to use the correct naming strategies for Hibernate. To do this, we will want to update our application.yml and set the following properties:

- `spring.jpa.hibernate.naming.implicit-strategy: "org.hibernate.boot.model.naming.ImplicitNamingStrategyLegacyJpaImpl"`

- `spring.jpa.hibernate.naming.physical-strategy: "org.hibernate.boot.model.naming.PhysicalNamingStrategyStandardImpl"`

With our data object mapped, we need to customize the TransactionManager we'll use for this batch job. By default, Spring Batch will give you a DataSourceTransactionManager. However we want a TransactionManager that will work across regular DataSource connections and a Hibernate Session. Spring just happens to have something we can use, the HibernateTransactionManager. That being said, we will need to configure it via a custom implementation of the BatchConfigurer. All we will need to do is override the DefaultBatchConfigurer.getTransactionManager() method and we should be good to go. Listing 7-51 shows the new BatchConfigurer.

Listing 7-51. HibernateBatchConfigurer

```
...
@Component
public class HibernateBatchConfigurer extends DefaultBatchConfigurer {

    private DataSource dataSource;
    private SessionFactory sessionFactory;
    private PlatformTransactionManager transactionManager;

    public HibernateBatchConfigurer(DataSource dataSource,
                EntityManagerFactory entityManagerFactory) {

        super(dataSource);
        this.dataSource = dataSource;
        this.sessionFactory = entityManagerFactory.unwrap(SessionFactory.class);
        this.transactionManager = new HibernateTransactionManager(this.sessionFactory);
    }
```

```
        @Override
        public PlatformTransactionManager getTransactionManager() {
                return this.transactionManager;
        }
}
```

You'll see that all we needed to do was create our own HibernateTransactionManager and return it via the overridden getTransactionManager() method. From here, Spring Batch will consume this where appropriate.[5]

With all of that configured, you need to actually configure the org.springframework.batch.item. database.HibernateCusorItemReader. Probably the simplest piece of the puzzle. Listing 7-52 shows using the HibernateCursorItemReaderBuilder to configure our reader. It takes a name, SessionFactory, query string, and any parameters for that query.

Listing 7-52. Configuring the HibernateCursorItemReader

```
...
@Bean
@StepScope
public HibernateCursorItemReader<Customer> customerItemReader(
            EntityManagerFactory entityManagerFactory,
            @Value("#{jobParameters['city']}") String city) {

    return new HibernateCursorItemReaderBuilder<Customer>()
                .name("customerItemReader")
                .sessionFactory(entityManagerFactory.unwrap(SessionFactory.class))
                .queryString("from Customer where city = :city")
                .parameterValues(Collections.singletonMap("city", city))
                .build();
    }
...
```

In this example, you used an HQL query as your method of querying the database. There are three other ways to specify the query to execute. Table 7-3 covers all three options.

Table 7-3. *Hibernate Query Options*

Option	Type	Description
queryName	String	This references a named Hibernate query as configured in your Hibernate configurations.
queryString	String	This is an HQL query specified in your Spring configuration.
queryProvider	HibernateQueryProvider	This provides the ability to programmatically build your Hibernate Query.
nativeQuery	String	Used to run a native SQL query and have Hibernate map the results.

---

[5]With this approach, the log messages will state that a DataSourceTransactionManager is being used. This is a bug in Spring Batch and is addressed in versions past 4.2.

That's all that is required to implement the Hibernate equivalent to `JdbcCursorItemReader`. Executing this job will output the same output as your previous job.

Paged Database Access with Hibernate

Hibernate, like JDBC, supports both cursor database access as well as paged database access. The only change required is to specify the `HibernatePagingItemReader` instead of the `HibernateCursorItemReader` in your job configuration class and specify a page size for your `ItemReader`. Listing 7-53 shows the updated `ItemReader` using paged database access with Hibernate.

Listing 7-53. Paging Database Access with Hibernate

```
...
@Bean
@StepScope
public HibernatePagingItemReader<Customer> customerItemReader (
      EntityManagerFactory entityManagerFactory,
        @Value("#{jobParameters['city']}") String city) {

    return new HibernatePagingItemReaderBuilder<Customer>()
            .name("customerItemReader")
            .sessionFactory(entityManagerFactory.unwrap(SessionFactory.class))
            .queryString("from Customer where city = :city")
            .parameterValues(Collections.singletonMap("city", city))
            .pageSize(10)
            .build();
}
...
```

Using Hibernate can speed up development of batch processing in situations where the mapping already exists as well as simplify the mapping of relational data to domain objects. However, Hibernate is not the only kid on the ORM block. The Java Persistence API (or JPA for short) is the native Java implementation of ORM persistence. You'll look at that next.

JPA

JPA or Java Persistence API brings a standardized approach to the object relational mapping (ORM) space. Hibernate was an early inspiration of JPA and currently implements the JPA specification. However, Hibernate is not a drop in replacement. JPA does not have a cursor based way to read items for example which is available in native Hibernate.[6] In this example, you will use JPA to provide paged database access similar to the Hibernate paged example you used previously.

Like most things that Spring Boot touches, configuring JPA when using Spring Boot is actually quite easy. In fact, we already did it in the Hibernate example previously. Any app that uses the spring-boot-starter-data-jpa will have all the necessary components required to use JPA with Spring Batch. In fact, we don't even need to create a custom `BatchConfigurer` implementation when using Spring Boot's starter because it handles the configuration of the `JpaTransactionManager` (similar to the Hibernate version) for us. Since our Hibernate examples used the JPA annotations, we actually don't need to do anything from a mapping perspective either.

---

[6]JPA 2.2 supports returning a `Stream`, but semantics make it difficult to use in a Spring Batch use case.

The only piece of the JPA puzzle we really need to be concerned with is to configure your ItemReader. As mentioned, JPA does not support cursor database access but it does support paging database access. The ItemReader will be the org.springframework.batch.item.database.JpaPagingItemReader which uses four dependencies: the name used as a prefix for entries in the ExecutionContext, the EntityManager provided by Spring Boot, a query to execute, and in your case, your query has a parameter, so you will inject the value of the parameter as well. Listing 7-54 shows the customerItemReader configured for JPA database access.

Listing 7-54. customerItemReader with JPA

```
...
@Bean
@StepScope
public JpaPagingItemReader<Customer> customerItemReader (
            EntityManagerFactory entityManagerFactory,
            @Value("#{jobParameters['city']}") String city) {

    return new JpaPagingItemReaderBuilder<Customer>()
                    .name("customerItemReader")
                    .entityManagerFactory(entityManagerFactory)
                    .queryString("select c from Customer c where c.city = :city")
                    .parameterValues(Collections.singletonMap("city", city))
                    .build();
}
...
```

Executing the job as it currently is configured will output all of the customers' names and addresses within the city specified at the command line. There's another way to specify queries in JPA: the Query object. To use JPA's Query API, you need to implement the org.springframework.batch.item.database. orm.JpaQueryProvider interface. The interface, which consists of a createQuery() method and a setEntityManager(EntityManager em) method is used by the JpaPagingItemReader to obtain the required Query to be executed. To make things easier, Spring batch provides an abstract base class for you to extend, the org.springframework.batch.item.database.orm.AbstractJpaQueryProvider. Listing 7-55 shows what the implementation to return the same query (as configured in Listing 7-54) looks like.

Listing 7-55. CustomerByCityQueryProvider

```
...
public class CustomerByCityQueryProvider extends AbstractJpaQueryProvider {

    private String cityName;

    public Query createQuery() {
        EntityManager manager = getEntityManager();

        Query query =
            manager.createQuery("select c from Customer " +
                                "c where c.city = :city");
        query.setParameter("city", cityName);

        return query;
    }
```

203

```
    public void afterPropertiesSet() throws Exception {
        Assert.notNull(cityName, "City name is required");
    }

    public void setCityName(String cityName) {
        this.cityName = cityName;
    }
}
```

For the `CustomerByCityQueryProvider`, you use the `AbstractJpaQueryProvider` base class to handle obtaining an `EntityManager` for you. From there, you create the JPA query, populate any parameters in the query and return it to Spring Batch for execution. To configure your `ItemReader` to use the `CustomerByCityQueryProvider` instead of the query string you provided previously, you simply swap the `queryString` parameter with the `queryProvider` parameter, as shown in Listing 7-56.

Listing 7-56. Using the JpaQueryProvider

```
...
@Bean
@StepScope
public JpaPagingItemReader<Customer> customerItemReader (
            EntityManagerFactory entityManagerFactory,
            @Value("#{jobParameters['city']}") String city) {

    CustomerByCityQueryProvider queryProvider =
                            new CustomerByCityQueryProvider();
    queryProvider.setCityName(city);

    return new JpaPagingItemReaderBuilder<Customer>()
                    .name("customerItemReader")
                    .entityManagerFactory(entityManagerFactory)
                    .queryProvider(queryProvider)
                    .parameterValues(Collections.singletonMap("city", city))
                    .build();
}
...
```

Using JPA can limit an application's dependencies on third party libraries while still providing many of the benefits of ORM libraries like Hibernate.

The last relational database topic we will take a look at in this chapter is how to read from the results of a stored procedure. This topic is up next.

Stored Procedures

In many enterprises, the relational database is not just a simple place with tables of data. It is an ecosystem of code that contains complex stored procedures that are used for all kinds of business purposes. While not the most friendly mechanism for working with a database for the average Java developer, stored procedures are a well-established tool found in many databases around the world.

What is a stored procedure? A stored procedure is a unit of database specific code that is saved in the database for future execution by a client of some kind. Not all databases support them, although they are available in most enterprise grade relational database options.

Before we can take a look at the configuration of our StoredProcedureItemReader, the component Spring Batch provides for reading data from a stored procedure, we should look at the stored procedure we will be using itself. In this case, since we are using MySQL as our database, we will be using MySQL's syntax for creating a stored procedure. For our procedure, we will do the same thing as the queries we have been executing up to now, find all customers by city. Listing 7-57 illustrates the code needed to create our stored procedure in MySQL.

Listing 7-57. customer_list stored procedure

```
DELIMITER //

CREATE PROCEDURE customer_list(IN cityOption CHAR(16))
  BEGIN
    SELECT * FROM CUSTOMER
    WHERE city = cityOption;
  END //

DELIMITER ;
```

To create the stored procedure we need to execute this code before we execute our job. The procedure we have defined takes one parameter in (cityOption), which is used in the query. By default, the query will return a ResultSet just like a regular SQL query would. It's important to note that you cannot just drop that code into your schema.sql file and expect Spring Boot to run it. You will need to execute the preceding code from the MySQL command line directly.

With the procedure created, we can take a look at how to configure our ItemReader. It should look similar to the JdbcCursorItemReader since the design of the class is based on it. They both have you configure a name, DataSource, RowMapper, and a PreparedStatementSettter. However, instead of defining the SQL for the query, we configure the StoredProcedureItemReader with the name of the procedure to call. Since stored procedures can handle more complex parameters we need to do a bit more mapping for our parameter definitions. The StoredProcedureItemReader takes an array of SqlParameter objects as the mechanism for defining the parameters the procedure takes. In our case, we will define a single parameter called cityOption of type VARCHAR. Listing 7-58 illustrates the configuration of the StoredProcedureItemReader.

Listing 7-58. StoredProcedureItemReader

```
...
@Bean
@StepScope
public StoredProcedureItemReader<Customer> customerItemReader(DataSource dataSource,
            @Value("#{jobParameters['city']}") String city) {

    return new StoredProcedureItemReaderBuilder<Customer>()
                .name("customerItemReader")
                .dataSource(dataSource)
                .procedureName("customer_list")
                .parameters(new SqlParameter[]{
                        new SqlParameter("cityOption", Types.VARCHAR)})
                .preparedStatementSetter(
                        new ArgumentPreparedStatementSetter(new Object[] {city}))
```

```
            .rowMapper(new CustomerRowMapper())
            .build();
}
...
```

Once this application is built and executed, the results are the same as we've seen from our previous relational database examples.

Relational databases hold the lion's share of the market when it comes to data stores. However, a new category of data stores has come on strong and that is the NoSQL variants. Spring Data, a project designed to simplify and provide a consistent programming model across these various datasources forms the foundation of what we are going to look at next.

Spring Data

According to the Spring Data website, it's "mission is to provide a familiar and consistent, Spring-based programming model for data access while still retaining the special traits of the underlying data store." Spring Data is not a single project, it represents a portfolio of projects that all have a set of consistent abstractions (the Repository) while still allowing users to access the unique features of each of the NoSQL and SQL data stores they support. In this section, we will look at a number of NoSQL data stores and how Spring Batch can consume data from them with the same declarative I/O style we've seen up to this point. We will begin by looking at MongoDB.

MongoDB

MongoDB's roots began at a company called 10gen back in 2007. 10gen developed it as part of a platform product they were working on at the time. The company later released MongoDB as a separate product and changed their name to MongoDB Inc.

MongoDB was one of the first NoSQL data stores to catch on, but it's difference from databases that most enterprise developers were used to and its lack of some traditional enterprise features (like ACID transactions), made it be considered more of a toy at the time. However, MongoDB has evolved into a popular document database for the enterprise.

The key feature for MongoDB is that it doesn't use tables. Instead, each database is made up of one or more collections. Each of these collections is a grouping of documents (typically JSON or BSON in format). These collections and documents can be traversed and queried via JavaScript or a JSON-based query language. This design allows for MongoDB to both be very fast and dynamic in that the schema can change based on the data you are looking at. Other features MongoDB has include

- *High availability and scaling*: MongoDB has great support for both high availability via replication and scalability via sharding natively.

- *Geospatial support*: MongoDB's query language supports queries that are attempting to determine things like if a point is within particular boundaries.

The MongoItemReader is a paging-based ItemReader in that it takes a query and returns data from the MongoDB servers in pages. The MongoItemReader takes a few dependencies as requirements:

- MongoOperations implementation: A MongoTemplate is required to execute the queries with.

- name: If saveState is true, just like all other stateful Spring Batch components.

- targetType: The Java class to be returned. This is what the document that is returned will be mapped to.

- Either a JSON-based query or a Query instance: The query to be executed.

Other items you can configure include sorts, hints, what fields to include, the MongoDB collection to query as well as any parameters needed for the query.

For our example, we will take a look at tweet data. Twitter conveniently uses JSON for all its communication mechanisms, so getting tweet data in JSON format is rather easy. We can obtain a simple dataset from Github here: https://github.com/ozlerhakan/mongodb-json-files. If we download the Tweets dataset from this git repository, we can import it into MongoDB using the command mongorestore -d tweets -c tweets_collection <PATH_TO_YOUR_UNZIPPED_FILE>/dump\ 2/twitter/tweets.bson. This will create a database called tweets with a collection called tweets_collection. The format of the JSON is such that within the root is an object called entities. Within that object are four fields: hashtags, symbols, user_mentions, and urls. The field we care about is hashtags. We will write a query to look for any element within that array to have the text value of the job parameter we pass in.

The next thing we need to do is add the appropriate dependency for working with MongoDB to our pom.xml. The dependency we will need is the spring-boot-starter-data-mongodb dependency as shown in Listing 7-59.

Listing 7-59. Spring Boot Starter for MongoDB

```
...
<dependency>
      <groupId>org.springframework.boot</groupId>
      <artifactId>spring-boot-starter-data-mongodb</artifactId>
</dependency>
...
```

With our data loaded and dependencies added, we can query it for a particular hash tag. To query MongoDB, we will need to configure our MongoItemReader with the following items:

- name: used for storing state in the ExecutionContext for restartability. Required if storeState is true.

- targetType: This is the class that each document returned will be deserialized into. For our use case, a Map will do.

- jsonQuery: We want to find any hashtag that is equal to the value we pass in via a job parameter.

- collection: The collection we want to query against. In this case, it will be tweets_collection.

- parameterValues: The values of any parameters in our query (the value of the hash tag to search for in our case).

- sorts: What fields to sort by and in what order. Since MongoItemReader is a paged ItemReader, output must be sorted.

- template: The MongoOperations implementation used to execute the queries against.

Listing 7-60 illustrates the configuration for the ItemReader and Step used in this job.

Listing 7-60. MongoItemReader

```
...
@Bean
@StepScope
public MongoItemReader<Map> tweetsItemReader(MongoOperations mongoTemplate,
            @Value("#{jobParameters['hashTag']}") String hashtag) {

    return new MongoItemReaderBuilder<Map>()
                    .name("tweetsItemReader")
                    .targetType(Map.class)
                    .jsonQuery("{ \"entities.hashtags.text\": { $eq: ?0 }}")
                    .collection("tweets_collection")
                    .parameterValues(Collections.singletonList(hashtag))
                    .pageSize(10)
                    .sorts(Collections.singletonMap("created_at", Sort.Direction.ASC))
                    .template(mongoTemplate)
                    .build();
}

@Bean
public Step copyFileStep() {
    return this.stepBuilderFactory.get("copyFileStep")
                    .<Map, Map>chunk(10)
                    .reader(tweetsItemReader(null, null))
                    .writer(itemWriter())
                    .build();
}
...
```

We need to configure one other item before running our job. When we imported our data into MongoDB, we did so into the tweets database. However, we haven't told our application that yet. In our application.yml, we will need to add the property `spring.data.mongodb.database: tweets`.

Once our application is built, we can run it with the job parameter hashTag=nodejs. On the dataset we imported, we should see 20 records printed into the console like the ones seen in Figure 7-7.

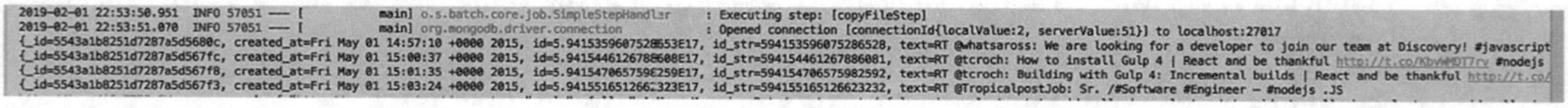

Figure 7-7. *MongoDB output*

MongoDB is one of the most popular NoSQL data stores on the market today. However it is not the only one. MongoDB serves a specific purpose with its document oriented approach. However there are other approaches to NoSQL data stores. The next reader we will explore is one that allows you to utilize any of Spring Data's `Repository` abstractions.

Spring Data Repository

Earlier in this section, I quoted the Spring Data's website as saying that their goal is to provide a consistent programming model. The main point of this consistency is the Repository abstraction. This abstraction allows you to do basic create/read/update/delete (CRUD) operations by simply defining an interface that extends one of the ones provided by Spring Data. This opens up a world of possibilities in that any data store that has a Spring Data project associated with it can be read via Spring Batch as long as they support the types of repository mechanisms needed. So, for example, if you needed to read from Apache Cassandra which does not have native support in Spring Batch, since it does have support in Spring Data for Apache Cassandra you can just by creating a repository for it.

What is a Spring Data repository though? Spring Data provides the feature where you can define an interface that extends one of their specific interfaces (the PagingAndSortingRepository for example) and Spring Data will handle the implementation for you. What is even more interesting is Spring Data provides a query language for these repositories that is based on the name of the method you define on your interface. For example, if you wanted to query our Customer table by the city attribute, you would create a method with the signature public List<Customer> findByCity(String city); This method will be interpreted by Spring Data and it will generate the appropriate query you need based on the data store you are using.

How Spring Batch fits into this is that Spring Batch takes advantage of Spring Data's PagingAndSortingRepository. This Repository interface defines repositories that have the ability to do both paging and sorting on the data in a standardized way. The RepositoryItemReader utilizes this to perform paged queries just like it does in the JdbcPagingItemReader or the HibernatePagingItemReader. The difference here is that this ItemReader can be utilized to query any data store that there is a Spring Data project for with repository support.

To take a look, we will go back to our JPA example. In that previous example, we queried a CUSTOMER table by city. To do that, we needed to define our query as a String that we passed into the JpaPagingItemReader. This time, we will create a repository interface and let Spring Data do the work. The domain object we will use is the same as shown in Listing 7-50.

Before we define our ItemReader, we need to create a repository that extends PagingAndSortingRepository. In our case, it will contain a single method for querying our database by city. The method signature mentioned previously in this section won't quite work because it does not utilize the paging mechanisms Spring Data provides. In order to get that functionality to be applied, we need to modify that signature in two small ways. First we need to add an additional parameter to the method: org.springframework.data.domain.Pageable. Implementations of this interface encapsulate the parameters related to requesting a page of data. Specifically, it includes the page size, the offset, the page number, the sort options, and other mechanisms needed to build the paged query. The other change we need to make to that method signature is that instead of returning a List<Customer>, we need to return a org.springframework.data.domain.Page<Customer>. A Page object contains the values from the Pageable we passed in, as well as the actual data for the specific page. It also contains some metadata about the dataset based on the query you ran like the total number of elements in the data set (not just the ones in the current page), an indicator if this page is the first page or the last page, and more. Listing 7-61 shows the repository interface we will use.

Listing 7-61. CustomerRepository

```
...
public interface CustomerRepository extends PagingAndSortingRepository<Customer, Long> {

    Page<Customer> findByCity(String city, Pageable pageRequest);
}
```

■ **Note** If more than one Spring Data starter is included in your project, you may need to use the data store specific interface like JpaRepository to indicate which data store your repository belongs to.

With our repository defined, we can now define our RepositoryItemReader to read data from it. To do so, we will use the RepositoryItemReaderBuilder passing it the name (for restartability), any arguments our method requires besides the Pageable parameter, the name of the method to call, the repository implementation itself, and define any sort we need (we will sort by the customer's last name in this example). Listing 7-62 illustrates this configuration.

Listing 7-62. RepositoryItemReader

```
...
@Bean
@StepScope
public RepositoryItemReader<Customer> customerItemReader(CustomerRepository repository,
            @Value("#{jobParameters['city']}") String city) {

    return new RepositoryItemReaderBuilder<Customer>()
                    .name("customerItemReader")
                    .arguments(Collections.singletonList(city))
                    .methodName("findByCity")
                    .repository(repository)
                    .sorts(Collections.singletonMap("lastName", Sort.Direction.ASC))
                    .build();
}
...
```

With this code in place, we can execute our job and view the results as shown in Listing 7-63.

Listing 7-63. RepositoryItemReader Job Results

```
...
2019-02-04 17:17:07.333  INFO 8219 --- [           main] o.s.batch.core.job.
SimpleStepHandler       : Executing step: [copyFileStep]
2019-02-04 17:17:07.448  INFO 8219 --- [           main] o.h.h.i.QueryTranslatorFactoryIniti
ator   : HHH000397: Using ASTQueryTranslatorFactory
Customer{id=831, firstName='Oren', middleInitial='Y', lastName='Benson', address='P.O. Box
201, 1204 Sed St.', city='Chicago', state='IL', zipCode='91416'}
Customer{id=297, firstName='Hermione', middleInitial='K', lastName='Kirby',
address='599-9125 Et St.', city='Chicago', state='IL', zipCode='95546'}
2019-02-04 17:17:07.657  INFO 8219 --- [           main] o.s.b.c.l.support.
SimpleJobLauncher       : Job: [SimpleJob: [name=job]] completed with the following
parameters: [{city=Chicago}] and the following status: [COMPLETED]
...
```

Up to this point you have covered various file and database input sources and the variety of ways you can obtain your input data from them. However, in a world of microservices and APIs direct access to the data is not always guaranteed. In the next section, you will cover how to obtain data from your existing Java services.

Existing Services

Many companies have Java applications (web or otherwise) currently in production. These applications have gone through strenuous amounts of analysis, development, testing, and bug fixing. The code that comprises these applications is battle tested and proven to work.

So why can't you use that code in your batch processes? Let's use the example that your batch process requires you to read in `Customer` objects. However, instead of a `Customer` object mapping to a single table or file like it has been up to now, your customer data is spread across multiple tables in multiple databases. Also, you never physically delete customers; instead you flag them as being deleted. A service to retrieve the customer objects already exists in your web-based application. How do you use that in your batch process? In this section you will look at how to call existing Spring services to provide data for your `ItemReader`.

Back in Chapter 4, you learned about a few adapters that Spring Batch provides for tasklets to be able to do different things, specifically the `org.springframework.batch.core.step.tasklet.CallableTaskletAdapter`, `org.springframework.batch.core.step.tasklet.MethodInvokingTaskletAdapter` and the `org.springframework.batch.core.step.tasklet.SystemCommandTasklet`. All three of these were used to wrap some other element in a way that Spring Batch could interact with it. To use an existing service within Spring Batch, the same pattern is used.

In the case of reading input, you will be using the `org.springframework.batch.item.adapter.ItemReaderAdapter`. Similarly to the way the `RepositoryItemReader` takes a reference to the `Repository` and the method to call on it, this `ItemReader` also takes a reference to the service to call and the name of the method to call as dependencies. You need to keep the following two things in mind when using the `ItemReaderAdapter`:

1. The object returned from each call is the object that will be returned by the `ItemReader`. If your service returns a single `Customer`, then that single `Customer` object will be the object passed onto the `ItemProcessor` and finally the `ItemWriter`. If a collection of `Customer` objects is returned by the service, it will be passed as a single item to the `ItemProcessor` and `ItemWriter` and it will be your responsibility to iterate over the collection.

2. Once the input is exhausted, the service method must return a `null`. This indicates to Spring Batch that the input is exhausted for this step.

For this example, you will use a service hardcoded to return a `Customer` object for each call until the `List` is exhausted. Once the `List` is exhausted, `null` will be returned for every call after. The `CustomerService` in Listing 7-64 generates a random list of `Customer` objects for your use.

Listing 7-64. `CustomerService`

```
...
@Component
public class CustomerService {

    private List<Customer> customers;
    private int curIndex;

    private String [] firstNames = {"Michael", "Warren", "Ann", "Terrence",
                                "Erica", "Laura", "Steve", "Larry"};
    private String middleInitial = "ABCDEFGHIJKLMNOPQRSTUVWXYZ";
    private String [] lastNames = {"Gates", "Darrow", "Donnelly", "Jobs",
                                "Buffett", "Ellison", "Obama"};
```

```
    private String [] streets = {"4th Street", "Wall Street", "Fifth Avenue",
                                 "Mt. Lee Drive", "Jeopardy Lane",
                                 "Infinite Loop Drive", "Farnam Street",
                                 "Isabella Ave", "S. Greenwood Ave"};
    private String [] cities = {"Chicago", "New York", "Hollywood", "Aurora",
                                "Omaha", "Atherton"};
    private String [] states = {"IL", "NY", "CA", "NE"};

    private Random generator = new Random();

    public CustomerService() {
        curIndex = 0;

        customers = new ArrayList<>();

        for(int i = 0; i < 100; i++) {
            customers.add(buildCustomer());
        }
    }

    private Customer buildCustomer() {
        Customer customer = new Customer();

        customer.setId((long) generator.nextInt(Integer.MAX_VALUE));
        customer.setFirstName(
            firstNames[generator.nextInt(firstNames.length - 1)]);
        customer.setMiddleInitial(
            String.valueOf(middleInitial.charAt(
                generator.nextInt(middleInitial.length() - 1))));
        customer.setLastName(
            lastNames[generator.nextInt(lastNames.length - 1)]);
        customer.setAddress(generator.nextInt(9999) + " " +
                            streets[generator.nextInt(streets.length - 1)]);
        customer.setCity(cities[generator.nextInt(cities.length - 1)]);
        customer.setState(states[generator.nextInt(states.length - 1)]);
        customer.setZip(String.valueOf(generator.nextInt(99999)));

        return customer;
    }

    public Customer getCustomer() {
        Customer cust = null;

        if(curIndex < customers.size()) {
            cust = customers.get(curIndex);
            curIndex++;
        }

        return cust;
    }
}
```

Finally, to use the service you have developed in Listing 7-64, using the `ItemReaderAdapter`, you configure your `customerItemReader` to call the `getCustomer` method for each item. Listing 7-65 shows the configuration for this.

Listing 7-65. Configuring the `CustomerService` and the `ItemReaderAdapter` to Call It

```
...
@Bean
public ItemReaderAdapter<Customer> itemReader(CustomerService customerService) {
    ItemReaderAdapter<Customer> adapter = new ItemReaderAdapter<>();

    adapter.setTargetObject(customerService);
    adapter.setTargetMethod("getCustomer");

    return adapter;
}
...
```

That's all that is required to use one of your existing services as the source of data for your batch job. Using existing services can allow you to reuse code that is tested and proven instead of running the risk of introducing new bugs by rewriting existing processes.

Spring Batch provides a wide array of `ItemReader` implementations, many of which you have covered up to now. However, there is no way the developers of the framework can plan for every possible scenario. Because of this, they provide the facilities for you to create your own `ItemReader` implementations. The next section will look at how to implement your own custom `ItemReader`.

Custom Input

Spring Batch provides readers for just about every type of input Java applications normally face, however if you are using a form of input that Spring Batch provides an `ItemReader`, you will need to create one yourself. Implementing the `ItemReader` interface's `read()` method is the easy part. However, what happens when you need to be able to restart your reader? How do you maintain state across executions? This section will look at how to implement an `ItemReader` that can handle state across executions.

As mentioned, implementing Spring Batch's `ItemReader` interface is actually quite simple. In fact, with a small tweak, you can convert the `CustomerService` you used in the previous section to an `ItemReader`. All you need to do is implement the interface and rename the method `getCustomer()` to `read()`. Listing 7-66 shows the updated code.

Listing 7-66. `CustomerItemReader`

```
...
public class CustomerItemReader implements ItemReader<Customer> {

    private List<Customer> customers;
    private int curIndex;

    private String [] firstNames = {"Michael", "Warren", "Ann", "Terrence",
                                    "Erica", "Laura", "Steve", "Larry"};
    private String middleInitial = "ABCDEFGHIJKLMNOPQRSTUVWXYZ";
    private String [] lastNames = {"Gates", "Darrow", "Donnelly", "Jobs",
                                   "Buffett", "Ellison", "Obama"};
```

213

```java
    private String [] streets = {"4th Street", "Wall Street", "Fifth Avenue",
                                "Mt. Lee Drive", "Jeopardy Lane",
                                "Infinite Loop Drive", "Farnam Street",
                                "Isabella Ave", "S. Greenwood Ave"};
    private String [] cities = {"Chicago", "New York", "Hollywood", "Aurora",
                                "Omaha", "Atherton"};
    private String [] states = {"IL", "NY", "CA", "NE"};

    private Random generator = new Random();

    public CustomerItemReader () {
        curIndex = 0;

        customers = new ArrayList<Customer>();

        for(int i = 0; i < 100; i++) {
            customers.add(buildCustomer());
        }
    }

    private Customer buildCustomer() {
        Customer customer = new Customer();

        customer.setFirstName(
            firstNames[generator.nextInt(firstNames.length - 1)]);
        customer.setMiddleInitial(
            String.valueOf(middleInitial.charAt(
                generator.nextInt(middleInitial.length() - 1))));
        customer.setLastName(
            lastNames[generator.nextInt(lastNames.length - 1)]);
        customer.setAddress(generator.nextInt(9999) + " " +
                            streets[generator.nextInt(streets.length - 1)]);
        customer.setCity(cities[generator.nextInt(cities.length - 1)]);
        customer.setState(states[generator.nextInt(states.length - 1)]);
        customer.setZip(String.valueOf(generator.nextInt(99999)));

        return customer;
    }

    @Override
    public Customer read() {
        Customer cust = null;

        if(curIndex < customers.size()) {
            cust = customers.get(curIndex);
            curIndex++;
        }

        return cust;
    }
}
```

Even if you ignore the fact that your CustomerItemReader builds a new list with each run, the CustomerItemReader as it is written in Listing 7-66 will restart at the beginning of your list each time the job is executed. Although this will be the behavior you want in many cases, it will not always be the case. Instead, if there is an error after processing half a million records out of a million, you will want to start over again in that same chunk.

To provide the ability for Spring Batch to maintain the state of your reader in the JobRepository and restart your reader where you left off, you need to implement an additional interface, the ItemStream interface. Shown in Listing 7-67, the ItemStream interface consists of three methods: open, update, and close.

Listing 7-67. The ItemStream Interface

```
package org.springframework.batch.item;

public interface ItemStream {

  void open(ExecutionContext executionContext) throws ItemStreamException;
  void update(ExecutionContext executionContext) throws ItemStreamException;
  void close() throws ItemStreamException;
}
```

Each of the three methods of the ItemStream interface are called by Spring Batch during the execution of a step. open is called to initialize any required state within your ItemReader. This includes the opening of any files or database connections as well as restoring state when restarting a job. For example, the open method could be used to reload the number of records that had been processed so they could be skipped during the second execution. update is used by Spring Batch as processing occurs to update that state. Keeping track of how many records or chunks have been processed is a use for the update method. Finally, the close method is used to close any required resources (close files, etc.).

You will notice that the open and update provide access to the ExecutionContext that you did not have a handle on in your ItemReader implementation. This is because Spring Batch will use the ExecutionContext in the open method to provide the previous state of the reader when a job is restarted. It will also use the update method to learn the current state of the reader (which record you are currently on) as each item is processed. Finally, the close method is used to clean up any resources used in the ItemStream.

Now you may be wondering how you can use the ItemStream interface for your ItemReader if it doesn't have the read method. Short answer: you don't. Instead we will extend a utility class called org.springframework.batch.item.ItemStreamSuport. ItemStreamSupport implements ItemStream as well as provides a utility method getExecutionContextKey(String key) that makes the key unique based on the name of the component. Listing 7-68 shows your CustomerItemReader updated to extend the ItemStreamSupport base class.[7]

Listing 7-68. CustomerItemReader extending ItemStreamSupport

```
...
public class CustomerItemReader extends ItemStreamSupport implements ItemReader<Customer> {

    private List<Customer> customers;
    private int curIndex;
    private String INDEX_KEY = "current.index.customers";
```

---

[7]For this use case, extending AbstractItemCountingItemStreamItemReader would be more efficient, however this illustrates the use of the ItemStreamReader interface.

```
    private String [] firstNames = {"Michael", "Warren", "Ann", "Terrence",
                                "Erica", "Laura", "Steve", "Larry"};
    private String middleInitial = "ABCDEFGHIJKLMNOPQRSTUVWXYZ";
    private String [] lastNames = {"Gates", "Darrow", "Donnelly", "Jobs",
                                "Buffett", "Ellison", "Obama"};
    private String [] streets = {"4th Street", "Wall Street", "Fifth Avenue",
                                "Mt. Lee Drive", "Jeopardy Lane",
                                "Infinite Loop Drive", "Farnam Street",
                                "Isabella Ave", "S. Greenwood Ave"};
    private String [] cities = {"Chicago", "New York", "Hollywood", "Aurora",
                                "Omaha", "Atherton"};
    private String [] states = {"IL", "NY", "CA", "NE"};

    private Random generator = new Random();

    public CustomerItemReader() {
        customers = new ArrayList<>();

        for(int i = 0; i < 100; i++) {
            customers.add(buildCustomer());
        }
    }

    private Customer buildCustomer() {
        Customer customer = new Customer();

        customer.setFirstName(
            firstNames[generator.nextInt(firstNames.length - 1)]);
        customer.setMiddleInitial(
            String.valueOf(middleInitial.charAt(
                generator.nextInt(middleInitial.length() - 1))));
        customer.setLastName(
            lastNames[generator.nextInt(lastNames.length - 1)]);
        customer.setAddress(generator.nextInt(9999) + " " +
                        streets[generator.nextInt(streets.length - 1)]);
        customer.setCity(cities[generator.nextInt(cities.length - 1)]);
        customer.setState(states[generator.nextInt(states.length - 1)]);
        customer.setZip(String.valueOf(generator.nextInt(99999)));

        return customer;
    }

    public Customer read() {
        Customer cust = null;

        if(curIndex == 50) {
            throw new RuntimeException("This will end your execution");
        }

        if(curIndex < customers.size()) {
            cust = customers.get(curIndex);
```

```
            curIndex++;
        }

        return cust;
    }

    public void close() throws ItemStreamException {
    }

    public void open(ExecutionContext executionContext) throws ItemStreamException {
        if(executionContext.containsKey(getExecutionContextKey(INDEX_KEY))) {
            int index = executionContext.getInt(getExecutionContextKey(INDEX_KEY));
            if(index == 50) {
                curIndex = 51;
            } else {
                curIndex = index;
            }
        } else {
            curIndex = 0;
        }
    }

    public void update(ExecutionContext executionContext) throws ItemStreamException {
        executionContext.putInt(getExecutionContextKey(INDEX_KEY), curIndex);
    }
}
```

The bold sections of Listing 7-68 show the updates to the CustomerItemReader. First, the class was changed to extend the ItemStreamSupport class. Then the close(), open(ExecutionContext executionContext), and update(ExecutionContext executionContext) methods were added. In the update method, you add a key value pair to the ExecutionContext that indicates the current record being processed. The open method will check to see if that value has been set. If it has been set, that means that this is the restart of your job. In the run method, to force the job to end, you added code to throw a RuntimeException after the 50th customer. In the open method, if the index being restored is 50, you'll know it was due to your previous code so you will just skip that record. Otherwise, you'll try again. You'll notice that all references to the key used in the ExecutionContext are passed through the getExecutionContextKey method provided by the ItemStreamSupport class as well.

The other piece you need to do is configure your new ItemReader implementation. In this case, your ItemReader has no dependencies, so all you will need to do is define the bean with the correct name (so it is referred to in your existing copyJob). Listing 7-69 shows the configuration of the CustomerItemReader.

Listing 7-69. CustomerItemReader Configuration

```
...
@Bean
public CustomerItemReader customerItemReader() {
        CustomerItemReader customerItemReader = new CustomerItemReader();

        customerItemReader.setName("customerItemReader");

        return customerItemReader;
}
...
```

That really is it. Now if you execute your job, after you process 50 records, your `CustomerItemReader` will throw an `Exception` causing your job to fail. However, if you look in the `BATCH_STEP_EXECUTION_CONTEXT` table of your job repository, you will be happy to see what is listed in Listing 7-70.

Listing 7-70. *The Step Execution Context*

```
mysql> select * from BATCH_STEP_EXECUTION_CONTEXT where STEP_EXECUTION_ID = 8495;
+------------------+--------------------------------------------------------------------
-----+------------------------+
| STEP_EXECUTION_ID | SHORT_CONTEXT
| SERIALIZED_CONTEXT |
+------------------+--------------------------------------------------------------------
-----+------------------------+
|              8495 | {"customerItemReader.current.index.customers":50,"batch.
taskletType":"org.springframework.batch.core.step.item.ChunkOrientedTasklet","batch.
stepType":"org.springframework.batch.core.step.tasklet.TaskletStep"} |
NULL              |
```

Although a bit hard to read, you'll notice that Spring Batch has saved your commit count in the job repository. Because of this and your logic to skip the 50th customer the second time around, you can re-execute your job knowing that Spring Batch will start back where it left off and your writer will skip the item that caused the error.

Files, databases, services and even your own custom `ItemReaders`—Spring Batch provides you with a wide array of input options of which you have truly only scratched the surface here. Unfortunately, not all of the data you work with in the real world is as pristine as the data you have been working with here. However, not all errors are ones that need to stop processing. In the next section you will look at some of the ways that Spring Batch allows you to deal with input errors.

Error Handling

Things can go wrong in any part of a Spring Batch application—on startup, when reading input, processing input, or writing output. In this section, you will look at ways to handle different errors that can occur during batch processing.

Skipping Records

When there is an error reading a record from your input, you have a couple different options. First, an `Exception` can be thrown that causes processing to stop. Depending on how many records need to be processed and the impact of not processing this single record, this may be a drastic resolution. Instead, Spring Batch provides the ability to skip a record when a specified `Exception` is thrown. This section will look at how to use this technique to skip records based upon specific `Exceptions`.

There are two pieces involved in choosing when a record is skipped. The first is under what conditions to skip the record, specifically what exceptions you will ignore. When any error occurs during the reading process, Spring Batch throws an exception. In order to determine what to skip, you need to identify what exceptions to skip.

The second part of skipping input records is how many records you will allow the step to skip before considering the step execution failed. If you skip one or two records out of a million, not a big deal; however, skipping half a million out of a million is probably wrong. It's your responsibility to determine the threshold.

To actually skip records, all you need to do is tell Spring Batch the exceptions you want to skip and how many times it's okay to do so. Say you want to skip the first 10 records that throw any `org.springframework.batch.item.ParseException`. Listing 7-71 shows the configuration for this scenario.

Listing 7-71. Configuring to Skip 10 ParseExceptions

```
@Bean
public Step copyFileStep() {

        return this.stepBuilderFactory.get("copyFileStep")
                        .<Customer, Customer>chunk(10)
                        .reader(itemReader())
                        .writer(outputWriter(null))
                        .faultTolerant()
                        .skip(ParseException.class)
                        .skipLimit(10)
                        .build();
}
```

In this scenario, you have a single exception that you want to be able to skip. However, sometimes this can be a rather exhaustive list. The configuration in Listing 7-71 allows the skipping of a specific exception, but it might be easier to configure the ones you don't want to skip instead of the ones you do. To do this, you use a combination of the `skip(Class exception)` method like Listing 7-71 did and the `noSkipMethod(Class exception)` method. Listing 7-72 shows how to configure the opposite of your previous example (skipping all exceptions except for the `ParseException`).

Listing 7-72. Configuring to Skip All Exceptions Except the `ParseException`

```
@Bean
public Step copyFileStep() {

        return this.stepBuilderFactory.get("copyFileStep")
                        .<Customer, Customer>chunk(10)
                        .reader(itemReader())
                        .writer(outputWriter(null))
                        .faultTolerant()
                        .skip(Exception.class)
                        .noSkip(ParseException.class)
                        .skipLimit(10)
                        .build();
}
```

The configuration in Listing 7-72 specifies that any `Exception` that extends `java.lang.Exception` except for `org.springframework.batch.item.ParseException` will be skipped up to 10 times.

There is a third way to specify what `Exceptions` to skip and how many times to skip them. Spring Batch provides an interface called `org.springframework.batch.core.step.skip.SkipPolicy`. This interface, with its single method `shouldSkip`, takes the `Exception` that was thrown and the number of times records have been skipped. From there, any implementation can determine what `Exceptions` they should skip and how many times. Listing 7-73 shows a `SkipPolicy` implementation that will not allow a `java.io.FileNotFoundException` to be skipped but 10 `ParseExceptions` to be skipped.

Listing 7-73. `FileVerificationSkipper`

```
...
public class FileVerificationSkipper implements SkipPolicy {

    public boolean shouldSkip(Throwable exception, int skipCount)
        throws SkipLimitExceededException {

        if(exception instanceof FileNotFoundException) {
            return false;
        } else if(exception instanceof ParseException && skipCount <= 10) {
            return true;
        } else {
            return false;
        }
    }
}
```

Skipping records is common practice in batch processing. It allows what is typically a much larger process than a single record to continue with minimal impact. Once you can skip a record that has an error, you may want to do something additional like log it for future evaluation. The next section discusses an approach for just that.

Logging Invalid Records

While skipping problematic records is a useful tool, by itself it can raise an issue. In some scenarios, the ability to skip a record is okay. Say you are mining data and come across something you can't resolve; it's probably okay to skip it. However, when you get into situations where money is involved, say when processing transactions, just skipping a record probably will not be a robust enough solution. In cases like these, it is helpful to be able to log the record that was the cause of the error. In this section, you will look at using an ItemListener to record records that were invalid.

The ItemReadListener interface consists of three methods: beforeRead(), afterRead(T item), and onReadError(Exception ex). For the case of logging invalid records as they are read in, you can use the ItemListenerSupport class and override the onReadError to log what happened or use a POJO with a method annotated with @OnReadError. It's important to point out that Spring Batch does a good job building its Exceptions for file parsing to inform you of what happened and why. On the database side, things are a little less in the framework's hands as most of the actual database work is done by other frameworks (Spring itself, Hibernate, etc.). It is important that as you develop your own processing (custom ItemReaders, RowMappers, etc.) that you include enough detail for you to diagnose the issue from the Exception itself.

In this example, you will read data in from the Customer file from the beginning of the chapter. When an Exception is thrown during input, you will log the record that caused the exception and the exception itself. To do this, the CustomerItemListener will take the exception thrown and if it is a FlatFileParseException, you will have access to the record that caused the issue and information on what went wrong. Listing 7-74 shows the CustomerItemListener.

Listing 7-74. `CustomerItemListener`

```
...
public class CustomerItemListener  {

        private static final Log logger = LogFactory.getLog(CustomerItemListener.class);

        @OnReadError
        public void onReadError(Exception e) {
                if(e instanceof FlatFileParseException) {
                        FlatFileParseException ffpe = (FlatFileParseException) e;

                        StringBuilder errorMessage = new StringBuilder();
                        errorMessage.append("An error occured while processing the " +
                                        ffpe.getLineNumber() +
                                        " line of the file.  Below was the faulty " +
                                        "input.\n");
                        errorMessage.append(ffpe.getInput() + "\n");

                        logger.error(errorMessage.toString(), ffpe);
                } else {
                        logger.error("An error has occurred", e);
                }
        }
}
```

Configuring your listener requires you to update the Step reading the file. In your case, you have only one step in your copyJob. Listing 7-75 shows the configuration for this listener.

Listing 7-75. Configuring the `CustomerItemListener`

```
...
@Bean
public CustomerItemListener customerListener() {
      return new CustomerItemListener ();
}

@Bean
public Step copyFileStep() {
    return this.stepBuilderFactory.get("copyFileStep")
                .<Customer, Customer>chunk(10)
                .reader(customerFileReader(null))
                .writer(outputWriter(null))
                .faultTolerant()
                .skipLimit(100)
                .skip(Exception.class)
                .listener(customerListener())
                .build();
}
 ...
```

If you use the fixed-length record job as an example and execute it with a file that contains an input record longer than 63 characters, an exception will be thrown. However, since you have configured your job to skip all exceptions that extend `Exception`, the exception will not affect your job's results, yet your `customerItemLogger` will be called and log the item as required. When you execute this job, you see two things. The first is a `FlatFileParseException` for each record that is invalid. The second are your log messages. Listing 7-76 shows an example of the log messages your job generates on error.

Listing 7-76. Output of the `CustomerItemLogger`

```
2011-05-03 23:49:22,148 ERROR main [com.apress.springbatch.chapter7.CustomerItemListener] -
<An error occured while processing the 1 line of the file.  Below was the faulty input.
Michael   TMinella  123   4th Street         Chicago  IL60606ABCDE
>
```

Using nothing more than a good logging framework, you can get the input that failed to parse from the `FlatFileParseException` and log it to your log file. However, this by itself does not accomplish your goal of logging the error record to a file and continuing on. In this scenario, your job will log the record that caused the issue and fail. In the last section, you will look at how to handle having no input when your jobs run.

Dealing with No Input

A SQL query that returns no rows is not an uncommon occurrence. Empty files exist in many situations. But do they make sense for your batch process? In this section, you will look at how Spring Batch handles reading input sources that have no data.

When a reader attempts to read from an input source and a `null` is returned the first time, by default this is treated like any other time a reader receives a `null`; it considers the step complete. While this approach may work in the majority of the scenarios, you may need to know when a given query returns zero rows or a file is empty.

If you want to cause your step to fail or take any other action (send an e-mail, etc.) when no input has been read, you use a `StepExecutionListner`. In Chapter 4, you used a `StepExecutionListner` to log the beginning and end of your step. In this case, you can use the `StepExecutionListner`'s `@AfterStep` method to see how many records were read and react accordingly. Listing 7-77 shows how you would mark a step failed if no records were read.

Listing 7-77. `EmptyInputStepFailer`

```
...
public class EmptyInputStepFailer {

    @AfterStep
    public ExitStatus afterStep(StepExecution execution) {
        if(execution.getReadCount() > 0) {
            return execution.getExitStatus();
        } else {
            return ExitStatus.FAILED;
        }
    }
}
```

To configure your listener, you configure it like you would any other `StepExecutionListener`. Listing 7-78 covers the configuration in this instance.

Listing 7-78. Configuring the `EmptyInputStepFailer`

```
...
@Bean
public EmptyInputStepFailer emptyFileFailer() {
      return new EmptyInputStepFailer();
}

@Bean
public Step copyFileStep() {
    return this.stepBuilderFactory.get("copyFileStep")
                .<Customer, Customer>chunk(10)
                .reader(customerFileReader(null))
                .writer(outputWriter(null))
                .listener(emptyFileFailer())
                .build();
}
...
```

By running a job with this step configured, instead of your job ending with the status `COMPLETED` if no input was found, the job will fail, allowing you to obtain the expected input and rerun the job.

Summary

Reading and writing takes up the vast majority of a batch process and, as such, is one of the most important pieces of the Spring Batch framework. In this chapter, you took a thorough (but not exhaustive) look at the `ItemReader` options within the framework. Now that you can read in an item, you need to be able to do something with it. `ItemProcessors`, which make things happen, are covered in the next chapter.

CHAPTER 8

■ ■ ■

ItemProcessors

In the previous chapter, you learned how to read various types of input using the components of Spring Batch. While obtaining the input for any piece of software is an important aspect of the project, it doesn't mean much if you don't do something with it. ItemProcessors are the component within Spring Batch where you do something with your input. In this chapter, you will look at the ItemProcessor interface and see how you can use it to develop your own processing of batch items.

- In the "Introduction to ItemProcessors" section, you will start with a quick overview of what an ItemProcessor is and how it fits into the flow of a step.

- Spring Batch provides utility ItemProcessor implementations like the ItemProcessorAdapter, which uses existing services as your ItemProcessor implementation. In the "Using Spring Batch's ItemProcessors" section, you'll take an in-depth look at each of the processors the framework provides.

- In many cases, you will want to develop your own ItemProcessor implementation. In the "Writing Your Own ItemProcessors" section, you will look at different considerations as you implement an example ItemProcessor.

- A common use of an ItemProcessor is to filter out items that were read in by an ItemReader from being written by the step's ItemWriter. In the "Filtering Items" section, you'll look at an example of how this is accomplished.

Introduction to ItemProcessors

In Chapter 7, you looked at ItemReaders, the input facility you use within Spring Batch. Once you have received your input, you have two options. The first is to just write it back out as you did in the examples in that chapter. There are many times when that will make sense. Migrating data from one system to another or loading data into a database initially are two examples of where reading input and writing it directly without any additional processing makes sense.

However, in most scenarios, you are going to need to do something with the data you read in. Spring Batch has broken up the pieces of a step to allow a good separation of concerns between reading, processing, and writing. Doing this allows you the opportunity to do a couple unique things, such as the following:

- *Validate input*: In the original version of Spring Batch, validation occurred at the ItemReader by subclassing the ValidatingItemReader class. The issue with this approach is that none of the provided readers subclassed the ValidatingItemReader class so if you wanted validation, you couldn't use any of the included readers. Moving the validation step to the ItemProcessor allows validation to occur on an object before processing, regardless of the input method. This makes much more sense from a division-of-concerns perspective.

- *Reuse existing services*: Just like the ItemReaderAdapter you looked at in Chapter 7 to reuse services for your input, Spring Batch provides an ItemProcessorAdapter for the same reason.

- *Execute scripts*: ItemProcessors can be a great opportunity to plug in logic from other developers or other teams. However, those other teams may not use the same Spring based toolset that you and your team does. The ScriptItemProcessor allows you to execute a script of some kind as an ItemProcessor, providing the script the item as the input and taking what the script returns as the output.

- *Chain ItemProcessors*: There are situations where you will want to perform multiple actions on a single item within the same transaction. Although you could write your own custom ItemProcessor to do all of the logic in a single class, that couples your logic to the framework, which is something you want to avoid. Instead, Spring Batch allows you to create a list of ItemProcessors that will be executed in order against each item.

To accomplish all of this, the org.springframework.batch.item.ItemProcessor interface consists of a single method process shown in Listing 8-1. It takes an item as read from your ItemReader and returns another item.

Listing 8-1. ItemProcessor Interface

```
package org.springframework.batch.item;

public interface ItemProcessor<I, O> {

    O process(I item) throws Exception;
}
```

It's important to note that the type the ItemProcessor receives as input does not need to be the same type it returns. The framework allows for you to read in one type of object and pass it to an ItemProcessor and have the ItemProcessor return a different type for writing. With this feature, you should note that the type the final ItemProcessor returns is required to be the type the ItemWriter takes as input. You should also be aware that if an ItemProcessor returns null, all processing of the item will stop. In other words, any further ItemProcessors for that item will not be called nor shall the ItemWriter for the step. However, unlike returning null from an ItemReader, which indicates to Spring Batch that all input has been exhausted, processing of other items will continue when an ItemProcessor returns null.

■ **Note** An ItemProcessor must be idempotent. An item may be passed through it more than once in fault tolerant scenarios.

Let's take a look at how to use ItemProcessors for your jobs. To start, you'll dig into the ones provided by the framework.

Using Spring Batch's `ItemProcessors`

When you looked at the `ItemReaders` previously, there was a lot of ground to cover regarding what was provided from Spring Batch because input and output are two relatively standard things. Reading from a file is the same in most cases. Writing to a database works the same with most databases. However, what you do to each item differs based on your business requirements. This is really what makes each job different. Because of this, the framework can only provide you with the facility to either implement your own or wrap existing logic. This section will cover the `ItemProcessor` implementations that are included in the Spring Batch framework.

ValidatingItemProcessor

You'll start your look at Spring Batch's `ItemProcessor` implementations with where you left off in Chapter 7. Previously, you handled obtaining input for your jobs; however, just because you can read it doesn't mean it's valid. Data validation with regards to types and format can occur within an `ItemReader`; however, validation via business rules is best left once the item has been constructed. That's why Spring Batch provides an `ItemProcessor` implementation for validating input called the `ValidatingItemProcessor`. In this section, you will look at how to use it to validate your input.

Input Validation

The `org.springframework.batch.item.validator.ValidatingItemProcessor` is an implementation of the `ItemProcessor` interface that allows you to set an implementation of Spring Batch's `Validator` interface[1] to be used to validate the incoming item prior to processing. If the item passes validation, it will be processed. If not, an `org.springframework.batch.item.validator.ValidationException` is thrown, causing normal Spring Batch error handling to kick in.

JSR 303 is the Java specification for bean validation. This specification provides a widely accepted method for validation in the Java ecosystem. The validation performed via the `javax.validation.*` code is configured via annotations. There is a collection of annotations that predefine validation functions out of the box; you also have the ability to create your own validation functions. Let's start by looking at how you would validate a `Customer` class like the one in Listing 8-2.

Listing 8-2. Customer Class

```
...
public class Customer {
    private String firstName;
    private String middleInitial;
    private String lastName;
    private String address;
    private String city;
    private String state;
    private String zip;

    // Getters & setters go here
...
}
```

---

[1]Although Spring does have a `Validator` interface of its own, the `ValidatingItemProcessor` uses one from Spring Batch instead.

If you look at the `Customer` class in Listing 8-2, you can quickly determine some basic validation rules:

- Not null: `firstName`, `lastName`, `address`, `city`, `state`, `zip`.

- Alphabetic: `firstName`, `middleInitial`, `lastName`, `city`, `state`.

- Numeric: `zip`.

- Size: `middleInitial` should be no longer than one character; `state` should be no longer than two characters; and `zip` should be no longer than five characters.

There are further validations you can perform on the data provided `zip` is a valid ZIP code for the city and state. However, this provides you with a good start. Now that you have identified the things you want to validate, you can describe them to your validator via annotations on the `Customer` object. Specifically, you will be using the `@NotNull`, `@Size`, and `@Pattern` annotations for these rules. To use these, we will need to use a new starter within our projects, spring-boot-starter-validation. This starter brings in the Hibernate implementation of the JSR-303 validation tooling.

We can begin our work with validation by creating a new project with the dependencies batch, jdbc, MySQL, and validation from Spring Initalizr. With our new project created, we can add the code from Listing 8-2 to it as our domain object. Next, we want to apply the preceding validation rules we mentioned using the JSR-303 annotations. Listing 8-3 shows their use on the `Customer` object.

Listing 8-3. Customer Object with Validation Annotations

```
...
public class Customer {

    @NotNull(message="First name is required")
    @Pattern(regexp="[a-zA-Z]+", message="First name must be alphabetical")
    private String firstName;

    @Size(min=1, max=1)
    @Pattern(regexp="[a-zA-Z]", message="Middle initial must be alphabetical")
    private String middleInitial;

    @NotNull(message="Last name is required")
    @Pattern(regexp="[a-zA-Z]+", message="Last name must be alphabetical")
    private String lastName;

    @NotNull(message="Address is required")
    @Pattern(regexp="[0-9a-zA-Z\\. ]+")
    private String address;

    @NotNull(message="City is required")
    @Pattern(regexp="[a-zA-Z\\. ]+")
    private String city;

    @NotNull(message="State is required")
    @Size(min=2,max=2)
    @Pattern(regexp="[A-Z]{2}")
    private String state;
```

```
@NotNull(message="Zip is required")
@Size(min=5,max=5)
@Pattern(regexp="\\d{5}")
private String zip;

// Accessors go here
...
}
```

A quick look at the rules defined in Listing 8-3 may make you ask why use both the @Size annotation and the @Pattern one when the regular expression defined in the @Pattern would satisfy both. You are correct. However, each annotation allows you to specify a unique error message (if you want); moreover, being able to identify if the field was the wrong size vs. the wrong format may be helpful in the future.

At this point, you have defined the validation rules you will use for your Customer item. In order to put this functionality to work, we need to provide a mechanism for Spring Batch to validate each of our items. The org.springframework.batch.item.validator.BeanValidatingItemProcessor will do that for us. This ItemProcessor is an extension of the ValidatingItemProcessor that specifically utilizes JSR-303 to provide the validation.

The ValidatingItemProcessor's validation capabilities come from its use of a org.springframework.batch.item.validator.Validator implementation. This interface has a single method, void validate(T value). This method does nothing if the item is valid and throws an org.springframework.batch.item.validator.ValidationException if the validation fails. The BeanValidatingItemProcessor is a special version of this in that it handles the creation of a Validator object for you that is JSR-303 specific.

■ **Note** The Validator interface included in the Spring Batch framework is not the same as the Validator interface that is part of the core Spring framework. Spring Batch provides an adapter class, SpringValidator, to handle the differences.

Let's see how all of this works together by creating a job to put them to use. Your job will read a comma-delimited file into your Customer object, which will then be validated as part of the BeanValidatingItemProcessor and written out to a csv, as you did in Chapter 7. To start, Listing 8-4 shows an example of the input you will process.

Listing 8-4. customer.csv

```
Richard,N,Darrow,5570 Isabella Ave,St. Louis,IL,58540
Barack,G,Donnelly,7844 S. Greenwood Ave,Houston,CA,38635
Ann,Z,Benes,2447 S. Greenwood Ave,Las Vegas,NY,55366
Laura,9S,Minella,8177 4th Street,Dallas,FL,04119
Erica,Z,Gates,3141 Farnam Street,Omaha,CA,57640
Warren,L,Darrow,4686 Mt. Lee Drive,St. Louis,NY,94935
Warren,M,Williams,6670 S. Greenwood Ave,Hollywood,FL,37288
Harry,T,Smith,3273 Isabella Ave,Houston,FL,97261
Steve,O,James,8407 Infinite Loop Drive,Las Vegas,WA,90520
Erica,Z,Neuberger,513 S. Greenwood Ave,Miami,IL,12778
Aimee,C,Hoover,7341 Vel Avenue,Mobile,AL,35928
Jonas,U,Gilbert,8852 In St.,Saint Paul,MN,57321
Regan,M,Darrow,4851 Nec Av.,Gulfport,MS,33193
Stuart,K,Mckenzie,5529 Orci Av.,Nampa,ID,18562
Sydnee,N,Robinson,894 Ornare. Ave,Olathe,KS,25606
```

Note that on line 4 of your input the middle initial field is 9S, which is invalid. This should cause your validation to fail at this point. With your input file defined, you can configure the job. The job you will be running will consist of a single step that reads in the input, passes it to an instance of the ValidatingItemProcessor, and then writes it to standard out. Listing 8-5 shows the configuration for the job.

Listing 8-5. ValidationJob

```
...
@EnableBatchProcessing
@SpringBootApplication
public class ValidationJob {

    @Autowired
    public JobBuilderFactory jobBuilderFactory;

    @Autowired
    public StepBuilderFactory stepBuilderFactory;

    @Bean
    @StepScope
    public FlatFileItemReader<Customer> customerItemReader(
                @Value("#{jobParameters['customerFile']}")Resource inputFile) {

        return new FlatFileItemReaderBuilder<Customer>()
                        .name("customerItemReader")
                        .delimited()
                        .names(new String[] {"firstName",
                                        "middleInitial",
                                        "lastName",
                                        "address",
                                        "city",
                                        "state",
                                        "zip"})
                        .targetType(Customer.class)
                        .resource(inputFile)
                        .build();
    }

    @Bean
    public ItemWriter<Customer> itemWriter() {
        return (items) -> items.forEach(System.out::println);
    }

    @Bean
    public BeanValidatingItemProcessor<Customer> customerValidatingItemProcessor() {
        return new BeanValidatingItemProcessor<>();
    }

    @Bean
    public Step copyFileStep() {
```

```
        return this.stepBuilderFactory.get("copyFileStep")
                    .<Customer, Customer>chunk(5)
                    .reader(customerItemReader(null))
                    .processor(customerValidatingItemProcessor())
                    .writer(itemWriter())
                    .build();
    }

    @Bean
    public Job job() throws Exception {

        return this.jobBuilderFactory.get("job")
                    .start(copyFileStep())
                    .build();
    }

    public static void main(String[] args) {
        SpringApplication.run(ValidationJob.class, "customerFile=/input/customer.csv");
    }
}
```

To walk through the ValidationJob class listed in Listing 8-5, let's start with definitions of the input file and the reader. This reader is a simple delimited file reader that maps the fields of the file to your Customer object. Next is the output configuration, which consists of defining a lambda for our ItemWriter to write to standard out. With the input and output defined, the bean customerValidatingItemProcessor will serve as your ItemProcessor. By default, the BeanValidatingItemProcessor just passes the item through from the ItemReader to the ItemWriter, which will work for this example.

With all of the beans defined, you can build your Step, which is the next piece of the file. All you need for your Step is to define the reader, processor, and writer. With your Step defined, you finish the file by configuring the Job itself.

To run the job, use the command in Listing 8-6 from the target directory of your project.

Listing 8-6. Running the copyJob

```
java -jar itemProcessors-0.0.1-SNAPSHOT.jar customerFile=/input/customer.csv
```

As mentioned, you have some bad input that will not pass validation. When you run the job, it fails due to the ValidationException that is thrown. To get the job to complete successfully, you have to fix your input to pass validation. Listing 8-7 shows the results of your job when the input fails validation.

Listing 8-7. copyJob Output

```
2019-02-05 17:19:35.287  INFO 39336 --- [            main] o.s.batch.core.job.SimpleStep
Handler     : Executing step: [copyFileStep]
2019-02-05 17:19:35.462 ERROR 39336 --- [            main] o.s.batch.core.step.Abstract
Step         : Encountered an error executing step copyFileStep in job job
```

```
org.springframework.batch.item.validator.ValidationException: Validation failed for Customer
{firstName='Laura', middleInitial='9S', lastName='Minella', address='8177 4th Street',
city='Dallas', state='FL', zip='04119'}:
Field error in object 'item' on field 'middleInitial': rejected value [9S]; codes [Size.
item.middleInitial,Size.middleInitial,Size.java.lang.String,Size]; arguments [org.
springframework.context.support.DefaultMessageSourceResolvable: codes [item.
middleInitial,middleInitial]; arguments []; default message [middleInitial],1,1]; default
message [size must be between 1 and 1]
Field error in object 'item' on field 'middleInitial': rejected value [9S]; codes [Pattern.
item.middleInitial,Pattern.middleInitial,Pattern.java.lang.String,Pattern]; arguments
[org.springframework.context.support.DefaultMessageSourceResolvable: codes [item.
middleInitial,middleInitial]; arguments []; default message [middleInitial],[Ljavax.
validation.constraints.Pattern$Flag;@3fd05b3e,org.springframework.validation.beanvalidation.
SpringValidatorAdapter$ResolvableAttribute@4eb9f2af]; default message [Middle initial must
be alphabetical]
        at org.springframework.batch.item.validator.SpringValidator.validate(SpringValidator.
        java:54) ~[spring-batch-infrastructure-4.1.1.RELEASE.jar:4.1.1.RELEASE]
...
```

That is all that is required to add item validation to your jobs in Spring Batch using JSR-303. However, what if you wanted to implement your own validation? To do that, we would change our ItemProcessor from the BeanValidatingItemProcessor to the ValidatingItemProcessor and inject our own implementation of the Validator interface.

Let's say that the lastName field must be unique across the data set. To validate that the records conform to that, we would implement a stateful Validator that kept track of the last names seen. If we wanted the state to be persisted across restarts, the Validator would also need to implement the ItemStream interface by extending ItemStreamSupport and store those last names in the ExecutionContext with each commit. Listing 8-8 illustrates the code for the new Validator that handles this logic.

Listing 8-8. Validating Last Name Is Unique in the Data Set

```
...
public class UniqueLastNameValidator extends ItemStreamSupport
        implements Validator<Customer> {

    private Set<String> lastNames = new HashSet<>();

    @Override
    public void validate(Customer value) throws ValidationException {
        if(lastNames.contains(value.getLastName())) {
            throw new ValidationException("Duplicate last name was found: "
                    + value.getLastName());
        }

        this.lastNames.add(value.getLastName());
    }
```

```
        @Override
        public void open(ExecutionContext executionContext) {
                String lastNames = getExecutionContextKey("lastNames");

                if(executionContext.containsKey(lastNames)) {
                        this.lastNames = (Set<String>) executionContext.get(lastNames);
                }
        }

        @Override
        public void update(ExecutionContext executionContext) {
                executionContext.put(getExecutionContextKey("lastNames"), this.lastNames);
        }
}
```

Starting from the top, the class extends `ItemStreamSupport` for the abilty to save the state from job execution to job execution and implements the `Validator` interface. We save the last name values in a `Set` defined next. The `validate(Customer customer)` method is required by the `Validator` interface and does the work for us. It confirms that if the last name isn't already saved in the `Set`, to save it and if it was put there by a previous record to throw the `ValidationExecption` identifying the bad data. The last two methods in the class are used to persist the state from execution to execution. The open method determines if the `lastNames` field was saved in a previous execution. If it was, it is restored before the step's processing begins. The update method (called once per chunk once the transaction commits) stores the current state in the `ExecutionContext` in case there is a failure in the next chunk.

Once we've created our `Validator` implementation, we need to configure it. There are three pieces to the configuration of this validation mechanism. First, defining the `UniqueLastNameValidator` as a bean, then injecting it into the `ValidatingItemProcessor`, and finally registering the `UniqueLastNameValidator` as a stream on our step so that Spring Batch will know to call the `ItemStream` related methods on it. Listing 8-9 illustrates this configuration.

Listing 8-9. UniqueLastNameValidator Configuration

```
...
@Bean
public UniqueLastNameValidator validator() {
    UniqueLastNameValidator uniqueLastNameValidator = new UniqueLastNameValidator();

    uniqueLastNameValidator.setName("validator");

    return uniqueLastNameValidator;
}

@Bean
public ValidatingItemProcessor<Customer> customerValidatingItemProcessor() {
    return new ValidatingItemProcessor<>(validator());
}

@Bean
public Step copyFileStep() {

return this.stepBuilderFactory.get("copyFileStep")
```

```
        .<Customer, Customer>chunk(5)
        .reader(customerItemReader(null))
        .processor(customerValidatingItemProcessor())
        .writer(itemWriter())
        .stream(validator())
        .build();
}
...
```

If you run these configuration updates with the data defined in Listing 8-4, it will take three attempts to get through the job:

1. The first attempt will commit the first chunk (five records), then fail on the second chunk.

2. Remove line 6 from the input file and run it again. This time, the records in the first chunk will be skipped, the second chunk will be committed, and then the third chunk will fail after finding the same last name (which was restored from the ExecutionContext previously).

3. Remove line 13 (12 after step 2 is done) and run it a final time to see it complete.

The ValidatingItemProcessor and its subclass, the BeanValidatingItemProcessor, are useful for being able to apply validation to your items as they are processed. However, these are only one of the three main areas Spring Batch provides implementations of the ItemProcessor interface. In the next section you will look at the ItemProcessorAdapter and how it allows you to use existing services as ItemProcessors.

ItemProcessorAdapter

In Chapter 7, you looked at the ItemReaderAdapter as a way to use existing services to provide input to your jobs. Spring Batch also allows you to put to use the various services you already have developed as ItemProcessors as well by using the org.springframework.batch.item.adapter.ItemProcessorAdapter. In this section, you will look at the ItemProcessorAdapter and see how it lets you use existing services as processors for your batch job items.

To take a look at this functionality, we will create a service that uppercases the customer's name (first name, middle initial, and last name). We can call this the UpperCaseNameService. It will have a single method that copies the Customer input object into a new Customer output object (making it idempotent), and then uppercases the name values on the new instance, returning that once it's done. Listing 8-10 shows the code for our UpperCaseNameService.

Listing 8-10. UpperCaseNameService

```
...
@Service
public class UpperCaseNameService {

    public Customer upperCase(Customer customer) {
        Customer newCustomer = new Customer(customer);

        newCustomer.setFirstName(newCustomer.getFirstName().toUpperCase());
        newCustomer.setMiddleInitial(newCustomer.getMiddleInitial().toUpperCase());
        newCustomer.setLastName(newCustomer.getLastName().toUpperCase());
```

```
            return newCustomer;
    }

}
```

With our service defined, we can replace the validation functionality in the previous job with the uppercase functionality in this new service. Listing 8-11 has the code to configure the new ItemProcessor and the updated step.

Listing 8-11. ItemProcessorAdapter Configuration

```
...
@Bean
public ItemProcessorAdapter<Customer, Customer> itemProcessor(UpperCaseNameService service)
{
    ItemProcessorAdapter<Customer, Customer> adapter = new ItemProcessorAdapter<>();

    adapter.setTargetObject(service);
    adapter.setTargetMethod("upperCase");

    return adapter;
}

@Bean
public Step copyFileStep() {

return this.stepBuilderFactory.get("copyFileStep")
            .<Customer, Customer>chunk(5)
            .reader(customerItemReader(null))
            .processor(itemProcessor(null))
            .writer(itemWriter())
            .build();
}
...
```

Looking at Listing 8-11, we begin by defining the bean for our ItemProcessorAdapter. We inject the UpperCaseNameService we defined in Listing 8-10. The adapter requires at least two values to be set with an optional third. The two required are the target object (the instance we are going to make the calls to) and the target method (the method on the instance to be called). There is another configuration option that enables you to provide an array of arguments, however any values passed to this on the ItemProcessorAdapter will be ignored.

With our adapter configured, the step needs to be updated to reference it as shown in Listing 8-11. With that configuration complete, when the job is run, the output shows the name fields in the Customer to be uppercased as shown in Listing 8-12.

Listing 8-12. `ItemProcessorAdapterJob` Output

```
...
2019-02-05 22:23:19.185  INFO 45123 --- [          main] o.s.batch.core.job.SimpleStep
Handler    : Executing step: [copyFileStep]
Customer{firstName='RICHARD', middleInitial='N', lastName='DARROW', address='5570 Isabella
Ave', city='St. Louis', state='IL', zip='58540'}
Customer{firstName='BARACK', middleInitial='G', lastName='DONNELLY', address='7844 S.
Greenwood Ave', city='Houston', state='CA', zip='38635'}
Customer{firstName='ANN', middleInitial='Z', lastName='BENES', address='2447 S. Greenwood
Ave', city='Las Vegas', state='NY', zip='55366'}
Customer{firstName='LAURA', middleInitial='9S', lastName='MINELLA', address='8177 4th
Street', city='Dallas', state='FL', zip='04119'}
Customer{firstName='ERICA', middleInitial='Z', lastName='GATES', address='3141 Farnam
Street', city='Omaha', state='CA', zip='57640'}
...
```

Scripting languages and their support on the JVM has had a big boost ever since Groovy came along.
Now you can run Ruby, JavaScript, Groovy, and a host of other scripting languages on the JVM. In the next
section, we will take a look at how you can use a script to implement your `ItemProcessor`.

ScriptItemProcessor

Scripting languages provide a whole host of unique opportunities. Scripts are typically easier to create
and modify so for frequently changing components, scripts can provide great flexibility. Prototyping is
another area where using a script instead of performing all the ceremony that a statically typed language
like Java requires. Spring Batch allows you to inject this flexibility into your batch jobs by executing
a script as an `ItemProcessor`. The `org.springframework.batch.item.support.ScriptItemProcessor`
allows you to specify a script that will receive the `ItemProcessor`'s input and return the object that
will be the `ItemProcessor`'s output. In this section, we will implement the same functionality as in the
`ItemProcessorAdapter` use case; however, instead of using a Java service to perform the logic, we will
put the logic in a JavaScript script.

To begin, we need to venture a bit out of our comfort zone and write some JavaScript. The script we
will create is actually quite simple. The `ScriptItemProcessor` binds the input from the `ItemProcessor` to
the variable item by default (this value is configurable if you want to change it). From there, we can perform
any JavaScript functionality on it and return it to our `ItemProcessor`. In our case, we will uppercase the first
name, middle initial, and last name as shown in Listing 8-13.

Listing 8-13. upperCase.js

```javascript
item.setFirstName(item.getFirstName().toUpperCase());
item.setMiddleInitial(item.getMiddleInitial().toUpperCase());
item.setLastName(item.getLastName().toUpperCase());
item;
```

With our script written, we can update our job to use the `ScriptItemProcessor`. All we need to provide
as a dependency for this `ItemProcessor` is a `Resource` that points to the script file we will use (you can also
define the script inline as a `String`). Listing 8-14 illustrates the configuration of the `ScriptItemProcessor` in
our job.

Listing 8-14. `ScriptItemProcessor` Configuration

```
...
@Bean
@StepScope
public ScriptItemProcessor<Customer, Customer> itemProcessor(
        @Value("#{jobParameters['script']}") Resource script) {

    ScriptItemProcessor<Customer, Customer> itemProcessor =
            new ScriptItemProcessor<>();

    itemProcessor.setScript(script);

    return itemProcessor;
}
...
```

To this configuration, we will need to add one additional job parameter to our job which is the location of the script file, making the command to execute this job: `java -jar copyJob.jar customerFile=/input/customer.csv script=/upperCase.js`. When you execute this command, the output from this job should match that from the `ItemProcessorAdapter` job illustrated in Listing 8-12.

The idea of applying a single action to an item within a transaction can be limiting in certain situations. For example, if you have a set of calculations that need to be done on some of the items, you may want to filter out the ones that don't need to be processed. In the next section, you will look at how to configure Spring Batch to execute a list of `ItemProcessor`s on each item within a step.

CompositeItemProcessor

You break up a step into three phases (reading, processing, and writing) to divide responsibilities between components. However, the business logic that needs to be applied to a given item may not make sense to couple into a single `ItemProcessor`. Spring Batch allows you to maintain that same division of responsibilities within your business logic by chaining `ItemProcessor`s within a step. In this section, you will look at how composition can be used to allow you to do more complex orchestration in the `ItemProcessor` phase of a step.

We will start with the `org.springframework.batch.item.support.CompositeItemProcessor`. This is an implementation of the `ItemProcessor` interface that delegates processing to each of a list of `ItemProcessor` implementations in order. As each processor returns its result, that result is passed onto the next processor until they all have been called. This pattern occurs regardless of the types returned, so if the first `ItemProcessor` takes a `String` as input, it can return a `Product` object as output as long as the next `ItemProcessor` takes a `Product` as input. At the end, the result is passed to the `ItemWriter` configured for the step. It is important to note that just like any other `ItemProcessor`, if any of the processors this one delegates to returns `null`, the item will not be processed further. Figure 8-1 shows how the processing within the `CompositeItemProcessor` occurs.

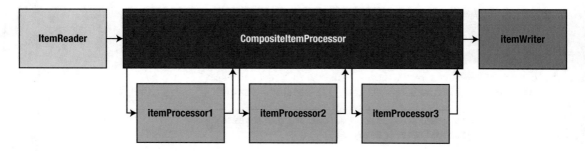

Figure 8-1. *CompositeItemProcessor processing*

As Figure 8-1 shows, the CompositeItemProcessor serves as a wrapper for multiple ItemProcessors, calling them in order. As one completes, the next one is called with the item returned from the previous one. Let's take a look at how this looks in practice.

In this example, we will take the previous examples of this chapter and apply them all at once. We will create a chain of ItemProcessors that do the following in order:

1. Validate the input, filtering out any bad records.

2. Uppercase the name using the UpperCaseNameService.

3. Lowercase the address, city, and state fields using a JavaScript script.

Let's begin by configuring the ItemProcessor we will use to validate the Customer objects passed received by the ItemProcessor. To do this, we will use the same configuration we did in Listing 8-9, with one small change. The ValidatingItemProcessor, by default, throws a ValidationException when an item is not valid. However, that may be too drastic of a measure for some use cases (like ours). Because of this, you can also configure that ItemProcessor to filter out the items that do not pass validation. This is the way we want to apply the logic this time around. Listing 8-15 shows the updated configuration for this ItemProcessor.

Listing 8-15. ValidatingItemProcessor Configured to Filter Items

```
...
@Bean
public UniqueLastNameValidator validator() {
    UniqueLastNameValidator uniqueLastNameValidator = new UniqueLastNameValidator();

    uniqueLastNameValidator.setName("validator");

    return uniqueLastNameValidator;
}

@Bean
public ValidatingItemProcessor<Customer> customerValidatingItemProcessor() {
    ValidatingItemProcessor<Customer> itemProcessor =
        new ValidatingItemProcessor<>(validator());

    itemProcessor.setFilter(true);

    return itemProcessor;
}
...
```

One down, three more to go. The second ItemProcessor we will configure is the ItemProcessorAdapter we configured earlier to uppercase the customer's name. This configuration is identical to the ItemProcessAdapter configuration we used previously as shown in Listing 8-16.

Listing 8-16. upperCaseItemProcessor

```
...
@Bean
public ItemProcessorAdapter<Customer, Customer> upperCaseItemProcessor(
    UpperCaseNameService service) {

    ItemProcessorAdapter<Customer, Customer> adapter = new ItemProcessorAdapter<>();

    adapter.setTargetObject(service);
    adapter.setTargetMethod("upperCase");

    return adapter;
}
...
```

For our third ItemProcessor, we will need to write a new script first. This one, instead of capitalizing the name related fields on the customer, will lowercase all the address related fields on the customer. Listing 8-17 shows the JavaScript for our new script.

Listing 8-17. lowerCase.js

```
item.setAddress(item.getAddress().toLowerCase());
item.setCity(item.getCity().toLowerCase());
item.setState(item.getState().toLowerCase());
item;
```

To put our script to use, we will configure the ScriptItemProcessor using the configuration in Listing 8-18.

Listing 8-18. lowerCaseItemProcessor

```
...
@Bean
@StepScope
public ScriptItemProcessor<Customer, Customer> lowerCaseItemProcessor(
            @Value("#{jobParameters['script']}") Resource script) {

    ScriptItemProcessor<Customer, Customer> itemProcessor =
                new ScriptItemProcessor<>();

    itemProcessor.setScript(script);

    return itemProcessor;
}
...
```

Finally, to put it all together, we need to configure our `CompositeItemProcessor`. This `ItemProcessor` takes a list of `ItemProcessors` to execute, so order is important. Listing 8-19 shows the configuration of the `CompositeItemProcessor` with its chain of delegates.

Listing 8-19. `CompositeItemProcessor` Configuration

```
...
@Bean
public CompositeItemProcessor<Customer, Customer> itemProcessor() {
    CompositeItemProcessor<Customer, Customer> itemProcessor =
                new CompositeItemProcessor<>();

    itemProcessor.setDelegates(Arrays.asList(
                customerValidatingItemProcessor(),
                upperCaseItemProcessor(null),
                lowerCaseItemProcessor(null)));

    return itemProcessor;
}
...
```

When our job is run with these updates, the output consists of all name fields uppercased, all address related fields lowercased, and two records being filtered out (the one for Warren M. Darrow and Regan M. Darrow). Listing 8-20 shows a sample of the output form the job.

Listing 8-20. `CompositeItemProcessor` Job Output

```
...
2019-02-05 23:46:49.884  INFO 46774 --- [           main] o.s.batch.core.job.SimpleStep
Handler      : Executing step: [copyFileStep]
Customer{firstName='RICHARD', middleInitial='N', lastName='DARROW', address='5570 isabella
ave', city='st. louis', state='il', zip='58540'}
Customer{firstName='BARACK', middleInitial='G', lastName='DONNELLY', address='7844 s.
greenwood ave', city='houston', state='ca', zip='38635'}
Customer{firstName='ANN', middleInitial='Z', lastName='BENES', address='2447 s. greenwood
ave', city='las vegas', state='ny', zip='55366'}
...
```

The `CompositeIemProcessor` allows for you to compose an `ItemProcessor` via other components for good design of your software. However, what happens if you don't want to send every item to every `ItemProcessor` on the list? What if you want some items to go to `ItemProcessorA` and some to go to `ItemProcessorB`? The `ClassifierCompositeItemProcessor` is the component you will want.

The `ClassifierCompositeItemProcessor` uses an `org.springframework.classify.Classifier` implementation to choose what `ItemProcessor` to use. The `Classifier`'s implementation of the `classify(C classifiable)` method must take an item as input and return the appropriate `ItemProcessor` to use. To look at this, we will create a `Classifier` that returns the `ItemProcessor` to uppercase the name if the zip code is odd, and the `ItemProcessor` to lowercase the address if the zip code is even.

The key piece of code we will need to write for this is our `Classifier` implementation. It will determine if the zip code is even or odd and return the correct delegate. Listing 8-21 lists the code for the `ZipCodeClassifier`.

Listing 8-21. ZipCodeClassifier

```
,,,
public class ZipCodeClassifier implements Classifier<Customer, ItemProcessor<Customer,
Customer>> {

    private ItemProcessor<Customer, Customer> oddItemProcessor;
    private ItemProcessor<Customer, Customer> evenItemProcessor;

    public ZipCodeClassifier(ItemProcessor<Customer, Customer> oddItemProcessor,
                ItemProcessor<Customer, Customer> evenItemProcessor) {

        this.oddItemProcessor = oddItemProcessor;
        this.evenItemProcessor = evenItemProcessor;
    }

    @Override
    public ItemProcessor<Customer, Customer> classify(Customer classifiable) {
        if(Integer.parseInt(classifiable.getZip()) % 2 == 0) {
            return evenItemProcessor;
        }
        else {
            return oddItemProcessor;
        }
    }
}
```

With our Classifier implementation created, we can configure it and the
ClassifierCompositeItemProcessor in our job. To begin, we'll configure the ZipCodeClassifier by
injecting the upperCaseItemProcessor and the lowerCaseItemProcessor into its constructor respectively.
This will be followed by the configuration for the ClassifierCompositeItemProcessor where we set the
Classifier as its only dependency. Listing 8-22 has the code for this configuration.

Listing 8-22. ClassifierCompositeItemProcessor Configuration

```
...
@Bean
public Classifier classifier() {
    return new ZipCodeClassifier(upperCaseItemProcessor(null),
                lowerCaseItemProcessor(null));
}

@Bean
public ClassifierCompositeItemProcessor<Customer, Customer> itemProcessor() {
    ClassifierCompositeItemProcessor<Customer, Customer> itemProcessor =
                new ClassifierCompositeItemProcessor<>();

    itemProcessor.setClassifier(classifier());

    return itemProcessor;
}
...
```

241

After we build and execute the job with this configuration as its ItemProcessor, the output will look like Listing 8-23.

Listing 8-23. Output from the ClassifierCompositeItemProcessor Job

```
...
2019-02-06 00:17:11.833  INFO 47362 --- [            main] o.s.b.c.l.support.SimpleJob
Launcher      : Job: [SimpleJob: [name=job]] launched with the following parameters:
[{customerFile=/input/customer.csv, script=/lowerCase.js}]
2019-02-06 00:17:11.882  INFO 47362 --- [            main] o.s.batch.core.job.
SimpleStepHandler     : Executing step: [copyFileStep]
Customer{firstName='Richard', middleInitial='N', lastName='Darrow', address='5570 isabella
ave', city='st. louis', state='il', zip='58540'}
Customer{firstName='BARACK', middleInitial='G', lastName='DONNELLY', address='7844 S.
Greenwood Ave', city='Houston', state='CA', zip='38635'}
Customer{firstName='Ann', middleInitial='Z', lastName='Benes', address='2447 s. greenwood
ave', city='las vegas', state='ny', zip='55366'}
...
```

The ClassifierCompositeItemProcessor is another way to build complex flows within an ItemProcessor. Combined with the CompositeItemProcessor, the options are nearly limitless as to how complex a chain of processors you can build within the ItemProcessor phase of a step.

In the next section, you will look at writing your own ItemProcessor to filter items from the ItemWriter. Although you have filtered records out in an earlier section, we have not implemented our own ItemProcessor yet. In the next section, you will look at how to change that.

Writing Your Own ItemProcessor

The ItemProcessor is really the easiest piece of the Spring Batch framework to implement yourself. This is by design. Input and output is standard across environments and business cases. Reading a file is the same regardless of whether it contains financial data or scientific data. Writing to a database works the same regardless of what the object looks like. However, the ItemProcessor is where the business logic of your process exists. Because of this, you will virtually always need to create custom implementations of them. In this section, you will look at how to create a custom ItemProcessor implementation that filters certain items that were read from begin written.

Filtering Items

In the previous section, we performed different logic based on if the zip code was even or odd. In this section, we will write an ItemProcessor that filters the even zip codes out, leaving only the odd ones to be written.

So how do we filter out records when they are going through an ItemProcessor? Spring Batch makes this simple by ensuring that any item that results in the ItemProcessor returning null is filtered out. Not only is it filtered out from the downstream impacts (other ItemProcessors or any ItemWriters involved in the step), but Spring Batch keeps a count of the number of records that are filtered and stores it in the job repository.

To implement our ItemProcessor, we will need to create a class that implements the ItemProcessor interface. We will put the logic in the process method returning null if the zip code is even, and returning the input parameter unmodified if the zip code is odd. Listing 8-24 has the code as described.

Listing 8-24. `EvenFilteringItemProcessor`

```
...
public class EvenFilteringItemProcessor implements ItemProcessor<Customer, Customer> {

    @Override
    public Customer process(Customer item)  {
        return Integer.parseInt(item.getZip()) % 2 == 0 ? null: item;
    }
}
```

The only thing left to do is to configure our job to use this `ItemProcessor`. Listing 8-25 has the configuration of our new `ItemProcessor`.

Listing 8-25. Custom `ItemProcessor` Configuration

```
...
@Bean
public EvenFilteringItemProcessor itemProcessor() {
    return new EvenFilteringItemProcessor();
}
...
```

With our new `ItemProcessor` configured, we can run the job and see that nine of the records are filtered out by our `EvenFilteringItemProcessor` leaving six written out to standard out as shown in Listing 8-26.

Listing 8-26. Output from the Custom `ItemProcessor` Job

```
...
2019-02-06 00:31:30.808  INFO 47626 --- [            main] o.s.batch.core.job.SimpleStep
Handler      : Executing step: [copyFileStep]
Customer{firstName='Barack', middleInitial='G', lastName='Donnelly', address='7844 S.
Greenwood Ave', city='Houston', state='CA', zip='38635'}
Customer{firstName='Laura', middleInitial='9S', lastName='Minella', address='8177 4th
Street', city='Dallas', state='FL', zip='04119'}
Customer{firstName='Warren', middleInitial='L', lastName='Darrow', address='4686 Mt. Lee
Drive', city='St. Louis', state='NY', zip='94935'}
Customer{firstName='Harry', middleInitial='T', lastName='Smith', address='3273 Isabella
Ave', city='Houston', state='FL', zip='97261'}
Customer{firstName='Jonas', middleInitial='U', lastName='Gilbert', address='8852 In St.',
city='Saint Paul', state='MN', zip='57321'}
Customer{firstName='Regan', middleInitial='M', lastName='Darrow', address='4851 Nec Av.',
city='Gulfport', state='MS', zip='33193'}
2019-02-06 00:31:30.949  INFO 47626 --- [            main] o.s.b.c.l.support.
SimpleJobLauncher      : Job: [SimpleJob: [name=job]] completed with the following
parameters: [{customerFile=/input/customer.csv}] and the following status: [COMPLETED]
```

If we are going to filter out records, it would be good to know how many in an easy way. Spring Batch records the count of items filtered in the job repository. If we look, we can see that using the data in Listing 8-4, 15 records were read, 9 were filtered out, and 6 were written. Listing 8-27 shows that data from the job repository.

Listing 8-27. Filter Count in the Job Repository

```
mysql> select step_execution_id as id, step_name, status, commit_count, read_count, filter_
count, write_count from SPRING_BATCH.BATCH_STEP_EXECUTION;
+----+--------------+-----------+--------------+------------+--------------+-------------+
| id | step_name    | status    | commit_count | read_count | filter_count | write_count |
+----+--------------+-----------+--------------+------------+--------------+-------------+
|  1 | copyFileStep | COMPLETED |            4 |         15 |            9 |           6 |
+----+--------------+-----------+--------------+------------+--------------+-------------+
1 row in set (0.01 sec)
```

In Chapter 4, you learned about skipping items, which used exceptions to identify records that were not to be processed. The difference between these two approaches is that this approach is intended for records that are technically valid records. Your customer had a zip code in the data. Instead, your business rules prevented you from being able to process this record, so you decided to filter it out of the steps results.

Although a simple concept, `ItemProcessors` are a piece of the Spring Batch framework that any batch developer will spend large amounts of time in. This is where the business logic lives and is applied to the items being processed.

Summary

`ItemProcessors` are where business logic can be applied to the items being processed in your jobs. Spring Batch, instead of trying to help you, does what it should do for this piece of the framework: it gets out of your way and lets you determine how to apply the logic of your business as needed. In the next chapter, you will finish your look at the core components of Spring Batch by taking a deep dive into `ItemWriters`.

CHAPTER 9

ItemWriters

It's amazing what computers can do. The numbers they can crunch. The images they can process. Yet it doesn't mean a thing unless the computer can communicate what it has done via its output. ItemWriters are the output facility for Spring Batch. And when you need a format to output the results of the Spring Batch process, Spring Batch delivers. In this chapter, you look at the different types of ItemWriters provided by Spring Batch as well as how to develop ItemWriters for situations that are more specific to your needs. Topics discussed include the following:

- *Introduction to ItemWriters:* Similar to the ItemReaders at the other end of step execution, ItemWriters have their own special nuances. This chapter talks about how ItemWriters work from a high level.

- *File-based ItemWriters:* File-based output is the easiest method to set up and is one of the most common forms used in batch processing. Because of this, you begin your exploration of ItemWriter implementations by looking at writing to flat files as well as XML files.

- *Database ItemWriters:* The relational database is king in the enterprise when it comes to data storage. However, databases create their own unique challenges when you're working with high volumes of data. You look at how Spring Batch handles these challenges with its unique architecture.

- *NoSQL ItemWriters:* While the relational database may be king, all the cool kids are using NoSQL stores like MongoDB, Apache Geode, and Neo4j for their storage needs. This section will look at the ItemWriter implementations that support the various forms of NoSQL stores.

- *Alternative output destination ItemWriters:* Files and databases aren't the only media to which enterprise software outputs. Systems send e-mails, write to Java Messaging Service (JMS) endpoints, and save data via other systems. This section looks at some of the less common but still very useful output methods that Spring Batch supports.

- *Multipart ItemWriters:* Unlike reading, where data typically comes from a single source and enriched later via an ItemProcessor, it's common to send output to multiple sources. Spring Batch provides ways to write to multiple systems as well as structure a single ItemWriter as a collaborative effort of multiple ItemWriters. This section looks at ItemWriters tasked with working with either multiple resources or multiple output formats.

To start with ItemWriters, let's look at how they work and how they fit into a step.

© Michael T. Minella 2019
M. T. Minella, *The Definitive Guide to Spring Batch*, https://doi.org/10.1007/978-1-4842-3724-3_9

Introduction to ItemWriters

The ItemWriter is the output mechanism used in Spring Batch. When Spring Batch first came out, ItemWriters were essentially the same as ItemReaders. They wrote each item out as it was processed. However, with Spring Batch 2 and the introduction of chunk-based processing, the role of the ItemWriter changed. Writing out each item as it's processed no longer makes sense.

With chunked-based processing, an ItemWriter doesn't write a single item: it writes a chunk of items. Because of this, the org.springframework.batch.item.ItemWriter interface is slightly different than the ItemReader interface. Listing 9-1 shows that the ItemWriter's write(List<T> items) method takes a List<T> of items, whereas the ItemReader interface you looked at in Chapter 7 returns only a single item from the read() method.

Listing 9-1. ItemWriter

```
package org.springframework.batch.item;

import java.util.List;

public interface ItemWriter<T> {
    void write(List<? extends T> items) throws Exception;
}
```

To illustrate the flow of how an ItemWriter fits into the step, Figure 9-1 shows a sequence diagram that walks through the processing within a step. The step reads each item individually via the ItemReader and passes it to the ItemProcessor for processing. This interaction continues until the number of items in a chunk has been processed. With the processing of a chunk complete, the items are passed into the ItemWriter to be written accordingly.

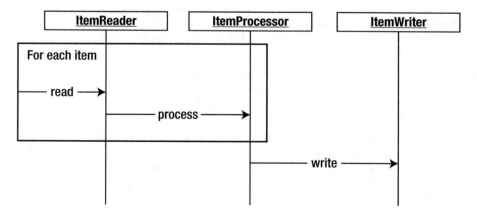

Figure 9-1. *Step interaction with an* ItemWriter

Since chunk-based processing was introduced, the number of calls to an ItemWriter is much less than it was. However, you need to handle things a bit differently. Take for example working with nontransactional resources like files. If a write to a file fails, there is no way to roll back what was already written. Because of that, extra safeguards must be put in place to limit your exposure to errors during the write.

Spring Batch provides a number of writers to handle the vast majority of output scenarios. Let's start with writers at the same place you started with readers: FlatFileItemWriter.

File-Based ItemWriters

Large amounts of data are moved via files in enterprise batch processing. There is a reason for this: files are simple and reliable. Backups are easy. So is recovery if you need to start over. This section looks at how to generate flat files in a variety of formats including formatted records (fixed width or other) and delimited files as well as how Spring Batch handles the issue of file creation.

FlatFileItemWriter

org.springframework.batch.item.file.FlatFileItemWriter is the ItemWriter implementation provided to generate text file output. Similar to FlatFileItemReader in many respects, this class addresses the issues with file-based output in Java with a clean, consistent interface for you to use. Figure 9-2 shows how the FlatFileItemWriter is constructed.

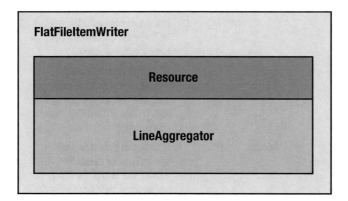

Figure 9-2. *FlatFileItemWriter pieces*

FlatFileItemWriter, as shown in Figure 9-2, consists of a Resource to write to and a LineAggregator implementation. The org.springframework.batch.item.file.transform.LineAggregator interface replaces the LineMapper of the FlatFileItemReader discussed in Chapter 7. Here, instead of parsing a String into an object as the LineMapper is responsible for doing, the LineAggregator is responsible for the generating of an output String based on an object.

FlatFileItemWriter has a number of interesting configuration options, which are reviewed in Table 9-1.

Table 9-1. *FlatFileItemWriter Configuration Options*

Option	Type	Default	Description
encoding	String	UTF-8	Character encoding for the file.
footerCallback	FlatFileFooterCallback	null	Executed after the last item of a file has been written.
headerCallback	FlatFileHeaderCallback	null	Executed before the first item of a file has been written.
lineAggregator	LineAggregator	null (required)	Used to convert an individual item to a String for output.
lineSeparator	String	System's line. separator	Generated file's newline character.
resource	Resource	null (required)	File or stream to be written to.
saveState	boolean	true	Determines if the state of the writer should be stored in the ExecutionContext as processing occurs.
shouldDeleteIfEmpty	boolean	false	If true and no records are written (not including header/footer records), the file is deleted on the close of the reader.
appendAllowed	boolean	false	If true and the file to be written to already exists, the output is appended to it instead of replacing the file. If true, shouldDeleteIfExists is automatically set to false.
shouldDeleteIfExists	boolean	true	If true and the file to be written to exists prior to the run of the job, the file is deleted and a new file is created.
transactional	boolean	true	If true and a transaction is currently active, the writing of the data to the file is delayed until the transaction is committed.

Unlike the `LineMapper` of `FlatFileItemReader`, the `LineAggregator` doesn't have any hard dependencies. However, a related interface to be aware of is `org.springframework.batch.item.file.transform.FieldExtractor`. This interface is used in most of the provided `LineAggregator` implementations as a way to access the required fields from a given item. Spring Batch provides two implementations of the `FieldExtractor` interface: `org.springframework.batch.item.file.transform.BeanWrapperFieldExtractor`, which uses the getters on the class to access the properties per traditional Java Beans, and `org.springframework.batch.item.file.transform.PassThroughFieldExtractor`, which returns the item (useful for items that are just a `String`, for example).

You will look at a few of the `LineAggregator` implementations over the course of this section. However, before we get into how to format our files, let's take a minute and talk about transactions. Spring Batch's transaction model is baked into chunk-based processing, and while it doesn't typically have an impact on the reading side of a step (transactional queues excluded), it has a big impact on the write side of things. How do transactions work with flat files?

■ **Note** The `FlatFileItemWriter` delays the persistence of the output data until the last moment before the commit to limit exposure to writing data and needing to roll back.

The `FlatFileItemWriter` is designed in a way that pushes off the actual write as late as possible in the transaction cycle. It does this by using a `TransactionSynchronizationAdapter`'s `beforeCommit(boolean readOnly)` method to do the actual write. This means that all other processing is complete and the only thing left to do is for the `PlatformTransactionManager` to commit the transaction before the data is actually pushed to disk. This allows for any other interactions that may cause problems (persisting data to the database, etc.) to fail before the data is written to the disk because there is no practical way to roll back the data once it is flushed to disk. All nontransactional data stores use a mechanism similar to this for persisting data during a transaction.

Let's begin to see how the `FlatFileItemWriter` really works by generating formatted files in the next section.

Formatted Text Files

When you looked at text files from the input side, you had three different types: fixed width, delimited, and XML. From the output side of things, you still have delimited and XML, but fixed width isn't just fixed width. In this case, it's really a formatted record. This section looks at how to construct batch output as a formatted text file.

Why the difference between a fixed-width input file and a formatted output file? Well, technically there is no difference. They're both files that contain a fixed format record of some kind, so it does not matter if that format is defining column widths or another format. However, typically input files have records that contain nothing but data and are defined via columns, whereas output files can be either fixed width or more robust (as you see later in this chapter with the statement job).

This example generates a list of customers and where they live. To get started, we will create a new project from Spring Initializr with the usual suspects for dependencies: batch, jdbc, and MySQL. Once we have our project, we can look at the input you're working with. Listing 9-2 shows an example of the customer.csv file.

Listing 9-2. customer.csv

```
Richard,N,Darrow,5570 Isabella Ave,St. Louis,IL,58540
Warren,L,Darrow,4686 Mt. Lee Drive,St. Louis,NY,94935
Barack,G,Donnelly,7844 S. Greenwood Ave,Houston,CA,38635
```

```
Ann,Z,Benes,2447 S. Greenwood Ave,Las Vegas,NY,55366
Erica,Z,Gates,3141 Farnam Street,Omaha,CA,57640
Warren,M,Williams,6670 S. Greenwood Ave,Hollywood,FL,37288
Harry,T,Darrow,3273 Isabella Ave,Houston,FL,97261
Steve,O,Darrow,8407 Infinite Loop Drive,Las Vegas,WA,90520
```

As Listing 9-2 shows, you're working with a file similar to the customer files you've been using up to this point in the book. However, the output for this job will be slightly different. In this case, you want to output a full sentence for each customer: "Richard Darrow lives at 5570 Isabella Ave in St. Louis, IL." Listing 9-3 shows an example of what the output file looks like.

Listing 9-3. Formatted Customer Output

```
Richard N Darrow lives at 5570 Isabella Ave in St. Louis, IL.
Warren L Darrow lives at 4686 Mt. Lee Drive in St. Louis, NY.
Barack G Donnelly lives at 7844 S. Greenwood Ave in Houston, CA.
Ann Z Benes lives at 2447 S. Greenwood Ave in Las Vegas, NY.
Laura 9S Minella lives at 8177 4th Street in Dallas, FL.
Erica Z Gates lives at 3141 Farnam Street in Omaha, CA.
Warren M Williams lives at 6670 S. Greenwood Ave in Hollywood, FL.
Harry T Darrow lives at 3273 Isabella Ave in Houston, FL.
Steve O Darrow lives at 8407 Infinite Loop Drive in Las Vegas, WA.
Erica Z Minella lives at 513 S. Greenwood Ave in Miami, IL.
```

How do you do this? For this example, you'll use a single step job that reads in the input file and writes it to the output file; we won't be needing an `ItemProcessor` for this example. Because the only code you need to write is that for the `Customer` class, you can start there; see Listing 9-4.

Listing 9-4. Customer.java

```
...
public class Customer {
    private static final long serialVersionUID = 1L;

    private long id;
    private String firstName;
    private String middleInitial;
    private String lastName;
    private String address;
    private String city;
    private String state;
    private String zip;

    // Accessors go here
    ...
}
```

As you can see in Listing 9-4, the fields of the Customer object map to the fields in the customer.csv file.[1] With the item coded, you can begin configuring the Job. The input side should be familiar from Chapter 7. Listing 9-5 shows the configuration of the input file as a resource (the value is passed in via a job parameter), the FlatFileItemReader configuration, and the required reference to the Customer object.

Listing 9-5. Configuring the Format Job's Input

```
...
@Bean
@StepScope
public FlatFileItemReader<Customer> customerFileReader(
        @Value("#{jobParameters['customerFile']}")Resource inputFile) {

    return new FlatFileItemReaderBuilder<Customer>()
            .name("customerFileReader")
            .resource(inputFile)
            .delimited()
            .names(new String[] {"firstName",
                    "middleInitial",
                    "lastName",
                    "address",
                    "city",
                    "state",
                    "zip"})
            .targetType(Customer.class)
            .build();
}
...
```

There shouldn't be a lot of surprises in the configuration in Listing 9-4. We define a step scoped bean so that we can inject the location of the input file via a job parameter. From there, we use the FlatFileItemReaderBuilder to construct our FlatFileItemReader by defining the name, injecting the Resource, identifying that the file is a comma-delimited file, and providing the names of each column and the type of object to be returned. Spring Batch uses this information to create a DefaultLineMapper that uses the DelimitedLineTokenizer and a BeanWrapperFieldSetMapper to parse the lines and populate our domain objects, respectively.

For the output side of things, we will use a FlatFileItemWriter and a LineAggregator. This example uses the org.springframework.batch.itemfile.transform.FormatterLineAggregator provided by Spring Batch. Having to configure a FieldExtractor and a LineAggregator by hand is a lot of work, so the framework rolls the configuration of those components into the FlatFileItemWriterBuilder. Listing 9-6 shows the configuration for the job's output.

Listing 9-6. Output Configuration for Format Job

```
...
@Bean
@StepScope
```

---

[1]The Customer object has an id attribute that you use later; it has no data in the file.

```
public FlatFileItemWriter<Customer> customerItemWriter(
        @Value("#{jobParameters['outputFile']}") Resource outputFile) {

        return new FlatFileItemWriterBuilder<Customer>()
                        .name("customerItemWriter")
                        .resource(outputFile)
                        .formatted()
                        .format("%s %s lives at %s %s in %s, %s.")
                        .names(new String[] {"firstName",
                                "lastName",
                                "address",
                                "city",
                                "state",
                                "zip"})
                        .build();
}
...
```

As Listing 9-6 shows, we begin by configuring a step scoped bean, allowing us to inject via a job parameter where the output file will result. We can then use the FlatFileItemWriterBuilder to construct the ItemWriter itself. We specify the name and Resource to write to. From there, we indicate to Spring Batch that we want to generate a formatted output file which returns the FormattedBuilder for us to configure the format we want our output to be generated in and the names of the fields to extract in the order they will be presented in the format. We call build to actually create the ItemWriter instance.

With all of the input and output configured, all you need to do to complete the job is configure the Step and Job. Listing 9-7 shows the complete configuration of formatJob including the previous input and output.

Listing 9-7. FormattedTextFileJob.java

```
...
@EnableBatchProcessing
@Configuration
public class FormattedTextFileJob {

    private JobBuilderFactory jobBuilderFactory;

    private StepBuilderFactory stepBuilderFactory;

    public FormattedTextFileJob(JobBuilderFactory jobBuilderFactory,
            StepBuilderFactory stepBuilderFactory) {

        this.jobBuilderFactory = jobBuilderFactory;
        this.stepBuilderFactory = stepBuilderFactory;
    }

    @Bean
    @StepScope
    public FlatFileItemReader<Customer> customerFileReader(
            @Value("#{jobParameters['customerFile']}")Resource inputFile) {
```

```
        return new FlatFileItemReaderBuilder<Customer>()
                .name("customerFileReader")
                .resource(inputFile)
                .delimited()
                .names(new String[] {"firstName",
                        "middleInitial",
                        "lastName",
                        "address",
                        "city",
                        "state",
                        "zip"})
                .targetType(Customer.class)
                .build();
}

@Bean
@StepScope
public FlatFileItemWriter<Customer> customerItemWriter(
        @Value("#{jobParameters['outputFile']}") Resource outputFile) {
            return new FlatFileItemWriterBuilder<Customer>()
                            .name("customerItemWriter")
                            .resource(outputFile)
                            .formatted()
                            .format("%s %s lives at %s %s in %s, %s.")
                            .names(new String[] {"firstName",
                                    "lastName",
                                    "address",
                                    "city",
                                    "state",
                                    "zip"})
                            .build();
}

@Bean
public Step formatStep() {
    return this.stepBuilderFactory.get("formatStep")
            .<Customer, Customer>chunk(10)
            .reader(customerFileReader(null))
            .writer(customerItemWriter(null))
            .build();
}

@Bean
public Job formatJob() {
    return this.jobBuilderFactory.get("formatJob")
            .start(formatStep())
            .incrementer(new RunIdIncrementer())
            .build();
}
}
```

After you build the project using Maven's `./mvnw clean install` command, you can execute the example via Spring Boot with the command shown in Listing 9-8.

Listing 9-8. How to Execute `formatJob` from the Command Line

```
java -jar itemWriters-0.0.1-SNAPSHOT.jar
customerFile=/data/customer.csv outputFile=file:/output/formattedCustomers.txt
```

When you run the job with the input specified in Listing 9-2, the result is a new file, formattedCustomers.txt, with the contents listed in Listing 9-9.

Listing 9-9. formattedCustomers.txt

```
Richard Darrow lives at 5570 Isabella Ave St. Louis in IL, 58540.
Warren Darrow lives at 4686 Mt. Lee Drive St. Louis in NY, 94935.
Barack Donnelly lives at 7844 S. Greenwood Ave Houston in CA, 38635.
Ann Benes lives at 2447 S. Greenwood Ave Las Vegas in NY, 55366.
Erica Gates lives at 3141 Farnam Street Omaha in CA, 57640.
Warren Williams lives at 6670 S. Greenwood Ave Hollywood in FL, 37288.
Harry Darrow lives at 3273 Isabella Ave Houston in FL, 97261.
Steve Darrow lives at 8407 Infinite Loop Drive Las Vegas in WA, 90520.
```

This method of formatting output can be used for a number of different requirements. Whether it's formatting items into human-readable output as you did here or formatting them into a fixed-width file as you used for input in Chapter 7, all that needs to change is the format `String` you configure for the `LineAggregator`.

The other main type of flat file you see on a regular basis is the delimited file. customer.csv is a comma-delimited file, for example. The next section looks at how to output files that contain delimited output.

Delimited Files

Unlike the formatted files you looked at in the previous section, delimited files don't have a single predefined format. Instead, a delimited file consists of a list of values separated by a predefined separator character. This section looks at how to use Spring Batch to generate a delimited file.

To see how generating a delimited file works, you use the same input for this job. For the output, you refactor the `ItemWriter` to generate the new, delimited output. In this case, you change the order of the fields and change the delimiter from a comma (,) to a semicolon (;). Listing 9-10 shows some sample output with the updated `formatJob`.

Listing 9-10. Output for Delimited `formatJob`

```
58540;IL;St. Louis;5570 Isabella Ave;Darrow;Richard
94935;NY;St. Louis;4686 Mt. Lee Drive;Darrow;Warren
38635;CA;Houston;7844 S. Greenwood Ave;Donnelly;Barack
55366;NY;Las Vegas;2447 S. Greenwood Ave;Benes;Ann
57640;CA;Omaha;3141 Farnam Street;Gates;Erica
37288;FL;Hollywood;6670 S. Greenwood Ave;Williams;Warren
97261;FL;Houston;3273 Isabella Ave;Darrow;Harry
90520;WA;Las Vegas;8407 Infinite Loop Drive;Darrow;Steve
```

To generate the output in Listing 9-10, all you need to do is update the configuration of the LineAggregator. Instead of using FormatterLineAggregator, you use Spring Batch's org. springframework.batch.item.file.transform.DelimitedLineAggregator implementation. Using the same BeanWrapperFieldExtractor to extract an Object array, the DelimitedLineAggregator concatenates the elements of the array with the configured delimiter between each element. Again, Spring Batch's FlatFileItemWriterBuilder provides another builder for configuring the LineAggregator resources needed to generate this delimited file by calling delimited(). Listing 9-11 shows the updated configuration for the ItemWriter.

Listing 9-11. flatFileOutputWriter Configuration

```
...
@Bean
@StepScope
public FlatFileItemWriter<Customer> customerItemWriter(
        @Value("#{jobParameters['outputFile']}") Resource outputFile) {

    return new FlatFileItemWriterBuilder<Customer>()
                    .name("customerItemWriter")
                    .resource(outputFile)
                    .delimited()
                    .delimiter(";")
                    .names(new String[] {"zip",
                                "state",
                                "city",
                                "address",
                                "lastName",
                                "firstName"})
                    .build();
}
...
```

By changing the configuration of the FormatterLineAggregator to use Spring Batch's DelimitedLineAggregator, the only other change you have to make is removing the format dependency and including the definition of a delimiter character. After building the project with the same ./mvnw clean install you used previously, you can run the job with the command in Listing 9-12.

Listing 9-12. Running formatJob to Generate Delimited Output

```
java -jar itemWriters-0.0.1-SNAPSHOT.jar
customerFile=/input/customer.csv outputFile=file:/output/delimitedCustomers.txt
```

The results of the formatJob with the updated configuration are shown in Listing 9-13.

Listing 9-13. formatJob Results for Delimited File Writing

```
58540;IL;St. Louis;5570 Isabella Ave;Darrow;Richard
94935;NY;St. Louis;4686 Mt. Lee Drive;Darrow;Warren
38635;CA;Houston;7844 S. Greenwood Ave;Donnelly;Barack
55366;NY;Las Vegas;2447 S. Greenwood Ave;Benes;Ann
57640;CA;Omaha;3141 Farnam Street;Gates;Erica
```

37288;FL;Hollywood;6670 S. Greenwood Ave;Williams;Warren
97261;FL;Houston;3273 Isabella Ave;Darrow;Harry
90520;WA;Las Vegas;8407 Infinite Loop Drive;Darrow;Steve

It's easy to create flat files with Spring Batch. With zero lines of code outside of the domain object, you can read in a file and convert its format to either a formatted file or a delimited file. Both of the examples for flat-file processing have assumed that the file is a new file to be created each time. The next section looks at some of the more advanced options Spring Batch provides for handling what file to write to.

File Management Options

Unlike reading from an input file where the file must exist or it is typically considered an error condition, an output file may or may not exist at the time of processing, and that may or may not be okay. Spring Batch provides the ability to configure how to handle each of these scenarios based on your needs. This section looks at how to configure FlatFileItemWriter to handle multiple file creation scenarios.

In Table 9-1, there were two options for FlatFileItemWriter that pertain to file creation: shouldDeleteIfEmpty and shouldDeleteIfExists. shouldDeleteIfEmpty actually deals with what to do when a step is complete. It's set to false by default. If a step executes, no items were written (a header and footer may have been, but no item records were written), and shouldDeleteIfEmpty is set to true, the file is deleted on the completion of the step. By default, the file is created and left empty. You can look at this behavior with the formatJob you ran in the previous section. By updating the configuration of flatFileOutputWriter to set shouldDeleteIfEmpty to true as shown in Listing 9-14, you can process an empty file and see that no output file is left behind.

Listing 9-14. Configuring formatJob to Delete the Output File If No Items Are Written

```
...
@Bean
@StepScope
public FlatFileItemWriter<Customer> delimitedCustomerItemWriter(
            @Value("#{jobParameters['outputFile']}") Resource outputFile) {

    return new FlatFileItemWriterBuilder<Customer>()
                    .name("customerItemWriter")
                    .resource(outputFile)
                    .delimited()
                    .delimiter(";")
                    .names(new String[] {"zip",
                                "state",
                                "city",
                                "address",
                                "lastName",
                                "firstName"})
                    .shouldDeleteIfEmpty(true)
                    .build();
}
 ...
```

If you execute formatJob with the updated file and pass it an empty customer.csv file as input, no output is left behind. It's important to note that the file is still created, opened, and closed. In fact, if the step is configured to write a header and/or footer in the file, that is written as well. However, if the number of items written to the file is zero, the file is deleted at the end of the step.

The next configuration parameter related to file creation/deletion is the shouldDeleteIfExists flag. This flag, set to true by default, deletes a file that has the same name as the output file the step intends to write to. For example, if you're going to run a job that writes to a file /output/jobRun.txt, and that file already exists when the job starts, Spring Batch deletes the file and creates a new one. If this file exists and the flag is set to false, an org.springframework.batch.item.ItemStreamException is thrown when the step attempts to create the new file. Listing 9-15 shows formatJob's flatFileOutputWriter configured to not delete the output file if it exists.

Listing 9-15. Configuring formatJob to Not Delete the Output File If It Already Exists

```
...
@Bean
@StepScope
public FlatFileItemWriter<Customer> delimitedCustomerItemWriter(
                @Value("#{jobParameters['outputFile']}") Resource outputFile) {

        return new FlatFileItemWriterBuilder<Customer>()
                        .name("customerItemWriter")
                        .resource(outputFile)
                        .delimited()
                        .delimiter(";")
                        .names(new String[] {"zip",
                                        "state",
                                        "city",
                                        "address",
                                        "lastName",
                                        "firstName"})
                        .shouldDeleteIfExists(false)
                        .build();
}
 ...
```

By running the job as it's configured in Listing 9-15, you receive the previously mentioned ItemStreamException as shown in Listing 9-16.

Listing 9-16. Results of a Job That Writes to an Existing File That Shouldn't Be There

```
2018-04-16 15:38:55.269  INFO 76152 --- [            main] o.s.batch.core.job.SimpleStep
Handler    : Executing step: [delimitedStep]
2018-04-16 15:38:55.316 ERROR 76152 --- [            main] o.s.batch.core.step.
AbstractStep          : Encountered an error executing step delimitedStep in job delimitedJob

org.springframework.batch.item.ItemStreamException: File already exists: [/Users/mminella/
Documents/IntelliJWorkspace/def-guide-spring-batch/Chapter9/target/formattedCustomers.txt]
```

```
    at org.springframework.batch.item.util.FileUtils.setUpOutputFile(FileUtils.java:56)
    ~[spring-batch-infrastructure-4.0.1.RELEASE.jar:4.0.1.RELEASE]
    at org.springframework.batch.item.file.FlatFileItemWriter$OutputState.initializeBuff
    eredWriter(FlatFileItemWriter.java:572) ~[spring-batch-infrastructure-4.0.1.RELEASE.
    jar:4.0.1.RELEASE]
    at org.springframework.batch.item.file.FlatFileItemWriter$OutputState.
    access$000(FlatFileItemWriter.java:414) ~[spring-batch-infrastructure-4.0.1.RELEASE.
    jar:4.0.1.RELEASE]
    at org.springframework.batch.item.file.FlatFileItemWriter.doOpen(FlatFileItemWriter.
    java:348) ~[spring-batch-infrastructure-4.0.1.RELEASE.jar:4.0.1.RELEASE]
    at org.springframework.batch.item.file.FlatFileItemWriter.open(FlatFileItemWriter.
    java:338) ~[spring-batch-infrastructure-4.0.1.RELEASE.jar:4.0.1.RELEASE]
```

The use of this parameter is a good idea in an environment where you want to preserve the output of each run. This prevents an accidental overwrite of your old file.

The final option related to file creation is the appendAllowed parameter. When this flag (which defaults to false) is set to true via the .append(boolean value) method call, Spring Batch automatically sets the shouldDeleteIfExists flag to false, creates a new file if one doesn't exist, and appends the data if it does. This option can be useful if you have an output file that you need to write to from multiple steps. Listing 9-17 shows formatJob configured to append data if the file exists.

Listing 9-17. Appending Data If the Output File Exists

```
...
@Bean
@StepScope
public FlatFileItemWriter<Customer> delimitedCustomerItemWriter(
                @Value("#{jobParameters['outputFile']}") Resource outputFile) {

        return new FlatFileItemWriterBuilder<Customer>()
                        .name("customerItemWriter")
                        .resource(outputFile)
                        .delimited()
                        .delimiter(";")
                        .names(new String[] {"zip",
                                        "state",
                                        "city",
                                        "address",
                                        "lastName",
                                        "firstName"})
                        .append(true)
                        .build();
}
...
```

With this configuration, you can run the job multiple times using the same output file (with different input files), and Spring Batch appends the output of the current job to the end of the existing output file.

As you can see, there are a number of options available to handle generation of flat file-based output, from being able to format your records any way you want to generating delimited files and even providing options for how Spring Batch handles files that already exist. However, flat files aren't the only type of file output. XML is the other type of file output that Spring Batch provides for, and you look at it next.

StaxEventItemWriter

When you looked at reading XML back in Chapter 7, you explored how Spring Batch views XML documents in fragments. Each of these fragments is the XML representation of a single item to be processed. On the `ItemWriter` side, the same concept exists. Spring Batch generates an XML fragment for each of the items the `ItemWriter` receives and writes the fragment to the file. This section looks at how Spring Batch handles XML as an output medium.

To handle writing XML using Spring Batch, you use `org.springframework.batch.item.xml.StaxEventItemWriter`. Just like the `ItemReader`, the Streaming API for XML (StAX) implementation allows Spring Batch to write fragments of XML as each chunk is processed. Just like `FlatFileItemWriter`, `StaxEventItemWriter` generates the XML a chunk at a time and writes it to the file just before the local transaction has been committed; this prevents rollback issues if there is an error writing to the file.

The configuration of the `StaxEventItemReader` consists of a `Resource` (file to read from), a root element name (the root tag for each fragment), and an `Unmarshaller` to be able to convert the XML input into an object. The configuration for `StaxEventItemWriter` is almost identical, with a `Resource` to write to, a root element name (the root tag for each fragment you generate), and a `Marshaller` to convert each item into an XML fragment.

`StaxEventItemWriter` has a collection of configurable attributes that are covered in Table 9-2.

Table 9-2. *Attributes Available in `StaxEventItemWriter`*

Option	Type	Default	Description
encoding	String	UTF-8	Character encoding for the file.
footerCallback	StaxWriterCallback	null	Executed after the last item of a file has been written.
headerCallback	StaxWriterCallback	null	Executed before the first item of a file has been written.
marshaller	Marshaller	null (required)	Used to convert an individual item to an XML fragment for output.
overwriteOutput	boolean	true	By default, the file is replaced if the output file already exists. If this is set to `true` and the file exists, an `ItemStreamException` is thrown.
resource	Resource	null (required)	File or stream to be written to.
rootElementAttributes	Map<String, String>	null	This key/value pairing is appended to the root tag of each fragment with the keys as the attribute names and value as their values.
rootTagName	String	null (required)	Defines the root XML tag the XML document.

(*continued*)

Table 9-2. (*continued*)

Option	Type	Default	Description
saveState	boolean	true	Determines if Spring Batch keeps track of the state of the ItemWriter (number of items written, and so on).
transactional	boolean	true	If true, the writing of the output is delayed until the transaction is committed, to prevent rollback issues.
version	String	"1.0"	Version of XML the file is written in.

To look at how StaxEventItemWriter works, let's update formatJob to output the customer output in XML. Using the same input from the previous examples, Listing 9-18 shows the new output you create when you update the job.

Listing 9-18. customer.xml

```xml
<?xml version="1.0" encoding="UTF-8"?>
<customers>
  <customer>
    <id>0</id>
    <firstName>Richard</firstName>
    <middleInitial>N</middleInitial>
    <lastName>Darrow</lastName>
    <address>5570 Isabella Ave</address>
    <city>St. Louis</city>
    <state>IL</state>
    <zip>58540</zip>
  </customer>
    ...
</customers>
```

In order to generate the output shown in Listing 9-18, you reuse the formatJob configuration but replace flatFileOutputWriter with a new xmlOutputWriter that uses the StaxEventItemWriter ItemWriter implementation. To configure the new ItemWriter, you provide three dependencies as shown in Listing 9-19: a Resource to write to, a reference to an org.springframework.oxm.Marshaller implementation, and a root tag name (customer in this case).

Listing 9-19. Configuration for formatJob with StaxEventItemWriter

```java
...
@Configuration
public class XmlFileJob {

        private JobBuilderFactory jobBuilderFactory;

        private StepBuilderFactory stepBuilderFactory;
```

```java
public XmlFileJob(JobBuilderFactory jobBuilderFactory,
                StepBuilderFactory stepBuilderFactory) {

        this.jobBuilderFactory = jobBuilderFactory;
        this.stepBuilderFactory = stepBuilderFactory;
}

@Bean
@StepScope
public FlatFileItemReader<Customer> customerFileReader(
                @Value("#{jobParameters['customerFile']}")Resource inputFile) {

        return new FlatFileItemReaderBuilder<Customer>()
                        .name("customerFileReader")
                        .resource(inputFile)
                        .delimited()
                        .names(new String[] {"firstName",
                                        "middleInitial",
                                        "lastName",
                                        "address",
                                        "city",
                                        "state",
                                        "zip"})
                        .targetType(Customer.class)
                        .build();
}

@Bean
@StepScope
public StaxEventItemWriter<Customer> xmlCustomerWriter(
        @Value("#{jobParameters['outputFile']}") Resource outputFile) {

        Map<String, Class> aliases = new HashMap<>();
        aliases.put("customer", Customer.class);

        XStreamMarshaller marshaller = new XStreamMarshaller();

        marshaller.setAliases(aliases);

        marshaller.afterPropertiesSet();

        return new StaxEventItemWriterBuilder<Customer>()
                        .name("customerItemWriter")
                        .resource(outputFile)
                        .marshaller(marshaller)
                        .rootTagName("customers")
                        .build();
}

@Bean
public Step xmlFormatStep() throws Exception {
```

```
                    return this.stepBuilderFactory.get("xmlFormatStep")
                            .<Customer, Customer>chunk(10)
                            .reader(customerFileReader(null))
                            .writer(xmlCustomerWriter(null))
                            .build();
        }

        @Bean
        public Job xmlFormatJob() throws Exception {
                return this.jobBuilderFactory.get("xmlFormatJob")
                            .start(xmlFormatStep())
                            .build();
        }
}
```

Of the 80 or so lines of Java code that it took to configure the original formatJob as shown in Listing 9-7, the formatJob in Listing 9-19 has changed only the delimitedCustomerItemWriter(Resource file) method. The changes begin with the definition of a new ItemWriter, xmlOutputWriter. This bean is a reference to the StaxEventItemWriter the section has been talking about and defines three dependencies: the resource to write to, the Marshaller implementation, and the root tag name for each XML fragment the Marshaller will generate.

We configure the Marshaller inline. This object is used to generate an XML fragment for each item the job processes. Using Spring's org.springframework.oxm.xtream.XStreamMarshaller class, the only further configuration you're required to provide is a Map of aliases to use for each type the Marshaller comes across. By default, the Marshaller uses the attribute's name as the tag name, but you provide an alias for the Customer class because the XStreamMarshaller uses the fully qualified name for the class by default as the root tag of each fragment (com.apress.springbatch.chatper8.Customer instead of just customer).

In order for the job to be able to compile and run, you need to make one more update. The POM file needs a new dependency to handle the XML processing, a reference to Spring's Object/XML Mapping (OXM) library as well as the XStream library we are using for XML processing. Listing 9-20 shows the update to the POM that is required.

Listing 9-20. Spring's OXM Library Maven Dependency

```
...
<dependency>
        <groupId>org.springframework</groupId>
        <artifactId>spring-oxm</artifactId>
</dependency>
<dependency>
        <groupId>com.thoughtworks.xstream</groupId>
        <artifactId>xstream</artifactId>
        <version>1.4.10</version>
</dependency>
 ...
```

With the POM updated and the job configured, you're ready to build and run formatJob to generate XML as the output. After running a ./mvnw clean install from the command line, you can use the command listed in Listing 9-21 to execute the job.

Listing 9-21. Executing formatJob to Generate XML

```
java -jar itemWriters-0.0.1-SNAPSHOT.jar
customerFile=/input/customer.csv outputFile=file:/output/xmlCustomer.xml
```

When you look at the results of the XML, notice that it was obviously generated by a library in that there is no formatting applied. However using your IDE or your favorite text editor, you can see clearly that the output is what you expected. Listing 9-22 shows a sample of the generated output XML.

Listing 9-22. formatJob XML Results

```
<?xml version="1.0" encoding="UTF-8"?>
<customers>
        <customer>
                <id>0</id>
                <firstName>Richard</firstName>
                <middleInitial>N</middleInitial>
                <lastName>Darrow</lastName>
                <address>5570 Isabella Ave</address>
                <city>St. Louis</city>
                <state>IL</state>
                <zip>58540</zip>
        </customer>
    ...
</customers>
```

With not much more than a couple lines of Java, you can easily generate XML output with the full power of any Spring-supported XML Marshaller.

The ability to process XML as both input and output is important in today's enterprise environment, as is the ability to process flat files. However, although files play a large part in batch processing, they aren't as prevalent in other processing in today's enterprise. Instead, the relational database has taken over. As such, the batch process must be able to not only read from a database (as you saw in Chapter 7) but write to it as well. The next section looks at the more common ways to handle writing to a database using Spring Batch.

Database-Based ItemWriters

Writing to a database offers a different set of constraints than file-based output. First, databases are transactional resources, unlike files. Because of this, you can include the physical write as part of the transaction instead of segmenting it as file-based processing does. Also, there are many different options for how to access a database. JDBC, Java Persistence API (JPA), and Hibernate all offer unique yet compelling models for handling writing to a database. This section looks at how to use JDBC, Hibernate, and JPA to write the output of a batch process to a database.

JdbcBatchItemWriter

The first way you can write to the database is the way most people learn how to access a database with Spring, via JDBC. Spring Batch's JdbcBatchItemWriter uses the JdbcTemplate and its batch SQL execution capabilities to execute all of the SQL for a single chunk at once. This section looks at how to use JdbcBatchItemWriter to write a step's output to a database.

org.springframework.batch.item.database.JdbcBatchItemWriter isn't much more than a thin wrapper around Spring's org.springframework.jdbc.support.JdbcTemplate, using the JdbcTemplate. batchUpdate() or JdbcTemplate.execute() method depending on whether named parameters are used in the SQL to execute mass database insert/updates. The important thing to note about this is that Spring uses PreparedStatement's batch-update capabilities to execute all the SQL statements for a single chunk at once instead of using multiple calls. This greatly improves performance while still allowing all the executions to execute within the current transaction.

To see how the JdbcBatchItemWriter works, again you work with the same input you used with the file-based writers, but you use it to populate a CUSTOMER database table instead of writing a file. Figure 9-3 shows the design of the table into which you're inserting the customer information.

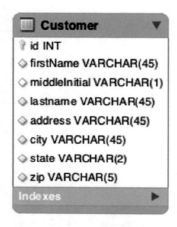

Figure 9-3. *CUSTOMER table design*

As you can see in Figure 9-3, the columns of the CUSTOMER table match up virtually one-to-one with the elements in the customer.csv file. The only difference is the id field, which you let the database populate for you. In order to insert the values into the table, you need to build the SQL in either of two ways: using question marks (?) as placeholders for the values or using named parameters (:name, for example) as placeholders. Each of these two options requires a slightly different approach in populating the values. You start with the question mark as shown in the sample SQL statement in Listing 9-23.

Listing 9-23. Prepared Statement for Inserting into the CUSTOMER Table

```
insert into CUSTOMER (firstName, middleInitial, lastName, address, city,
state, zip) values (?, ?, ?, ?, ?, ?, ?)
```

As you can see, there is nothing unusual about the prepared statement. However, providing the SQL statement is only one of the configuration options for JdbcBatchItemWriter. Table 9-3 lists all the configuration options.

Table 9-3. *JdbcBatchItemWriter Configuration Options*

Option	Type	Default	Description
assertUpdates	boolean	true	If true, causes JdbcBatchItemWriter to validate that every item resulted in an insert or update. If any item didn't trigger an insert or an update of a record, an EmptyResultDataAccessException is thrown.
dataSource	DataSource	null (required)	Provides access to the required database.
itemPrepared StatementSetter	ItemPrepared Statement Setter	null	If a standard PreparedStatement is provided (using ? for parameters), JdbcBatchItemWriter uses this class to populate the parameter values.
itemSqlParameter SourceProvider	ItemSqlParameter SourceProvider	null	If named parameters are used in the SQL provided, the JdbcBatchItemWriter uses this class to populate the parameter values.
simpleJdbcTemplate	SimpleJdbc Template	null	Allows you to inject an implementation of the SimpleJdbcOperations interface.
sql	String	null (required)	SQL to be executed for each item.

To use JdbcBatchItemWriter in formatJob, you replace xmlOutputWriter with a new jdbcBatchWriter bean. Because you begin with a standard PreparedStatement syntax for the query (using question marks), you need to provide it with a DataSource, the SQL to be executed, and an implementation of the org.springframework.batch.item.database.ItemPreparedStatementSetter interface. Yes, you're correct if you realized that you're going to have to write some code to make this one work.

ItemPreparedStatementSetter is a simple interface used to abstract the extraction of values from each item and set them on the PreparedStatement. It contains a single method, as shown in Listing 9-24.

Listing 9-24. ItemPreparedStatementSetter Interface

```
package org.springframework.batch.item.database;

import java.sql.PreparedStatement;
import java.sql.SQLException;

public interface ItemPreparedStatementSetter<T> {
    void setValues(T item, PreparedStatement ps) throws SQLException;
}
```

To implement the ItemPreparedStatementSetter interface, you create your own CustomerItemPreparedStatementSetter. This class implements the single setValues(T item, PreparedStatement ps) method that is required by the ItemPreparedStatementSetter interface by using the normal PreparedStatement API to populate each value of the PreparedStatement with the appropriate value from the item. Listing 9-25 shows the code for CustomerItemPreparedStatementSetter.

Listing 9-25. `CustomerItemPreparedStatementSetter.java`

```java
...
public class CustomerItemPreparedStatementSetter implements
        ItemPreparedStatementSetter<Customer> {

    public void setValues(Customer customer, PreparedStatement ps)
            throws SQLException {

        ps.setString(1, customer.getFirstName());
        ps.setString(2, customer.getMiddleInitial());
        ps.setString(3, customer.getLastName());
        ps.setString(4, customer.getAddress());
        ps.setString(5, customer.getCity());
        ps.setString(6, customer.getState());
        ps.setString(7, customer.getZip());
    }
}
```

As Listing 9-25 shows, there is no magic involved in setting the values for each `PreparedStatement`. With this code, you can update `formatJob`'s configuration to write its output to the database. Listing 9-26 shows the configuration for the new ItemWriter.

Listing 9-26. `jdbcBatchWriter`'s Configuration

```java
...
@Bean
@StepScope
public JdbcBatchItemWriter<Customer> jdbcCustomerWriter(DataSource dataSource) throws
Exception {
        return new JdbcBatchItemWriterBuilder<Customer>()
                        .dataSource(dataSource)
                        .sql("INSERT INTO CUSTOMER (first_name, " +
                                        "middle_initial, " +
                                        "last_name, " +
                                        "address, " +
                                        "city, " +
                                        "state, " +
                                        "zip) VALUES (?, ?, ?, ?, ?, ?, ?)")
                        .itemPreparedStatementSetter(new
                                CustomerItemPreparedStatementSetter())
                        .build();
}
...
```

As you can see in Listing 9-26, the new `jdbcBatchItemWriter` references the `dataSource` bean from the Spring Boot (the `CUSTOMER` table is in the same schema as the Spring Batch tables you use for the `JobRepository`). The SQL value is the same as the SQL statement you previously defined in Listing 9-23. The last dependency you provide is the reference to the `CustomerItemPreparedStatementSetter`.

The final piece of the puzzle to configure the new ItemWriter is to update the configuration for the step to reference the new ItemWriter. To do this, all you need to do is update `formatStep`'s configuration to reference the `jdbcBatchWriter` bean in place of its current reference to the `xmlOutputWriter` from the previous section. Listing 9-27 shows the full listing of JdbcFormatJob.java configured to write to the database.

Listing 9-27. `formatJob` Configured for JDBC Database Writing

```
...
@Configuration
public class JdbcFormatJob {

        private JobBuilderFactory jobBuilderFactory;

        private StepBuilderFactory stepBuilderFactory;

        public JdbcFormatJob(JobBuilderFactory jobBuilderFactory,
                        StepBuilderFactory stepBuilderFactory) {

                this.jobBuilderFactory = jobBuilderFactory;
                this.stepBuilderFactory = stepBuilderFactory;
        }

        @Bean
        @StepScope
        public FlatFileItemReader<Customer> customerFileReader(
                        @Value("#{jobParameters['customerFile']}")Resource inputFile) {

                return new FlatFileItemReaderBuilder<Customer>()
                                .name("customerFileReader")
                                .resource(inputFile)
                                .delimited()
                                .names(new String[] {"firstName",
                                                "middleInitial",
                                                "lastName",
                                                "address",
                                                "city",
                                                "state",
                                                "zip"})
                                .targetType(Customer.class)
                                .build();
        }

        @Bean
        @StepScope
        public JdbcBatchItemWriter<Customer> jdbcCustomerWriter(DataSource dataSource)
        throws Exception {
                return new JdbcBatchItemWriterBuilder<Customer>()
                                .dataSource(dataSource)
                                .sql("INSERT INTO CUSTOMER (first_name, " +
                                                "middle_initial, " +
                                                "last_name, " +
```

```
                                          "address, " +
                                          "city, " +
                                          "state, " +
                                          "zip) VALUES (?, ?, ?, ?, ?, ?, ?)")
                            .itemPreparedStatementSetter(
                                    new CustomerItemPreparedStatementSetter())
                            .build();
        }

        @Bean
        public Step xmlFormatStep() throws Exception {
                return this.stepBuilderFactory.get("xmlFormatStep")
                                .<Customer, Customer>chunk(10)
                                .reader(customerFileReader(null))
                                .writer(jdbcCustomerWriter(null))
                                .build();
        }

        @Bean
        public Job xmlFormatJob() throws Exception {
                return this.jobBuilderFactory.get("xmlFormatJob")
                                .start(xmlFormatStep())
                                .build();
        }
}
```

Because you already have the JDBC drivers configured in the POM and the DataSource configured for the JobRepository, all you need to do is execute an ./mvnw clean install and execute the command in Listing 9-28 to see the results of the updated formatJob.

Listing 9-28. Command to Execute formatJob

```
java -jar itemWriters-0.0.1-SNAPSHOT.jar customerFile=/input/customer.csv
```

The output of this job isn't in a file this time but in the database. You can confirm the execution in two ways. The first is by going to the database to validate the input. Listing 9-29 shows the results of the job in the database.

Listing 9-29. Job Results with jdbcBatchWriter

```
mysql> select id, first_name, middle_initial as middle, last_name, address, city, state as
st, zip from SPRING_BATCH.CUSTOMER;
+----+------------+--------+-----------+-------------------------+-----------+------+-------+
| id | first_name | middle | last_name | address                 | city      | st   | zip   |
+----+------------+--------+-----------+-------------------------+-----------+------+-------+
|  1 | Richard    | N      | Darrow    | 5570 Isabella Ave       | St. Louis | IL   | 58540 |
|  2 | Warren     | L      | Darrow    | 4686 Mt. Lee Drive      | St. Louis | NY   | 94935 |
|  3 | Barack     | G      | Donnelly  | 7844 S. Greenwood Ave   | Houston   | CA   | 38635 |
|  4 | Ann        | Z      | Benes     | 2447 S. Greenwood Ave   | Las Vegas | NY   | 55366 |
|  5 | Erica      | Z      | Gates     | 3141 Farnam Street      | Omaha     | CA   | 57640 |
```

```
|  6 | Warren     | M      | Williams   | 6670 S. Greenwood Ave   | Hollywood | FL   | 37288 |
|  7 | Harry      | T      | Darrow     | 3273 Isabella Ave       | Houston   | FL   | 97261 |
|  8 | Steve      | O      | Darrow     | 8407 Infinite Loop Drive | Las Vegas | WA   | 90520 |
+----+------------+--------+------------+-------------------------+-----------+------+-------+
8 rows in set (0.00 sec)
```

The PreparedStatement notation is useful given most Java developers' familiarity with it. However, the named parameter approach provided by Spring's JdbcTemplate is a much safer way to go and is the preferred way to populate parameters in most Spring environments. With that in mind, you can put this feature to use by making two small updates to the configuration:

- Update the configuration to remove the ItemPreparedStatementSetter implementation you wrote and replace it with an implementation of the ItemSqlParameterSourceProvider interface.

- Update the SQL to use named parameters instead of question marks for parameters.

The org.springframework.batch.item.database.ItemSqlParameterSourceProvider interface is slightly different from the ItemPreparedStatementSetter interface in that it doesn't set the parameters on the statement to be executed. Instead, an implementation of the ItemSqlParameterSourceProvider's responsibility is to extract the parameter values from an item and return them as an org.springframework. jdbc.core.namedparam.SqlParameterSource object.

The nice thing about this approach is that not only is it the safer approach (no concerns about needing to keep the SQL in the configuration code in synch with the code of the ItemPreparedStatementSetter implementation) but Spring Batch provides implementations of this interface that allow you to use convention over code to extract the values from the items. In this example, you use Spring Batch's BeanPropertyItemSqlParameterSourceProvider (try saying that three times fast) to extract the values from the items to be populated in the SQL which the JdbcBatchItemWriterBuilder makes easy to use via the beanMapped() method. Listing 9-30 shows the updated jdbcBatchWriter configuration for this change.

Listing 9-30. jdbcBatchWriter Using BeanPropertyItemSqlParameterSourceProvider

```java
...
@Bean
public JdbcBatchItemWriter<Customer> jdbcCustomerWriter(DataSource dataSource)
        throws Exception {
        return new JdbcBatchItemWriterBuilder<Customer>()
                    .dataSource(dataSource)
                    .sql("INSERT INTO CUSTOMER (first_name, " +
                            "middle_initial, " +
                            "last_name, " +
                            "address, " +
                            "city, " +
                            "state, " +
                            "zip) VALUES (:firstName, " +
                            ":middleInitial, " +
                            ":lastName, " +
                            ":address, " +
                            ":city, " +
                            ":state, " +
                            ":zip)")
```

```
                          .beanMapped()
                          .build();
}
...
```

You can quickly note in Listing 9-30 that there is no reference to the `ItemPreparedStatementSetter` implementation. By using this configuration, you don't need any custom code. Yet the results are the same.

Although JDBC is known for its speed compared to other persistence frameworks that lie on top of it, other frameworks are popular in the enterprise. Next you look at how to use the most popular of those to do database writing: Hibernate.

HibernateItemWriter

When you have most of your database tables and applications already mapped with Hibernate, reusing all that is a logical choice to start. You saw how Hibernate works as a competent reader in Chapter 7. This section looks at how you can use `HibernateItemWriter` to write the changes to a database.

Like `JdbcBatchItemWriter`, `org.springframework.batch.item.database.HibernateItemWriter` serves as a thin wrapper to Hibernate's `Session` API. When a chunk completes, the list of items is passed to `HibernateItemWriter` where Hibernate's `Session.saveOrUpdate(Object item)` method is called for each item that is not already associated with the `Session`. When all the items have been saved or updated, `HibernateItemWriter` makes a single call to `Session#flush()` method, executing all the changes at once. This provides a batching functionality similar to `JdbcBatchItemWriter`'s implementation without dealing directly with the SQL.

Configuring `HibernateItemWriter` is simple. All but the configuration of the actual `ItemWriter` should be familiar, because it's the same as the configuration and coding you did for the Hibernate-supported `ItemReaders`. To modify `formatJob` to use Hibernate, you need to update the following:

- *The POM:* The POM needs to incorporate the Hibernate dependencies.

- *application.yml:* You need to configure the `CurrentSessionContext` class via the property `spring.jpa.properties.hibernate.current_session_context_class=org.springframework.orm.hibernate5.SpringSessionContext`.

- Customer.java: You use annotations to configure the mapping for the `Customer` object, so you need to add those to the `Customer` class.

- `SessionFactory`: You need to configure both the `SessionFactory` and a new `TransactionManager` to support Hibernate.

- `HibernateItemWriter`: You can configure the new `ItemWriter` using `HibernateItemWriter`.

Let's start with the POM updates. For Hibernate to work with Spring Batch, we'll use the Spring JPA starter since it provides everything we need. Listing 9-31 shows the additions you need to make to the POM.

Listing 9-31. POM Additions for Supporting Hibernate

```
...
<dependency>
        <groupId>org.springframework.boot</groupId>
        <artifactId>spring-boot-starter-data-jpa</artifactId>
</dependency>
...
```

Now you can begin updating formatJob. Let's begin with the only code you need to write: the annotations you add to the Customer class to map it to the database. Listing 9-32 shows the Customer class updated.

Listing 9-32. Customer.java Mapped to the CUSTOMER Table

```
...
@Entity
@Table(name = "CUSTOMER")
public class Customer implements Serializable {
        private static final long serialVersionUID = 1L;

        @Id
        @GeneratedValue(strategy = GenerationType.IDENTITY)
        private long id;
        private String firstName;
        private String middleInitial;
        private String lastName;
        private String address;
        private String city;
        private String state;
        private String zip;

    // Accessors go here
    ....
}
```

The annotations you use here are the same as the ones you used in the ItemReader example in Chapter 7. The mapping for the Customer class is pretty straightforward because the column names of the CUSTOMER table match those of the Customer class. The other thing to notice is that you aren't using any Hibernate-specific annotations. All the annotations used here are JPA-supported annotations, which allows you to switch from Hibernate to any JPA-supported implementation if you choose with no code changes required.

The next piece we need to add when using Hibernate is our custom HibernateBatchConfigurer. This will be the exact same one we used in Chapter 7. It provides a HibernateTransactionManager in place of the normally provided DataSourceTransactionManager. Listing 9-33 provides the HibernateBatchConfigurer.

Listing 9-33. HibernateBatchConfigurer.java

```
...
@Component
public class HibernateBatchConfigurer implements BatchConfigurer {

        private DataSource dataSource;
        private SessionFactory sessionFactory;
        private JobRepository jobRepository;
        private PlatformTransactionManager transactionManager;
        private JobLauncher jobLauncher;
        private JobExplorer jobExplorer;
```

```
public HibernateBatchConfigurer(DataSource dataSource,
        EntityManagerFactory entityManagerFactory) {

    this.dataSource = dataSource;
    this.sessionFactory = entityManagerFactory.unwrap(SessionFactory.class);
}

@Override
public JobRepository getJobRepository() throws Exception {
    return this.jobRepository;
}

@Override
public PlatformTransactionManager getTransactionManager() throws Exception {
    return this.transactionManager;
}

@Override
public JobLauncher getJobLauncher() throws Exception {
    return this.jobLauncher;
}

@Override
public JobExplorer getJobExplorer() throws Exception {
    return this.jobExplorer;
}

@PostConstruct
public void initialize() {

    try {
        HibernateTransactionManager transactionManager =
        new HibernateTransactionManager(sessionFactory);
        transactionManager.afterPropertiesSet();

        this.transactionManager = transactionManager;

        this.jobRepository = createJobRepository();
        this.jobExplorer = createJobExplorer();
        this.jobLauncher = createJobLauncher();

    }
    catch (Exception e) {
        throw new BatchConfigurationException(e);
    }
}

private JobLauncher createJobLauncher() throws Exception {
    SimpleJobLauncher jobLauncher = new SimpleJobLauncher();
```

```
                jobLauncher.setJobRepository(this.jobRepository);
                jobLauncher.afterPropertiesSet();

                return jobLauncher;
        }

        private JobExplorer createJobExplorer() throws Exception {
                JobExplorerFactoryBean jobExplorerFactoryBean = new JobExplorerFactoryBean();

                jobExplorerFactoryBean.setDataSource(this.dataSource);
                jobExplorerFactoryBean.afterPropertiesSet();

                return jobExplorerFactoryBean.getObject();
        }

        private JobRepository createJobRepository() throws Exception {
                JobRepositoryFactoryBean jobRepositoryFactoryBean = new
                JobRepositoryFactoryBean();

                jobRepositoryFactoryBean.setDataSource(this.dataSource);
                jobRepositoryFactoryBean.setTransactionManager(this.transactionManager);
                jobRepositoryFactoryBean.afterPropertiesSet();

                return jobRepositoryFactoryBean.getObject();
        }
}
```

Finally, you can configure HibernateItemWriter. It's probably the easiest ItemWriter to configure given that other components and the Hibernate framework do all the work. HibernateItemWriter requires a single dependency and has one optional dependency. The required dependency is a reference to the SessionFactory. The optional dependency (which you aren't using in this case) is an indicator if the writer should call clear() on the Session after calling flush() (defaults to true). Listing 9-34 show the configuration of the job complete with the new HibernateItemWriter configuration.

Listing 9-34. HibernateImportJob.java Using Hibernate

```
...
@Configuration
public class HibernateImportJob {

        private JobBuilderFactory jobBuilderFactory;

        private StepBuilderFactory stepBuilderFactory;

        public HibernateImportJob(JobBuilderFactory jobBuilderFactory,
                        StepBuilderFactory stepBuilderFactory) {

                this.jobBuilderFactory = jobBuilderFactory;
                this.stepBuilderFactory = stepBuilderFactory;
        }
```

```java
@Bean
@StepScope
public FlatFileItemReader<Customer> customerFileReader(
                @Value("#{jobParameters['customerFile']}")Resource inputFile) {

        return new FlatFileItemReaderBuilder<Customer>()
                        .name("customerFileReader")
                        .resource(inputFile)
                        .delimited()
                        .names(new String[] {"firstName",
                                        "middleInitial",
                                        "lastName",
                                        "address",
                                        "city",
                                        "state",
                                        "zip"})
                        .targetType(Customer.class)
                        .build();
}

@Bean
public HibernateItemWriter<Customer> hibernateItemWriter(
        EntityManagerFactory entityManager) {

        return new HibernateItemWriterBuilder<Customer>()
                        .sessionFactory(entityManager.unwrap(SessionFactory.class))
                        .build();
}

@Bean
public Step hibernateFormatStep() throws Exception {
        return this.stepBuilderFactory.get("jdbcFormatStep")
                        .<Customer, Customer>chunk(10)
                        .reader(customerFileReader(null))
                        .writer(hibernateItemWriter(null))
                        .build();
}

@Bean
public Job hibernateFormatJob() throws Exception {
        return this.jobBuilderFactory.get("hibernateFormatJob")
                        .start(hibernateFormatStep ())
                        .build();
}
}
```

The configuration for this job changes only with the configuration of hibernateBatchWriter and its reference in the hibernateFormatStep. As you saw previously, HibernateItemWriter requires only a reference to a SessionFactory, which is provided via Spring Boot. Executing this job returns the same results as the JdbcBatchItemWriter example previously.

When other frameworks do all of the heavy lifting, the Spring Batch configuration is quite simple, as this Hibernate example shows. Hibernate's official spec cousin, JPA, is the other database access framework you can use to do database writing.

JpaItemWriter

The Java Persistence API (JPA) provides very similar functionality and requires almost the exact same configuration as its Hibernate cousin. It, like Hibernate, does the heavy lifting in the case of writing to the database, so the Spring Batch piece of the puzzle is very small. This section looks at how to configure JPA to perform database writing.

When you look at the `org.springframework.batch.item.writer.JpaItemWriter`, it serves as a thin wrapper around JPA's `javax.persistence.EntityManager`. When a chunk completes, the list of items within the chunk is passed to `JpaItemWriter`. The writer loops over the items in the `List<T>`, calling the `EntityManager`'s `merge(T entity)` method on each item before calling `flush()` after all the items have been saved.

To see `JpaItemWriter` in action, you use the same customer input as earlier and insert it into the same `CUSTOMER` table. To hook JPA into the job, you need to do the following two things:

1. Create a `BatchConfigurer` implementation that creates a `JpaTransactionManager`. This acts the same as the Hibernate version in the previous section.

2. Configure the `JpaItemWriter`. The last step is to configure the new `ItemWriter` to save the items read in the job.

The rest is provided via Spring Boot. We can start the code we need to provide with the `JpaBatchConfigurer`. The code here is exactly the same as the Hibernate version in the previous section except for two minor things. First, we'll save off an `EntityManager` instead of a `SessionFactory` in the constructor. Second, instead of creating a `HibernateTransactionManager` in the `initialize()` method, we'll create a `JpaTransactionManager`. Listing 9-35 illustrates the configuration.

Listing 9-35. JpaBatchConfigurer.java

```
...
@Component
public class JpaBatchConfigurer implements BatchConfigurer {

        private DataSource dataSource;
        private EntityManagerFactory entityManagerFactory;
        private JobRepository jobRepository;
        private PlatformTransactionManager transactionManager;
        private JobLauncher jobLauncher;
        private JobExplorer jobExplorer;

        public JpaBatchConfigurer(DataSource dataSource,
                EntityManagerFactory entityManagerFactory) {
                this.dataSource = dataSource;
                this.entityManagerFactory = entityManagerFactory;
        }

        @Override
        public JobRepository getJobRepository() throws Exception {
                return this.jobRepository;
        }
```

```java
    @Override
    public PlatformTransactionManager getTransactionManager() throws Exception {
            return this.transactionManager;
    }

    @Override
    public JobLauncher getJobLauncher() throws Exception {
            return this.jobLauncher;
    }

    @Override
    public JobExplorer getJobExplorer() throws Exception {
            return this.jobExplorer;
    }

    @PostConstruct
    public void initialize() {

            try {
                    JpaTransactionManager transactionManager =
                            new JpaTransactionManager(entityManagerFactory);
                    transactionManager.afterPropertiesSet();

                    this.transactionManager = transactionManager;

                    this.jobRepository = createJobRepository();
                    this.jobExplorer = createJobExplorer();
                    this.jobLauncher = createJobLauncher();

            }
            catch (Exception e) {
                    throw new BatchConfigurationException(e);
            }
    }

    private JobLauncher createJobLauncher() throws Exception {
            SimpleJobLauncher jobLauncher = new SimpleJobLauncher();

            jobLauncher.setJobRepository(this.jobRepository);
            jobLauncher.afterPropertiesSet();

            return jobLauncher;
    }

    private JobExplorer createJobExplorer() throws Exception {
            JobExplorerFactoryBean jobExplorerFactoryBean =
                    new JobExplorerFactoryBean();

            jobExplorerFactoryBean.setDataSource(this.dataSource);
            jobExplorerFactoryBean.afterPropertiesSet();
```

```
                return jobExplorerFactoryBean.getObject();
        }

        private JobRepository createJobRepository() throws Exception {
                JobRepositoryFactoryBean jobRepositoryFactoryBean =
                        new JobRepositoryFactoryBean();

                jobRepositoryFactoryBean.setDataSource(this.dataSource);
                jobRepositoryFactoryBean.setTransactionManager(this.transactionManager);
                jobRepositoryFactoryBean.afterPropertiesSet();

                return jobRepositoryFactoryBean.getObject();
        }
}
```

Since we used the JPA annotations to map our `Customer` object in the previous section, there are no modifications needed to it for this example. The final aspect of configuring the job to use JPA is to configure `JpaItemWriter`. It requires only a single dependency—a reference to `EntityManagerFactory`—so that it can obtain an `EntityManager` to work with. Listing 9-36 shows the configuration for the new `ItemWriter` and the job updated to use it.

Listing 9-36. `formatJob` Configured to Use `JpaItemWriter`

```
...
@Configuration
public class JpaImportJob {

        private JobBuilderFactory jobBuilderFactory;

        private StepBuilderFactory stepBuilderFactory;

        public JpaImportJob(JobBuilderFactory jobBuilderFactory,
                        StepBuilderFactory stepBuilderFactory) {

                this.jobBuilderFactory = jobBuilderFactory;
                this.stepBuilderFactory = stepBuilderFactory;
        }

        @Bean
        @StepScope
        public FlatFileItemReader<Customer> customerFileReader(
                        @Value("#{jobParameters['customerFile']}")Resource inputFile) {

                return new FlatFileItemReaderBuilder<Customer>()
                                .name("customerFileReader")
                                .resource(inputFile)
                                .delimited()
                                .names(new String[] {"firstName",
                                                "middleInitial",
                                                "lastName",
                                                "address",
                                                "city",
```

```
                                              "state",
                                              "zip"})
                          .targetType(Customer.class)
                          .build();
        }

        @Bean
        public JpaItemWriter<Customer> jpaItemWriter(
                EntityManagerFactory entityManagerFactory) {

                JpaItemWriter<Customer> jpaItemWriter = new JpaItemWriter<>();

                jpaItemWriter.setEntityManagerFactory(entityManagerFactory);

                return jpaItemWriter;
        }

        @Bean
        public Step jpaFormatStep() throws Exception {
                return this.stepBuilderFactory.get("jpaFormatStep")
                                .<Customer, Customer>chunk(10)
                                .reader(customerFileReader(null))
                                .writer(jpaItemWriter(null))
                                .build();
        }

        @Bean
        public Job jpaFormatJob() throws Exception {
                return this.jobBuilderFactory.get("jpaFormatJob")
                                .start(jpaFormatStep())
                                .build();
        }
}
```

You can now build the job with a quick ./mvnw clean install. To execute the job, use the command in Listing 9-37, which returns the results you've seen in the other database examples.

Listing 9-37. Command to Execute formatJob with JPA Configured

```
java -jar itemWriters-0.0.1-SNAPSHOT.jar customerFile=/input/customer.csv
```

The relational database rules in the modern enterprise, for better or worse. As you can see, writing job results to a database is easy with Spring Batch. But relational databases aren't the only databases that are available both from Spring Batch or needed in an enterprise. The next section looks at other NoSql stores supported by Spring Data.

Spring Data ItemWriters

In Chapter 7 we learned about the Spring Data project and how it brings a common programming model to a number of different data stores. Spring Batch takes advantage of Spring Data's capabilities to offer the ability to write to a number of NoSql data stores, specifically MongoDB, Neo4J, Pivotal Gemfire, or Apache Geode, as well as offering support for any other Spring Data project with `CrudRepository` support. In this section, we will look at how each of these options can be integrated into our Spring Batch projects.

MongoDB

Mongo's features as a high performance, highly scalable datastore make it an attractive option for the enterprise. Spring Batch supports the use of storing objects as documents in a MongoDB collection via the `MongoItemWriter`.

In order to use MongoDB, we first need to update our `Customer` object in a few minor ways. First of all, MongoDB does not support `long` values for ids, it requires `String` ids. Secondly, we can remove the JPA annotations from the `Customer` object since there are no tables in MongoDB. This leaves us with a `Customer` domain object that looks like what is in Listing 9-38.

Listing 9-38. Customer.java for MongoDB

```
...
public class Customer implements Serializable {
        private static final long serialVersionUID = 1L;

        @Id
        private String id;
        private String firstName;
        private String middleInitial;
        private String lastName;
        private String address;
        private String city;
        private String state;
        private String zip;

        // Getters and setters removed

        @Override
        public String toString() {
                return "Customer{" +
                                "id=" + id +
                                ", firstName='" + firstName + '\" +
                                ", middleInitial='" + middleInitial + '\" +
                                ", lastName='" + lastName + '\" +
                                ", address='" + address + '\" +
                                ", city='" + city + '\" +
                                ", state='" + state + '\" +
                                ", zip='" + zip + '\" +
                                '}';
        }
}
```

With our domain object updated, we need to add the correct dependency to our pom.xml to bring in the MongoDB dependencies. Like most other instances when you need to bring in new functionality using Spring Boot, we will bring in the correct starter. In this case, it is the spring-boot-starter-data-mongodb as shown in Listing 9-39.

Listing 9-39. spring-boot-starter-data-mongodb

```
...
<dependency>
        <groupId>org.springframework.boot</groupId>
        <artifactId>spring-boot-starter-data-mongodb</artifactId>
</dependency>
...
```

The next thing we need to configure is our application.yml to point to our MongoDB database. By default, Spring Boot looks for it on localhost with standard login credentials, so we do not need to configure any of those; however, we do need to tell our application the name of the database to write to, which in our case is called customerdb. We set that via the spring.data.mongodb.database property.

The MongoDB ItemWriter works in a similar way to the Hibernate and JPA based ItemWriter implementations. The mapping is handled via the annotations on the domain object, so there is only minimal configuration required on the ItemWriter itself. In our case, we will need to configure the name of the collection within our database to write to (customers), provide an instance of MongoOperations, and call build(). The only other configuration option is a flag called delete that indicates if the ItemWriter should delete the matching items or it should save them. By default, it saves them. Listing 9-40 shows the configuration for our mongoFormatJob including the updated MongoItemWriter.

Listing 9-40. mongoFormatJob

```
...
@Configuration
public class MongoImportJob {

        private JobBuilderFactory jobBuilderFactory;

        private StepBuilderFactory stepBuilderFactory;

        public MongoImportJob(JobBuilderFactory jobBuilderFactory,
                        StepBuilderFactory stepBuilderFactory) {

                this.jobBuilderFactory = jobBuilderFactory;
                this.stepBuilderFactory = stepBuilderFactory;
        }

        @Bean
        @StepScope
        public FlatFileItemReader<Customer> customerFileReader(
                        @Value("#{jobParameters['customerFile']}")Resource inputFile) {

                return new FlatFileItemReaderBuilder<Customer>()
                                .name("customerFileReader")
                                .resource(inputFile)
                                .delimited()
```

```
                        .names(new String[] {"firstName",
                                    "middleInitial",
                                    "lastName",
                                    "address",
                                    "city",
                                    "state",
                                    "zip"})
                        .targetType(Customer.class)
                        .build();
    }

    @Bean
    public MongoItemWriter<Customer> mongoItemWriter(MongoOperations mongoTemplate) {
            return new MongoItemWriterBuilder<Customer>()
                            .collection("customers")
                            .template(mongoTemplate)
                            .build();
    }

    @Bean
    public Step mongoFormatStep() throws Exception {
            return this.stepBuilderFactory.get("mongoFormatStep")
                            .<Customer, Customer>chunk(10)
                            .reader(customerFileReader(null))
                            .writer(mongoItemWriter(null))
                            .build();
    }

    @Bean
    public Job mongoFormatJob() throws Exception {
            return this.jobBuilderFactory.get("mongoFormatJob")
                            .start(mongoFormatStep())
                            .build();
    }
}
```

One point of contention with MongoDB with many enterprises is that historically MongoDB has not supported ACID (Atomicity, Consistency, Isolation, Durability) transactions. Because of this, Spring Batch treats MongoDB like any other data store that does not support transactions and buffers the writes until just before the commit occurs, performing the write at the last moment.

With our job written, if we build and run it, we will find nine documents were inserted into our collection as illustrated via the GUI client Robot 3T[2] Figure 9-4.

---

[2]https://robomongo.org/

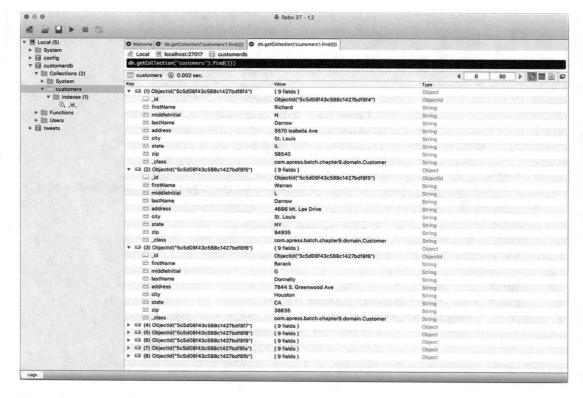

Figure 9-4. *Output of mongoFormatJob in Robot 3T*

While MongoDB is probably the most popular NoSQL data store on the market, the format of storing documents is not ideal for all use cases. That's the whole idea behind the NoSQL movement, to use the data store that is right for your data. In the next section, we will look at another NoSQL store that has a very different model, it's a graph database.

Neo4J

In our society where everything is connected, our lives have been turned into large graphs. Graphs of friends on Facebook, networks of connections on LinkedIn...the list goes on. Each node of these graphs has very different forms of relationships which make traditional relational data stores inefficient for storing this kind of data. Neo4J is the leading graph database on the market today with over three million downloads which is increasing by 50 thousand a month.[3] Spring Data brings support to the Spring portfolio for Neo4j. In this section, we will take a look at how to write records to a Neo4j database using the Neo4jItemWriter.

To get started with Neo4j, just like in JPA and MongoDB, we map our data to it's representation in the database via annotations. With this, we will need to update our Customer class with the appropriate annotations to map it in a Neo4j database. We will begin with the @NodeEntity annotation at the class level, indicating to the ItemWriter that this class represents a node in the graph. Neo4j also has an @Relationship

---

[3]https://neo4j.com/top-ten-reasons/

annotation that allows you to map the relationships between nodes in the graph. In our case, our node does not have any relationships, but we do need to identify an id for it. Using a relational database we used a long for the id. MongoDB required that we use a String. For Neo4j we are going to use a UUID. Listing 9-41 illustrates the updated Customer domain object, mapped for Neo4j.

Listing 9-41. Customer Mapped for Neo4j

```
...
@NodeEntity
public class Customer implements Serializable {
        private static final long serialVersionUID = 1L;

        @Id
        @GeneratedValue(strategy = UuidStrategy.class)
        private UUID id;
        private String firstName;
        private String middleInitial;
        private String lastName;
        private String address;
        private String city;
        private String state;
        private String zip;

        // Getters and setters removed
        @Override
        public String toString() {
                return "Customer{" +
                                "id=" + id +
                                ", firstName='" + firstName + '\" +
                                ", middleInitial='" + middleInitial + '\" +
                                ", lastName='" + lastName + '\" +
                                ", address='" + address + '\" +
                                ", city='" + city + '\" +
                                ", state='" + state + '\" +
                                ", zip='" + zip + '\" +
                                '}';
        }
}
```

In order to put Neo4j to use from our application, we will need to add the appropriate Spring Boot starter to our pom.xml. At this point in the book, assuming you have been following along, it should come as no surprise that the starter we are going to bring in is the spring-boot-starter-data-neo4j artifact as shown in Listing 9-42.

Listing 9-42. Neo4j Dependencies

```
...
<dependency>
        <groupId>org.springframework.boot</groupId>
        <artifactId>spring-boot-starter-data-neo4j</artifactId>
</dependency>
...
```

With the correct dependencies included in our application, we need to add the correct configuration to our application.yml. To get started, I recommend using the Community Edition of the Neo4j server.[4] To use it, we will need to configure the username, password, and uri as shown in Listing 9-43.

Listing 9-43. Neo4j Dependencies

```
spring:
  data:
...
    neo4j:
      username: neo4j
      password: password
      embedded:
        enabled: false
      uri: bolt://localhost:7687
```

The last piece of the puzzle for using Neo4j as our `ItemWriter` is to actually configure the `ItemWriter`. The Neo4jItemWriter simply requires one dependency, an `org.neo4j.orgm.session.SessionFactory` instance which is provided by Spring Boot via the starter we added earlier. Listing 9-44 illustrates the full job configured to write to Neo4j.

Listing 9-44. Neo4jImportJob

```
...
@Configuration
public class Neo4jImportJob {

        private JobBuilderFactory jobBuilderFactory;

        private StepBuilderFactory stepBuilderFactory;

        public Neo4jImportJob(JobBuilderFactory jobBuilderFactory,
                        StepBuilderFactory stepBuilderFactory) {

                this.jobBuilderFactory = jobBuilderFactory;
                this.stepBuilderFactory = stepBuilderFactory;
        }

        @Bean
        @StepScope
        public FlatFileItemReader<Customer> customerFileReader(
                        @Value("#{jobParameters['customerFile']}")Resource inputFile) {

                return new FlatFileItemReaderBuilder<Customer>()
                                .name("customerFileReader")
                                .resource(inputFile)
                                .delimited()
                                .names(new String[] {"firstName",
```

---

[4]https://neo4j.com/download-center/#releases

```
                                        "middleInitial",
                                        "lastName",
                                        "address",
                                        "city",
                                        "state",
                                        "zip"})
                        .targetType(Customer.class)
                        .build();
    }

    @Bean
    public Neo4jItemWriter<Customer> neo4jItemWriter(SessionFactory sessionFactory) {
            return new Neo4jItemWriterBuilder<Customer>()
                            .sessionFactory(sessionFactory)
                            .build();
    }

    @Bean
    public Step neo4jFormatStep() throws Exception {
            return this.stepBuilderFactory.get("neo4jFormatStep")
                            .<Customer, Customer>chunk(10)
                            .reader(customerFileReader(null))
                            .writer(neo4jItemWriter(null))
                            .build();
    }

    @Bean
    public Job neo4jFormatJob() throws Exception {
            return this.jobBuilderFactory.get("neo4jFormatJob")
                            .start(neo4jFormatStep())
                            .build();
    }
}
```

After building and running the Neo4jImportJob, you can verify that your results were successfully imported into the database by opening the Neo4j browser provided by the server. Once open, you can execute the cypher equivalent of SELECT firstName, lastName FROM customer; by executing the cypher query MATCH(c:Customer) RETURN c.firstName, c.lastName. The results you will receive are illustrated in Figure 9-5.

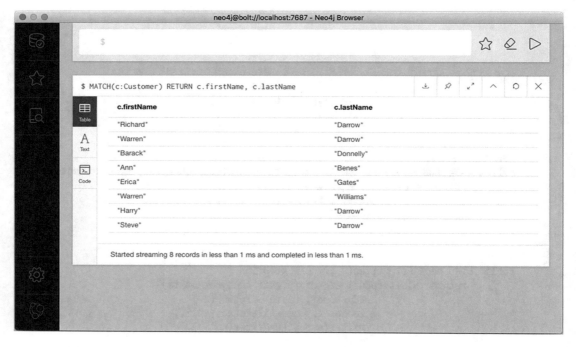

Figure 9-5. *Output of neo4jFormatJob in Neo4j's browser*

Graph databases provide powerful tools for solving the right problem; however, there is yet another data store that Spring Batch supports that utilizes another storage model. In the next section, we will look at how to use a particular key value store for high performance in memory use cases.

Pivotal Gemfire and Apache Geode

In the financial world, milliseconds can count. Fraud detection during a retail transaction does not have time to wait to retrieve data across multiple hops on a network or off of slow disk. Data must be cached in memory to be fast enough in these types of environments. This is the type of environment that Pivotal Gemfire was born in. In this section we will learn about Pivotal Gemfire and it's open source version Apache Geode as well as how to write to it using Spring Batch.

Pivotal Gemfire is an in memory data grid. At its core, it is a high performance, distributed `HashMap`. It is a key value store that keeps all data in memory and provides extremely fast read times due to it's network topology. So what does this have to do with Spring Batch? Simple. One of the main use cases for Gemfire is caching. And in order for a cache to be useful, it needs to be primed. Spring Batch provides an excellent facility for priming a cold cache on application startup in an efficient manner.

We will begin as we have in the past couple sections by updating the mapping of the domain object. For Gemfire, since we are working with a key value store, the key will be something derived from the domain object and the domain object itself will be the value. All we really need to do is to identify the Region (similar to a collection in MongoDB) where the values will be persisted to. To do this, we add the `@Region` annotation on the `Customer` class as shown in Listing 9-45.

Listing 9-45. Neo4jImportJob

```
...
@Region(value = 'Customers")
public class Customer implements Serializable {
        private static final long serialVersionUID = 1L;

        private long id;
        private String firstName;
        private String middleInitial;
        private String lastName;
        private String address;
        private String city;
        private String state;
        private String zip;

        // Accessors removed for brevity

        @Override
        public String toString() {
                return "Customer{" +
                                "id=" + id +
                                ", firstName='" + firstName + '\" +
                                ", middleInitial='" + middleInitial + '\" +
                                ", lastName='" + lastName + '\" +
                                ", address='" + address + '\" +
                                ", city='" + city + '\" +
                                ", state='" + state + '\" +
                                ", zip='" + zip + '\" +
                                '}';
        }
}
```

With our domain object defined, we need to update our pom.xml to bring in the Pivotal Gemfire dependencies. To do this, we need to make two changes to our POM. First, we need to add two new dependencies, one for Spring Data Gemfire and one for Spring Shell (Spring Data Gemfire requires it). The other thing we need to do is we need to exclude the logging dependency on our spring-boot-starter-batch due to a conflict between what Spring Boot uses by default and what Gemfire uses by default. Listing 9-46 has the POM updates we need to make.

Listing 9-46. pom.xml Updates for Pivotal Gemfire

```
...
<dependency>
        <groupId>org.springframework.boot</groupId>
        <artifactId>spring-boot-starter-batch</artifactId>
        <exclusions>
                <exclusion>
                        <groupId>org.springframework.boot</groupId>
                        <artifactId>spring-boot-starter-logging</artifactId>
                </exclusion>
        </exclusions>
</dependency>
```

```
...
<dependency>
        <groupId>org.springframework.data</groupId>
        <artifactId>spring-data-gemfire</artifactId>
</dependency>
<dependency>
        <groupId>org.springframework.shell</groupId>
        <artifactId>spring-shell</artifactId>
</dependency>
...
```

With our dependencies added, we can now configure Pivotal Gemfire. Unlike the other data stores that we have used up to this point, all of which were external to our application, we are going to run Pivotal Gemfire internal to our application. The reason for this is that it's a common practice so that it limits network hops.

To begin our configuration, we will add the annotation @PeerCacheApplication to our configuration class. This annotation is what bootstraps the Pivotal Gemfire service within our application. We will provide it with two options, a name of the Pivotal Gemfire application and the level of logging we want. From there, we need to configure a Region. As mentioned earlier, a Region for Gemfire is similar to a Collection in MongoDB. Pivotal Gemfire allows us to configure things like this directly via Spring instead of using external mechanisms to do so. Once we have our Region configured, we need to create a GemfireTemplate for our ItemWriter to use. It takes a single dependency (the Region). Once we have our template defined, we can create our ItemWriter. The GemfireItemWriter requires two items to be configured on it. The first is the template we just created. The second is an org.springframework.core.convert.converter.Converter. This Converter instance is used to convert the item being written to Pivotal Gemfire to the key it will use. In our case, we will use one of the implementations provided by Spring Batch for this purpose, the SpELItemKeyMapper. This implementation will take a SpEL expression to create a key from the current item. The final addition to our configuration compared to previous iterations of it is a new CommandLineRunner. We are going to use this to validate our job ran successfully since we do not have a GUI to view the results and the Pivotal Gemfire server shuts down when our application shuts down. This CommandLineRunner will query our Gemfire instance and list out the items in it. Listing 9-47 configures our Pivotal Gemfire instance and our Spring Batch job to load it.

Listing 9-47. GemfireImportJob

```
...
@Configuration
@PeerCacheApplication(name = "AccessingDataGemFireApplication", logLevel = "info")
public class GemfireImportJob {

        private JobBuilderFactory jobBuilderFactory;

        private StepBuilderFactory stepBuilderFactory;

        public GemfireImportJob(JobBuilderFactory jobBuilderFactory,
                        StepBuilderFactory stepBuilderFactory) {

                this.jobBuilderFactory = jobBuilderFactory;
                this.stepBuilderFactory = stepBuilderFactory;
        }
```

```
@Bean
@StepScope
public FlatFileItemReader<Customer> customerFileReader(
                @Value("#{jobParameters['customerFile']}")Resource inputFile) {

        return new FlatFileItemReaderBuilder<Customer>()
                        .name("customerFileReader")
                        .resource(inputFile)
                        .delimited()
                        .names(new String[] {"firstName",
                                        "middleInitial",
                                        "lastName",
                                        "address",
                                        "city",
                                        "state",
                                        "zip"})
                        .targetType(Customer.class)
                        .build();
}

@Bean
public GemfireItemWriter<Long, Customer> gemfireItemWriter(
        GemfireTemplate gemfireTemplate) {

        return new GemfireItemWriterBuilder<Long, Customer>()
                        .template(gemfireTemplate)
                        .itemKeyMapper(new SpELItemKeyMapper<>(
                                "firstName + middleInitial + lastName"))
                        .build();
}

@Bean
public Step gemfireFormatStep() throws Exception {
        return this.stepBuilderFactory.get("gemfireFormatStep")
                        .<Customer, Customer>chunk(10)
                        .reader(customerFileReader(null))
                        .writer(gemfireItemWriter(null))
                        .build();
}

@Bean
public Job gemfireFormatJob() throws Exception {
        return this.jobBuilderFactory.get("gemfireFormatJob")
                        .start(gemfireFormatStep())
                        .build();
}
```

```
    @Bean(name="customer")
    public Region<Long, Customer> getCustomer(final GemFireCache cache) throws Exception {
            LocalRegionFactoryBean<Long, Customer> customerRegion =
                    new LocalRegionFactoryBean<>();
            customerRegion.setCache(cache);
            customerRegion.setName("customer");
            customerRegion.afterPropertiesSet();
            Region<Long, Customer> object = customerRegion.getRegion();
            return object;
    }


    @Bean
    public GemfireTemplate gemfireTemplate() throws Exception {
            return new GemfireTemplate(getCustomer(null));
    }


    @Bean
    public CommandLineRunner validator(final GemfireTemplate gemfireTemplate) {
            return args -> {
                    List<Object> customers =
                            gemfireTemplate.find("select * from /customer").asList();

                    for (Object customer : customers) {
                            System.out.println(">> object: " + customer);
                    }
            };
    }
}
```

With this configuration we can build and run our job, the output of which will write to System.out all the items we loaded into our Gemfire Region as shown in Listing 9-48.

Listing 9-48. GemfireImportJob Output

```
...
[info 2019/02/09 12:59:40.617 CST <main> tid=0x1] Job: [SimpleJob: [name=gemfireFormatJob]]
completed with the following parameters: [{customerFile=/data/customer.csv}] and the
following status: [COMPLETED]

>> object: Customer{id=0, firstName='Ann', middleInitial='Z', lastName='Benes',
address='2447 S. Greenwood Ave', city='Las Vegas', state='NY', zip='55366'}
>> object: Customer{id=0, firstName='Warren', middleInitial='M', lastName='Williams',
address='6670 S. Greenwood Ave', city='Hollywood', state='FL', zip='37288'}
>> object: Customer{id=0, firstName='Erica', middleInitial='Z', lastName='Gates',
address='3141 Farnam Street', city='Omaha', state='CA', zip='57640'}
>> object: Customer{id=0, firstName='Warren', middleInitial='L', lastName='Darrow',
address='4686 Mt. Lee Drive', city='St. Louis', state='NY', zip='94935'}
>> object: Customer{id=0, firstName='Richard', middleInitial='N', lastName='Darrow',
address='5570 Isabella Ave', city='St. Louis', state='IL', zip='58540'}
>> object: Customer{id=0, firstName='Steve', middleInitial='O', lastName='Darrow',
address='8407 Infinite Loop Drive', city='Las Vegas', state='WA', zip='90520'}
```

```
>> object: Customer{id=0, firstName='Harry', middleInitial='T', lastName='Darrow',
address='3273 Isabella Ave', city='Houston', state='FL', zip='97261'}
>> object: Customer{id=0, firstName='Barack', middleInitial='G', lastName='Donnelly',
address='7844 S. Greenwood Ave', city='Houston', state='CA', zip='38635'}
[info 2019/02/09 12:59:40.660 CST <Distributed system shutdown hook> tid=0x21] VM is exiting
- shutting down distributed system
...
```

Pivotal Gemfire and Apache Geode both can provide best in class performance capabilities on a variety of workloads. Spring Batch provides efficient mechanisms for updating them as well. The last Spring Data related `ItemWriter` is the `RepositoryItemWriter`. Just like the `RepositoryItemReader` discussed in Chapter 7, this one takes advantage of Sping Data's `Repository` abstraction to write records to any data store Spring Data supports. We will look at it in the next section.

Repository

Spring Data's `Repository` abstraction provides a very useful way to construct an `ItemWriter`. In Chapter 7, we looked at how Spring Data's `PagingAndSortingRepository` can be used to create an `ItemReader` for any data store Spring Data has repository support for. The difference between the `ItemReader` use of repositories and the `ItemWriter` use of them is which repository we use. When reading, we used the `PagingAndSortingRepository`. However on the writing side of the coin, we really aren't worried about paging and sorting, so we use its super interface, `org.springframework.data.repository.CrudRepository`. In this section, we will look at using the `RepositoryItemWriter` to persist data to a data store supported by Spring Data.

To take a look at how this `ItemWriter` works, we will write data to the `CUSTOMER` table just like we did in the JPA example earlier in this chatper. However, instead of using the `JpaItemWriter`, we will use the `RepositoryItemWriter`. We can begin with the domain object. Since we are using JPA under the hood of this example, we can actually use the same domain object configuration we did in the JPA section of this chapter. Here it is in Listing 9-49 for your convenience.

Listing 9-49. Customer Mapped for JPA

```
...
@Entity
@Table(name = "CUSTOMER")
public class Customer implements Serializable {
        private static final long serialVersionUID = 1L;

        @Id
        @GeneratedValue(strategy = GenerationType.IDENTITY)
        private long id;
        private String firstName;
        private String middleInitial;
        private String lastName;
        private String address;
        private String city;
        private String state;
        private String zip;

        // Getters and setters removed for brevity
```

```
        @Override
        public String toString() {
                return "Customer{" +
                                "id=" + id +
                                ", firstName='" + firstName + '\" +
                                ", middleInitial='" + middleInitial + '\" +
                                ", lastName='" + lastName + '\" +
                                ", address='" + address + '\" +
                                ", city='" + city + '\" +
                                ", state='" + state + '\" +
                                ", zip='" + zip + '\" +
                                '}';
        }
}
```

With our domain object configured as needed, we can also create our repository definition. Since we are just storing the Customer objects into the database, we actually just need to create an interface that extends CrudRepository. Spring Data will do the rest. Listing 9-50 shows our CustomerRepository definition.

Listing 9-50. CustomerRepository

```
...
public interface CustomerRepository extends CrudRepository<Customer, Long> {
}
```

Since we already imported our JPA related dependencies earlier, we do not need to add them again leaving us only with the configuration of the job itself. The RepositoryItemWriter takes two dependencies, the repository we are going to use and the name of the method to call. The only other new item here is that to bootstrap the repository functionality, we need to tell Spring to do so and where to look for repositories. We do this via the @EnableJpaRepositories and specifying a class in the package where our repository lives. Listing 9-51 has the configuration of our job.

Listing 9-51. RepositoryImportJob

```
...
@Configuration
@EnableJpaRepositories(basePackageClasses = Customer.class)
public class RepositoryImportJob {

        private JobBuilderFactory jobBuilderFactory;

        private StepBuilderFactory stepBuilderFactory;

        public RepositoryImportJob(JobBuilderFactory jobBuilderFactory,
                        StepBuilderFactory stepBuilderFactory) {

                this.jobBuilderFactory = jobBuilderFactory;
                this.stepBuilderFactory = stepBuilderFactory;
        }
```

```
@Bean
@StepScope
public FlatFileItemReader<Customer> customerFileReader(
                @Value("#{jobParameters['customerFile']}")Resource inputFile) {

        return new FlatFileItemReaderBuilder<Customer>()
                        .name("customerFileReader")
                        .resource(inputFile)
                        .delimited()
                        .names(new String[] {"firstName",
                                        "middleInitial",
                                        "lastName",
                                        "address",
                                        "city",
                                        "state",
                                        "zip"})
                        .targetType(Customer.class)
                        .build();
}

@Bean
public RepositoryItemWriter<Customer> repositoryItemWriter(
        CustomerRepository repository) {

        return new RepositoryItemWriterBuilder<Customer>()
                        .repository(repository)
                        .methodName("save")
                        .build();
}

@Bean
public Step repositoryFormatStep() throws Exception {
        return this.stepBuilderFactory.get("repositoryFormatStep")
                        .<Customer, Customer>chunk(10)
                        .reader(customerFileReader(null))
                        .writer(repositoryItemWriter(null))
                        .build();
}

@Bean
public Job repositoryFormatJob() throws Exception {
        return this.jobBuilderFactory.get("repositoryFormatJob")
                        .start(repositoryFormatStep())
                        .build();
}
}
```

Once our job is built and ran, we can verify the results by looking in our CUSTOMER table in MySQL. Listing 9-52 shows the results.

Listing 9-52. Results of the RepositoryImportJob

```
mysql> select id, first_name, middle_initial as middle, last_name, address, city, state as
st, zip from SPRING_BATCH.CUSTOMER;
+----+------------+--------+-----------+-------------------------+-----------+------+-------+
| id | first_name | middle | last_name | address                 | city      | st   | zip   |
+----+------------+--------+-----------+-------------------------+-----------+------+-------+
|  1 | Richard    | N      | Darrow    | 5570 Isabella Ave       | St. Louis | IL   | 58540 |
|  2 | Warren     | L      | Darrow    | 4686 Mt. Lee Drive      | St. Louis | NY   | 94935 |
|  3 | Barack     | G      | Donnelly  | 7844 S. Greenwood Ave   | Houston   | CA   | 38635 |
|  4 | Ann        | Z      | Benes     | 2447 S. Greenwood Ave   | Las Vegas | NY   | 55366 |
|  5 | Erica      | Z      | Gates     | 3141 Farnam Street      | Omaha     | CA   | 57640 |
|  6 | Warren     | M      | Williams  | 6670 S. Greenwood Ave   | Hollywood | FL   | 37288 |
|  7 | Harry      | T      | Darrow    | 3273 Isabella Ave       | Houston   | FL   | 97261 |
|  8 | Steve      | O      | Darrow    | 8407 Infinite Loop Drive| Las Vegas | WA   | 90520 |
+----+------------+--------+-----------+-------------------------+-----------+------+-------+
8 rows in set (0.01 sec)
```

The list of databases Spring Batch can interact with is quite long as we have seen. However, databases and files are not the only thing you can call from a Spring Batch job to write to. Let's look at some other options.

Alternative Output Destination ItemWriters

Files and databases aren't the only ways you can communicate the end result of an item being processed. Enterprises use a number of other means to store an item after it has been processed. In Chapter 7, you looked at Spring Batch's ability to call an existing Spring service to obtain data. It should come as no surprise then that the framework offers similar functionality on the writing end. Spring Batch also exposes Spring's powerful JMS interactions with a `JmsItemWriter`. Finally, if you have a requirement to send e-mails from a batch process, Spring Batch can handle that too. This section looks at how to call existing Spring services, write to a JMS destination, and send e-mail using provided Spring Batch `ItemWriters`.

ItemWriterAdapter

In most enterprises that use Spring, there are a number of existing services already written and battle-tested in production. There is no reason they can't be reused in your batch processes. In Chapter 7, you looked at how to use them as sources of input for the jobs. This section looks at how the `ItemWriterAdapter` allows you to use existing Spring services as `ItemWriters` as well.

`org.springframework.batch.item.adapter.ItemWriterAdapter` is nothing more than a thin wrapper around the service you configure. As with any other `ItemWriter`, the `write` method receives a `List<T>` of items to be written. `ItemWriterAdapter` loops through the list calling the service method configured for each item in the list. It's important to note that the method being called by `ItemWriterAdapter` can only accept the item type being processed. For example, if the step is processing `Car` objects, the method being called must take a single argument of type `Car`.

To configure an `ItemWriterAdapter`, two dependencies are required:

- `targetObject`: The Spring bean that contains the method to be called
- `targetMethod`: The method to be called with each item

■ **Note** The method being called by `ItemWriterAdapter` must take a single argument of the type that is being processed by the current step.

Let's look at an example of `ItemWriterAdapter` in action. Listing 9-53 shows the code for a service that logs `Customer` items to `System.out`.

Listing 9-53. `CustomerService.java`

```java
package com.apress.springbatch.chapter9;

@Service
public class CustomerService {

public void logCustomer(Customer cust) {
        System.out.println("I just saved " + cust);
    }
}
```

As you can see in Listing 9-53, `CustomerService` is short, sweet, and to the point. But it serves the purpose for the example. To put this service to work in `formatJob`, you can configure it to be the target of a new `ItemWriterAdapter`. Using the same input configuration you've used in the other jobs this chapter, Listing 9-54 shows the configuration for the `ItemWriter` using the `CustomerServiceImpl`'s `logCustomer(Customer cust)` method and job referencing it.

Listing 9-54. `ItemWriterAdapter` Configuration

```java
    ...
@Bean
public ItemWriterAdapter<Customer> itemWriter(CustomerService customerService) {
        ItemWriterAdapter<Customer> customerItemWriterAdapter = new ItemWriterAdapter<>();

        customerItemWriterAdapter.setTargetObject(customerService);
        customerItemWriterAdapter.setTargetMethod("logCustomer");

        return customerItemWriterAdapter;
}

@Bean
public Step formatStep() throws Exception {
        return this.stepBuilderFactory.get("jpaFormatStep")
                        .<Customer, Customer>chunk(10)
                        .reader(customerFileReader(null))
                        .writer(itemWriter(null))
                        .build();
}
```

```
@Bean
public Job itemWriterAdapterFormatJob() throws Exception {
        return this.jobBuilderFactory.get("itemWriterAdapterFormatJob")
                        .start(formatStep())
                        .build();
}
...
```

Listing 9-54 starts with the configuration of the ItemWriter as the itemWriterAdapter. The two dependencies it uses are a reference to customerService and the name of the logCustomer method. Finally, you reference the itemWriterAdapter in the step to be used by the job.

To execute this job, you build it, like all jobs, with a ./mvnw clean install from the command line. With the job built, you can execute it by executing the jar file as you've done in the past. A sample of the output of this job is shown in Listing 9-55.

Listing 9-55. ItemWriterAdapter Output

```
2018-05-03 21:55:01.287  INFO 61906 --- [            main] o.s.b.c.l.support.
SimpleJobLauncher    : Job: [SimpleJob: [name=itemWriterAdapterFormatJob]] launched
with the following parameters: [{customerFile=/data/customer.csv, outputFile=file:/
Users/mminella/Documents/IntelliJWorkspace/def-guide-spring-batch/Chapter9/target/
formattedCustomers.xml}]
2018-05-03 21:55:01.299  INFO 61906 --- [            main] o.s.batch.core.job.
SimpleStepHandler    : Executing step: [jpaFormatStep]
Customer{id=0, firstName='Richard', middleInitial='N', lastName='Darrow', address='5570
Isabella Ave', city='St. Louis', state='IL', zip='58540'}
Customer{id=0, firstName='Warren', middleInitial='L', lastName='Darrow', address='4686 Mt.
Lee Drive', city='St. Louis', state='NY', zip='94935'}
Customer{id=0, firstName='Barack', middleInitial='G', lastName='Donnelly', address='7844 S.
Greenwood Ave', city='Houston', state='CA', zip='38635'}
Customer{id=0, firstName='Ann', middleInitial='Z', lastName='Benes', address='2447 S.
Greenwood Ave', city='Las Vegas', state='NY', zip='55366'}
Customer{id=0, firstName='Erica', middleInitial='Z', lastName='Gates', address='3141 Farnam
Street', city='Omaha', state='CA', zip='57640'}
Customer{id=0, firstName='Warren', middleInitial='M', lastName='Williams', address='6670 S.
Greenwood Ave', city='Hollywood', state='FL', zip='37288'}
Customer{id=0, firstName='Harry', middleInitial='T', lastName='Darrow', address='3273
Isabella Ave', city='Houston', state='FL', zip='97261'}
Customer{id=0, firstName='Steve', middleInitial='O', lastName='Darrow', address='8407
Infinite Loop Drive', city='Las Vegas', state='WA', zip='90520'}
2018-05-03 21:55:01.373  INFO 61906 --- [            main] o.s.b.c.l.support.
SimpleJobLauncher    : Job: [SimpleJob: [name=itemWriterAdapterFormatJob]] completed
with the following parameters: [{customerFile=/data/customer.csv, outputFile=file:/
Users/mminella/Documents/IntelliJWorkspace/def-guide-spring-batch/Chapter9/target/
formattedCustomers.xml}] and the following status: [COMPLETED]
```

As you would expect, calling an existing service with the item you've processed in your step is made easy with Spring Batch. However, what if your service doesn't take the same object you're processing? If you want to be able to extract values out of your item and pass them to your service, Spring Batch has you covered. PropertyExtractingDelegatingItemWriter (yes, that really is its name) is next.

PropertyExtractingDelegatingItemWriter

The use case for `ItemWriterAdapter` is pretty simple. Take the item being processed, and pass it to an existing Spring service. However, software is rarely that straightforward. Because of that, Spring Batch has provided a mechanism to extract values from an item and pass them to a service as parameters. This section looks at `PropertyExtractingDelegatingItemWriter` and how to use it with an existing service.

Although it has a long name, `org.springframework.batch.item.adapter.` `PropertyExtractingDelegatingItemWriter` is a lot like the `ItemWriterAdapter`. Just like `ItemWriterAdapter`, it calls a specified method on a referenced Spring service. The difference is that instead of blindly passing the item being processed by the step, `PropertyExtractingDelegatingItemWriter` passes only the attributes of the item that are requested. For example, if you have an item of type `Product` that contains fields for a database id, name, price, and SKU number, you're required to pass the entire `Product` object to the service method as with `ItemWriterAdapter`. But with `PropertyExtractingDelegatingItemWriter`, you can specify that you only want the database id and price to be passed as parameters to the service.

To look at this as an example, you can use the same customer input that you're familiar with by this point. You add a method to the `CustomerService` that allows you to log the address of the `Customer` item being processed and use `PropertyExtractingDelegatingItemWriter` to call the new method. Let's start by looking at the updated `CustomerService` (see Listing 9-56).

Listing 9-56. `CustomerService` with `logCustomerAddress()`

```
...
@Service
public class CustomerService {

        public void logCustomer(Customer customer) {
                System.out.println(customer);
        }

        public void logCustomerAddress(String address,
                        String city,
                        String state,
                        String zip) {
                System.out.println(
                                String.format("I just saved the address:\n%s\n%s, %s\n%s",
                                        address,
                                        city,
                                        state,
                                        zip));
        }
}
```

As you can see in Listing 9-56, we've added a new method `logCustomerAddress(String address, String city, String state, String zip)`; however, our new method doesn't take the `Customer` item. Instead it takes values that you have within it. To use this method, you use `PropertyExtractingDelegatingItemWriter` to extract the address fields (address, city, state, and zip) from each `Customer` item and call the service with the values it receives. To configure this `ItemWriter`, you pass in an ordered list of properties to extract from the item along with the target object and method to be called. The list you pass is in the same order as the parameters required for the property; Spring does support dot notation (`address.city`, for example) as well as index properties (`e-mail[5]`). Just like the `ItemWriterAdapter`, this `ItemWriter` implementation also exposes an `arguments` property that isn't used because the arguments are extracted by the writer dynamically. Listing 9-57 shows the job updated to call the `logCustomerAddress(String address, String city, String state, String zip)` method instead of handling the entire `Customer` item.

Listing 9-57. `formatJob` Configured to Call the `logCustomerAddress` Method on `CustomerService`

```
    ...
@Bean
public PropertyExtractingDelegatingItemWriter<Customer> itemWriter(CustomerService
customerService) {
        PropertyExtractingDelegatingItemWriter<Customer> itemWriter =
                    new PropertyExtractingDelegatingItemWriter<>();

        itemWriter.setTargetObject(customerService);
        itemWriter.setTargetMethod("logCustomerAddress");
        itemWriter.setFieldsUsedAsTargetMethodArguments(
                    new String[] {"address", "city", "state", "zip"});

        return itemWriter;
}

@Bean
public Step formatStep() throws Exception {
        return this.stepBuilderFactory.get("formatStep")
                        .<Customer, Customer>chunk(10)
                        .reader(customerFileReader(null))
                        .writer(itemWriter(null))
                        .build();
}

@Bean
public Job propertiesFormatJob() throws Exception {
        return this.jobBuilderFactory.get("propertiesFormatJob")
                        .start(formatStep())
                        .build();
}
...
```

When you run the job, the output of it consists of a sentence written to `System.out` with a formatted address. Listing 9-58 shows a sample of the output you can expect.

Listing 9-58. Output of `formatJob` Using `PropertyExtractingDelegatingItemWriter`

```
2018-05-03 22:15:06.509  INFO 62192 --- [           main] o.s.b.c.l.support.
SimpleJobLauncher      : Job: [SimpleJob: [name=propertiesFormatJob]] launched with the
following parameters: [{customerFile=/data/customer.csv, outputFile=file:/Users/mminella/
Documents/IntelliJWorkspace/def-guide-spring-batch/Chapter9/target/formattedCustomers.xml}]
2018-05-03 22:15:06.523  INFO 62192 --- [           main] o.s.batch.core.job.
SimpleStepHandler      : Executing step: [formatStep]
I just saved the address:
5570 Isabella Ave
St. Louis, IL
```

```
58540
I just saved the address:
4686 Mt. Lee Drive
St. Louis, NY
94935
I just saved the address:
7844 S. Greenwood Ave
Houston, CA
38635
I just saved the address:
2447 S. Greenwood Ave
Las Vegas, NY
55366
I just saved the address:
3141 Farnam Street
Omaha, CA
57640
I just saved the address:
6670 S. Greenwood Ave
Hollywood, FL
37288
I just saved the address:
3273 Isabella Ave
Houston, FL
97261
I just saved the address:
8407 Infinite Loop Drive
Las Vegas, WA
90520
2018-05-03 22:15:06.598  INFO 62192 --- [           main] o.s.b.c.l.support.
SimpleJobLauncher     : Job: [SimpleJob: [name=propertiesFormatJob]] completed with the
following parameters: [{customerFile=/data/customer.csv, outputFile=file:/Users/mminella/
Documents/IntelliJWorkspace/def-guide-spring-batch/Chapter9/target/formattedCustomers.xml}]
and the following status: [COMPLETED]
2018-05-03 22:15:06.599  INFO 62192 --- [       Thread-7] s.c.a.AnnotationConfigApplication
Context : Closing org.springframework.context.annotation.AnnotationConfigApplicationContext
@22635ba0: startup date [Thu May 03 22:15:04 CDT 2018]; root of context hierarchy
```

Spring Batch provides the ability to reuse just about any existing Spring service you've created as an ItemWriter, with good reason. The code your enterprise has is battle tested in production, and reusing it is less likely to introduce new bugs and also speeds up development time. The next section looks at using JMS resources as the destination of items processed within a step.

JmsItemWriter

Java Messaging Service (JMS) is a message-oriented method of communicating between two or more endpoints. By using either point-to-point communication (a JMS queue) or a publish-subscribe model (JMS topic), Java applications can communicate with any other technology that can interface with the messaging implementation. This section looks at how you can put messages on a JMS queue using Spring Batch's JmsItemWriter.

Spring has made great progress in simplifying a number of common Java concepts. JDBC and integration with the various ORM frameworks come to mind as examples. But Spring's work in simplifying interfacing with JMS resources is just as impressive. In order to work with JMS, you need to use a JMS broker. This example uses Apache's ActiveMQ. Apache ActiveMQ is a simple broker that we can utilize in memory for our examples.

Before you can work with ActiveMQ, you need to add its dependencies and Spring's JMS dependencies to the POM so that it's available. This example works with ActiveMQ version 5.15.3, which is the most current version as of this writing. Listing 9-5 shows the dependencies you need to add to the POM.

Listing 9-59. Dependencies for ActiveMQ and Spring JMS

```
...
<dependency>
        <groupId>org.springframework.boot</groupId>
        <artifactId>spring-boot-starter-activemq</artifactId>
</dependency>
<dependency>
        <groupId>org.apache.activemq</groupId>
        <artifactId>activemq-broker</artifactId>
</dependency>
...
```

Now you can begin to put ActiveMQ to work. Before you get into the code, however, let's look at the processing for this job because it's slightly different than before.

In previous examples in this chapter, you have had a single step that read in the `customer.csv` file and wrote it out using the appropriate `ItemWriter` for the example. For this example, however, that won't be enough. If you read in the items and write them to the JMS queue, you won't know if everything got onto the queue correctly because you can't see what is in the queue. Instead, as Figure 9-6 shows, you use two steps for this job. The first one reads the `customer.csv` file and writes it to the ActiveMQ queue. The second step reads from the queue and writes the records out to an XML file.

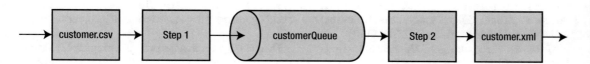

Figure 9-6. *Processing for* `jmsFormatJob`

It's important to note that you don't want to do this in an actual production environment because a message isn't pulled off the queue until all of them have been put on it. This could lead to running out of room in your queue depending on how it's configured and the resources available. However, for this example and given the small number of customers you're processing, this approach demonstrates the point.

- *A* `MessageConverter`*:* This will convert the message to JSON for transport over the wire.

- *A* `JmsTemplate`*:* While Spring Boot will provide a `JmsTemplate` autoconfigured, it does not provide one using a `CachingConnectionFactory`, which is the recommended approach when using `JmsTemplate`. Because of this, we'll configure our own `JmsTemplate`.

Let's start by looking at the `MessageConverter`. By default Spring Integration will handle the serialization of `Message` objects on it's own via Java serialization. However, that isn't very useful. Instead, we'll configure a `MessageConverter` to convert the `Message` passed to JSON for transport over the wire on ActiveMQ.

The `JmsTemplate` is the next bean we need to configure ourselves. Spring Boot provides a `JmsTemplate`; however, the `ConnectionFactory` it provides does not work well with a `JmsTemplate`. Because of this, we configure our own `JmsTemplate` utilizing a `CachingConnectionFactory`.

Listing 9-60 shows the configuration of the JMS resources in JmsJob.java.

Listing 9-60. JMS Resource Configuration

```
...
@Bean // Serialize message content to json using TextMessage
public MessageConverter jacksonJmsMessageConverter() {
        MappingJackson2MessageConverter converter = new MappingJackson2MessageConverter();
        converter.setTargetType(MessageType.TEXT);
        converter.setTypeIdPropertyName("_type");
        return converter;
}

@Bean
public JmsTemplate jmsTemplate(ConnectionFactory connectionFactory) {
        CachingConnectionFactory cachingConnectionFactory = new CachingConnectionFactory
        (connectionFactory);
        cachingConnectionFactory.afterPropertiesSet();

        JmsTemplate jmsTemplate = new JmsTemplate(cachingConnectionFactory);
        jmsTemplate.setDefaultDestinationName("customers");
        jmsTemplate.setReceiveTimeout(5000L);

        return jmsTemplate;
}
...
```

Now you can configure the job. You use the same reader you've used up to this point in the chapter for the first step and the same writer you used in the XML example earlier in the chapter for the writer in the second step. Their configuration can be found in Listing 9-61.

Listing 9-61. Input and Output of `jmsFormatJob`

```
...
@Bean
@StepScope
public FlatFileItemReader<Customer> customerFileReader(
                @Value("#{jobParameters['customerFile']}")Resource inputFile) {

        return new FlatFileItemReaderBuilder<Customer>()
                        .name("customerFileReader")
                        .resource(inputFile)
                        .delimited()
                        .names(new String[] {"firstName",
                                        "middleInitial",
                                        "lastName",
```

301

```
                                        "address",
                                        "city",
                                        "state",
                                        "zip"})
                        .targetType(Customer.class)
                        .build();
}

@Bean
@StepScope
public StaxEventItemWriter<Customer> xmlOutputWriter(
                @Value("#{jobParameters['outputFile']}") Resource outputFile) {

        Map<String, Class> aliases = new HashMap<>();
        aliases.put("customer", Customer.class);

        XStreamMarshaller marshaller = new XStreamMarshaller();
        marshaller.setAliases(aliases);

        return new StaxEventItemWriterBuilder<Customer>()
                        .name("xmlOutputWriter")
                        .resource(outputFile)
                        .marshaller(marshaller)
                        .rootTagName("customers")
                        .build();
}
...
```

JmsReader and JmsWriter are configured the same way. Both of them are basic Spring beans with a reference to the JmsTemplate configured in Listing 9-60. In Listing 9-62, you see the configuration of JmsItemReader, JmsItemWriter, and the job to put all the readers/writers to work.

Listing 9-62. JmsItemReader and JmsItemWriter and the Job That Uses Them

```
    ...
@Bean
public JmsItemReader<Customer> jmsItemReader(JmsTemplate jmsTemplate) {

        return new JmsItemReaderBuilder<Customer>()
                        .jmsTemplate(jmsTemplate)
                        .itemType(Customer.class)
                        .build();
}

@Bean
public JmsItemWriter<Customer> jmsItemWriter(JmsTemplate jmsTemplate) {

        return new JmsItemWriterBuilder<Customer>()
                        .jmsTemplate(jmsTemplate)
                        .build();
}
```

```
@Bean
public Step formatInputStep() throws Exception {
        return this.stepBuilderFactory.get("formatInputStep")
                        .<Customer, Customer>chunk(10)
                        .reader(customerFileReader(null))
                        .writer(jmsItemWriter(null))
                        .build();
}

@Bean
public Step formatOutputStep() throws Exception {
        return this.stepBuilderFactory.get("formatOutputStep")
                        .<Customer, Customer>chunk(10)
                        .reader(jmsItemReader(null))
                        .writer(xmlOutputWriter(null))
                        .build();
}

@Bean
public Job jmsFormatJob() throws Exception {
        return this.jobBuilderFactory.get("jmsFormatJob")
                        .start(formatInputStep())
                        .next(formatOutputStep())
                        .build();
}
...
```

That's all it takes! With all the resources configured, building and running this job is no different than any of the others you've executed. However, when you run this job, notice that nothing obvious is outputted from step 1 to tell you that anything happened besides looking into the JobRepository or browsing the queue before the second step executes. When you look at the XML generated in step 2, you can see that the messages have successfully been passed through the queue as expected. Listing 9-63 shows a sample of the XML generated by this job.

Listing 9-63. Sample Output from the JMS Version of formatJob

```
<?xml version="1.0" encoding="UTF-8"?>
<customers>
        <customer>
                <id>0</id>
                <firstName>Richard</firstName>
                <middleInitial>N</middleInitial>
                <lastName>Darrow</lastName>
                <address>5570 Isabella Ave</address>
                <city>St. Louis</city>
                <state>IL</state>
                <zip>58540</zip>
        </customer>
```

```
        <customer>
                <id>0</id>
                <firstName>Warren</firstName>
                <middleInitial>L</middleInitial>
                <lastName>Darrow</lastName>
                <address>4686 Mt. Lee Drive</address>
                <city>St. Louis</city>
                <state>NY</state>
                <zip>94935</zip>
        </customer>
        ...
</customers>
```

By using Spring's `JmsTemplate`, Spring Batch exposes the full power of Spring's JMS processing capabilities to the batch processes with minimal effort. The next section looks at a writer you may not have thought about: it lets you send e-mail from batch processes.

SimpleMailMessageItemWriter

The ability to send an e-mail may sound very useful. Heck, when a job completes, it might be handy to receive an e-mail that things ended nicely. However, that isn't what this `ItemWriter` is for. It's an `ItemWriter`, which means it's called once for each item processed in the step where it's used. If you want to run your own spam operation, this is the `ItemWriter` for you! This section looks at how to use Spring Batch's `SimpleMailMessageItemWriter` to send e-mails from jobs.

Although you probably won't be using this `ItemWriter` to write a spam-processing program, you can use it for other things as well. Let's say the customer file you've been processing up to this point is really a customer import file; after you import all the new customers, you want to send a welcome e-mail to each one. Using the `org.springframework.batch.item.mail.SimpleMailMessageItemWriter` is a perfect way to do that.

For this example, you have a two-step process as you did in the JMS example. The first step imports the customer.csv file into the `CUSTOMER` database table. The second step reads all the customers that have been imported and sends them the welcome e-mail. Figure 9-7 shows the flow for this job.

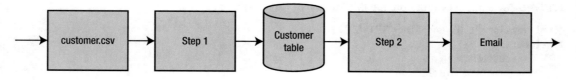

***Figure 9-7.** Flow for the `customerImport` job*

Before you begin coding, let's look at `SimpleMailMessageItemWriter`. Like all other `ItemWriters`, it implements the `ItemWriter` interface by executing a single `write(List<T> items)` method that takes a list of objects. However, unlike the `ItemWriters` you've looked at up to this point, `SimpleMailMessageItemWriter` doesn't take just any item. Sending an e-mail requires more information than the text of the e-mail. It needs a subject, a to address, and a from address. Because of this, `SimpleMailMessageItemWriter` requires that the list of objects it takes contain objects that extend Spring's `SimpleMailMessage`. By doing this, `SimpleMailMessageItemWriter` has all the information it needs to build the e-mail message.

But does that mean any item you read in must extend `SimpleMailMessage`? That seems like a poor job of decoupling e-mail functionality from business logic—which is why you don't have to do that. If you remember, Chapter 8 talked about how `ItemProcessors` don't need to return an object of the same type they receive. For example, you can receive a `Car` object but return an object of type `House`. In this case, you create an `ItemProcessor` that takes in the `Customer` object and returns the required `SimpleMailMessage`.

To make this work, you reuse the same input file format with a single field appended to the end: the customer's e-mail address. Listing 9-64 shows an example of the input file you're processing.

Listing 9-64. `customerWithEmail.csv`

```
Ann,A,Smith,2501 Mt. Lee Drive,Miami,NE,62935,ASmith@yahoo.com
Laura,B,Jobs,9542 Isabella Ave,Aurora,FL,62344,LJobs@yahoo.com
Harry,J,Williams,1909 4th Street,Seatle,TX,48548,HWilliams@hotmail.com
Larry,Y,Minella,7839 S. Greenwood Ave,Miami,IL,65371,LMinella@hotmail.com
Richard,Q,Jobs,9732 4th Street,Chicago,NV,31320,RJobs@gmail.com
Ann,P,Darrow,4195 Jeopardy Lane,Aurora,CA,24482,ADarrow@hotmail.com
Larry,V,Williams,3075 Wall Street,St. Louis,NY,34205,LWilliams@hotmail.com
Michael,H,Gates,3219 S. Greenwood Ave,Boston,FL,24692,MGates@gmail.com
Harry,H,Johnson,7520 Infinite Loop Drive,Hollywood,MA,83983,HJohnson@hotmail.com
Harry,N,Ellison,6959 4th Street,Hollywood,MO,70398,HEllison@gmail.com
```

To handle the need for an e-mail address per customer, you need to add an e-mail field to the `Customer` object as well. Listing 9-65 shows the updated `Customer` class.

Listing 9-65. `Customer.java` Updated with an E-mail Field

```java
package com.apress.springbatch.chapter9;

import java.io.Serializable;

import javax.persistence.Entity;
import javax.persistence.GeneratedValue;
import javax.persistence.GenerationType;
import javax.persistence.Id;
import javax.persistence.Table;

@Entity
@Table(name="CUSTOMER")
public class Customer implements Serializable {
    private static final long serialVersionUID = 1L;

    @Id
    @GeneratedValue(strategy = GenerationType.IDENTITY)
    private long id;
    private String firstName;
    private String middleInitial;
    private String lastName;
    private String address;
    private String city;
    private String state;
```

```
    private String zip;
    private String email;

    // Accessors go here
    ...
}
```

Because the job is storing the customer information in the database, let's take a quick look at how that interaction works. To start, Figure 9-8 has the data model for the CUSTOMER table you use in this example.

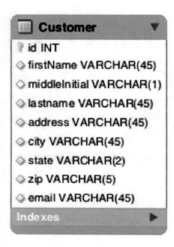

Figure 9-8. *CUSTOMER table*

Before we can get into any of the email pieces of the code, we need to add the correct dependencies to our POM file. Spring Boot makes adding email support to an application simple by requiring a single dependency, the mail starter as shown in Listing 9-66.

Listing 9-66. Java Mail Dependency

```
...
<dependency>
        <groupId>org.springframework.boot</groupId>
        <artifactId>spring-boot-starter-mail</artifactId>
</dependency>
 ...
```

With the starter added, we need to do a bit of configuration in order to send emails. For this example, I'll be using a personal Gmail account to send the emails via Google's SMTP server.[5] If you use Gmail, you can do the same within the bounds of their terms of service. To do so, Listing 9-67 shows the properties you need to configure in your application.properties.

---

[5]You may need to turn on allow less secure applications to use Google's SMTP server. Learn more about it here: https://support.google.com/accounts/answer/6010255?hl=en

Listing 9-67. Java Mail Dependency

```
spring.mail.host=smtp.gmail.com
spring.mail.port=587
spring.mail.username=<SOME_USERNAME>
spring.mail.password=<SOME_PASSWORD>
spring.mail.properties.mail.smtp.auth=true
spring.mail.properties.mail.smtp.starttls.enable=true
```

With the email configuration complete, we can configure the components of our job. The first three beans should all be familiar to you. They are two ItemReaders and one ItemWriter that we've used before. The customerEmailFileReader is the same as the other FlatFileItemReaders we've used in this chapter with an extra field (email) configured to be read in each record. Following that bean is the customerBatchWriter which is a JdbcBatchItemWriter that is configured to use JDBC to write to the database. Listing 9-68 shows how you wire that up for the first step in the job.

Listing 9-68. ItemReader and ItemWriter for Step 1

```
...
@Bean
@StepScope
public FlatFileItemReader<Customer> customerEmailFileReader(
                @Value("#{jobParameters['customerFile']}")Resource inputFile) {

        return new FlatFileItemReaderBuilder<Customer>()
                        .name("customerFileReader")
                        .resource(inputFile)
                        .delimited()
                        .names(new String[] {"firstName",
                                        "middleInitial",
                                        "lastName",
                                        "address",
                                        "city",
                                        "state",
                                        "zip",
                                        "email"})
                        .targetType(Customer.class)
                        .build();
}

@Bean
public JdbcBatchItemWriter<Customer> customerBatchWriter(DataSource dataSource) {

        return new JdbcBatchItemWriterBuilder<Customer>()
                        .namedParametersJdbcTemplate(new NamedParameterJdbcTemplate
                        (dataSource))
                        .sql("INSERT INTO CUSTOMER (first_name, middle_initial, last_name,
                        address, city, state, zip, email) " +
                                        "VALUES(:firstName, :middleInitial, :lastName,
                                        :address, :city, :state, :zip, :email)")
```

```
                              .beanMapped()
                              .build();
}
...
```

After the components for the first step are configured, customerCursorItemReader, the reader for the second step is configured. This ItemReader is a JdbcCursorItemReader that returns all of the data in the CUSTOMER table. You'll note that we aren't developing a custom RowMapper, but instead, using the BeanPropertyRowMapper to map column names with bean setters to map our database data.

None of the configuration up to this point should be new, because you've seen it previously. The new parts come when you configure the ItemWriter for step 2. For step 2, you're using a SimpleMailMessageItemWriter as the ItemWriter. Listing 9-69 shows the configuration of the beans required for step 2 along with the job configuration.

Listing 9-69. Step 2 and the Job Configuration

```
...
@Bean
public JdbcCursorItemReader<Customer> customerCursorItemReader(DataSource dataSource) {

        return new JdbcCursorItemReaderBuilder<Customer>()
                        .name("customerItemReader")
                        .dataSource(dataSource)
                        .sql("select * from customer")
                        .rowMapper(new BeanPropertyRowMapper<>(Customer.class))
                        .build();
}

@Bean
public SimpleMailMessageItemWriter emailItemWriter(MailSender mailSender) {

        return new SimpleMailMessageItemWriterBuilder()
                        .mailSender(mailSender)
                        .build();
}

@Bean
public Step importStep() throws Exception {
        return this.stepBuilderFactory.get("importStep")
                        .<Customer, Customer>chunk(10)
                        .reader(customerEmailFileReader(null))
                        .writer(customerBatchWriter(null))
                        .build();
}

@Bean
public Step emailStep() throws Exception {
        return this.stepBuilderFactory.get("emailStep")
                        .<Customer, SimpleMailMessage>chunk(10)
                        .reader(customerCursorItemReader(null))
                        .processor((ItemProcessor<Customer, SimpleMailMessage>) customer -> {
                                SimpleMailMessage mail = new SimpleMailMessage();
```

```
                           mail.setFrom("prospringbatch@gmail.com");
                           mail.setTo(customer.getEmail());
                           mail.setSubject("Welcome!");
                           mail.setText(String.format("Welcome %s %s,\nYou were
                           imported into the system using Spring Batch!",
                                          customer.getFirstName(), customer.
                                          getLastName()));

                           return mail;
                   })
                   .writer(emailItemWriter(null))
                   .build();
}

@Bean
public Job emailJob() throws Exception {
        return this.jobBuilderFactory.get("emailJob")
                      .start(importStep())
                      .next(emailStep())
                      .build();
}
...
```

You'll notice that the SimpleMailMessageItemWriter only requires one dependency, a MailSender which is provided by Spring Boot. Once that ItemWriter is configured, we can configure our steps. The importStep looks like all the others we've seen up to this point, specifying both an ItemReader and an ItemWriter. The second step, emailStep, also provides an ItemReader and an ItemWriter; however, we're using a lambda expression for the ItemProcessor that we need to convert a Customer item to a SimpleMailMessage (the input type for our ItemWriter). That's all there is to it! You can build this job with ./mvnw clean install from the command line and run it with the command listed in Listing 9-70 to process the input file and send out the e-mails.

Listing 9-70. Executing the E-mail Job

```
java -jar itemWriters-0.0.1-SNAPSHOT.jar customerFile=/input/customerWithEmail.csv
```

When the job is complete, you can check your e-mail inbox as shown in Figure 9-9 to see that the customers have successfully received their e-mails.

Figure 9-9. *The result of the e-mail job*

Spring Batch provides a full collection of ItemWriters to handle the vast majority of output handling that you need to be able to do. The next section looks at how you can use the individual features of each of these ItemWriters together to address more complex output scenarios, such as writing to multiple places based on a number of scenarios.

Multipart ItemWriters

As part of your new system, you have the requirement to extract customer data into two different formats. You need an XML file for the Sale's department's customer relationship management (CRM) application. You also need a CSV for the billing department's database import system. The issue is, you expect to extract one million customers.

Using the tools discussed up to this point, you would be stuck looping through the one million items twice (once for a step that outputs the XML file and once for the step that outputs the CSV file) or creating a custom ItemWriter implementation to write to each file as an item is processed. Neither option is what you're looking for. The first will take too long, tying up resources; and the other requires you to code and test something that the framework should already provide. Fortunately for you, it does. This section looks at how you can use the various composite ItemWriters available in Spring Batch to address more complex output scenarios.

MultiResourceItemWriter

Chapter 7 looked at Spring Batch's ability to read from multiple files with the same format in a single step. Spring Batch provides a similar feature on the ItemWriter side as well. This section looks at how to generate multiple resources based on the number of items written to a file.

Spring Batch offers the ability to create a new resource after a given number of records has been processed. Say you want to extract all the customer records and write them to XML files with only ten customers per file. To do that, you use MultiResourceItemWriter.

MultiResourceItemWriter dynamically creates output resources based on the number of records it has processed. It passes each item it processes to a delegate writer so that the actual writing piece is handled there. All MultiResourceItemWriter is responsible for is maintaining the current count and creating new resources as items are processed. Figure 9-10 shows the flow of a step using org.springframework.batch. item.file.MultiResourceItemWriter.

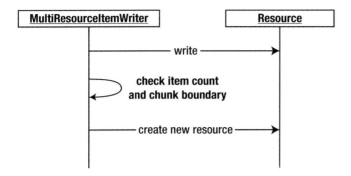

Figure 9-10. *Processing using a* `MultiResourceItemWriter`

When the `write(List<T>)` method on `MultiResourceItemWriter` is called, it verifies that the current `Resource` has been created and is open (if not, it creates and opens a new file) and passes the items to the delegate `ItemWriter`. Once the items have been written, it checks to see if the number of items written to the file has reached the configured threshold for a new resource. If it has, the current file is closed.

It's important to note that when `MultiResourceItemWriter` is processing, it doesn't create a new resource mid-chunk. It waits for the end of the chunk before creating a new resource. For example, if the writer is configured to roll the file after 15 items have been processed but the chunk size is configured to 20, `MultiResourceItemWriter` writes the 20 items in the chunk before creating a new resource.

`MultiResourceItemWriter` has five available dependencies you can configure. Table 9-4 shows each one and how they're used.

Table 9-4. `MultiResourceItemWriter` *Configuration Options*

Option	Type	Default	Description
delegate	ResourceAwareItem WriterItemStream	null (required)	The delegate `ItemWriter` that the `MultiResourceItemWriter` uses to write each item.
itemCountLimit PerResource	int	Integer.MAX_VALUE	The number of items to write to each resource.
resource	Resource	null (required)	A prototype of the resources to be created by `MultiResourceItemWriter`.
resourceSuffix Creator	ResourceSuffix Creator	null	Optionally, `MultiResourceItemWriter` can use this class to append a suffix to the end of the file names it creates.
saveState	boolean	true	If `false`, the state of the `ItemWriter` isn't maintained in the `JobRepository`.

To see how this works, you extract customers from the database and create XML files containing ten customers each. To make this work, you don't need to develop any new code (you created the XML). All you need to do is wire everything up. Let's start working with this example by looking at the configuration for the job.

Listing 9-71 shows the configuration for the ItemReader in this example. In this case, it's a simple JdbcCursorItemReader configured to select all customers. From there, you pass the customers you receive from the database to the ItemWriter you configure next.

Listing 9-71. The multiResource formatJob's ItemReader

```
...
@Bean
public JdbcCursorItemReader<Customer> customerJdbcCursorItemReader(DataSource dataSource) {

        return new JdbcCursorItemReaderBuilder<Customer>()
                    .name("customerItemReader")
                    .dataSource(dataSource)
                    .sql("select * from customer")
                    .rowMapper(new BeanPropertyRowMapper<>(Customer.class))
                    .build();
}
...
```

The configuration for this ItemWriter is in layers. First you configure the StaxEventItemWriter that you use for the XML generation. With that configured you layer MultiResourceItemWriter on top to generate multiple resources the StaxEventItemWriter writes to. Listing 9-72 shows the configuration of the output half of the job as well as the step and job configuration.

Listing 9-72. ItemWriters and Step and Job Configuration

```
...
@Bean
@StepScope
public StaxEventItemWriter<Customer> delegateItemWriter() throws Exception {

        Map<String, Class> aliases = new HashMap<>();
        aliases.put("customer", Customer.class);

        XStreamMarshaller marshaller = new XStreamMarshaller();

        marshaller.setAliases(aliases);

        marshaller.afterPropertiesSet();

        return new StaxEventItemWriterBuilder<Customer>()
                    .name("customerItemWriter")
                    .marshaller(marshaller)
                    .rootTagName("customers")
                    .build();
}
```

```java
@Bean
public MultiResourceItemWriter<Customer> multiCustomerFileWriter() throws Exception {

        return new MultiResourceItemWriterBuilder<Customer>()
                        .name("multiCustomerFileWriter")
                        .delegate(delegateItemWriter())
                        .itemCountLimitPerResource(25)
                        .resource(new FileSystemResource("Chapter9/target/customer"))
                        .build();
}

@Bean
public Step multiXmlGeneratorStep() throws Exception {
        return this.stepBuilderFactory.get("multiXmlGeneratorStep")
                        .<Customer, Customer>chunk(10)
                        .reader(customerJdbcCursorItemReader(null))
                        .writer(multiCustomerFileWriter())
                        .build();
}

@Bean
public Job xmlGeneratorJob() throws Exception {
        return this.jobBuilderFactory.get("xmlGeneratorJob")
                        .start(multiXmlGeneratorStep())
                        .build();
}
...
```

We start by can configuring `delegateItemWriter` to generate the XML as required. Although similar to the `StaxEventItemWriters` we've configured in other examples, it's important to note that `delegateItemWriter` doesn't have a direct reference to the output file. Instead, `multiResourceItemWriter` provides it when needed.

For this example, `multiCustomerFileWriter` uses three dependencies: a `Resource` to serve as the template of where to write the file to (directory and the file name), the `delegateItemWriter` that does the actual work of writing to the files it creates, and the number of customers that the `ItemWriter` writes per file (`itemCountLimitPerResource`)—25 in this case. The last piece for this job is configuring the step and job to put them to use. The configuration for the `Job` itself is straightforward, as Listing 9-72 shows. To use this job, you use the command listed in Listing 9-73.

Listing 9-73. Command Used to Execute the `multiResource` Job

```
java -jar itemWriters-0.0.1-SNAPSHOT.jar
```

When you look at the output of this job, you find in the `/output` directory one file for every ten customers currently loaded in the database. However, Spring Batch did something interesting. First, note that you didn't pass in a file extension on the `outputFile` parameter you passed into the job. This was for a reason. If you look at the directory listing shown in Listing 9-74, you see that `MultiResourceItemWriter` added a `.X` to each file, where X is the number of the file that was created.

Listing 9-74. File Names Created by the Job

```
michael-minellas-macbook-pro:temp mminella$ ls Chapter9/target/customer
customer.1 customer.2 customer.3 customer.4
```

Although it makes sense that you need to distinguish each file name from another, this may or may not be a workable solution for how to name the files (they don't exactly open nicely with your favorite editor by default). Because of that, Spring Batch lets you to configure the suffix for each file created. You do that by implementing the org.springframework.batch.item.file.ResourceSuffixCreator interface and adding that as a dependency to the multiResourceItemWriter bean. When the MultiResourceItemWriter is creating a new file, it uses ResourceSuffixCreator to generate a suffix that it tacks onto the end of the new file's name. Listing 9-75 shows the suffix creator for the example.

Listing 9-75. CustomerOutputFileSuffixCreator

```
...
@Component
public class CustomerOutputFileSuffixCreator implements ResourceSuffixCreator {

    @Override
    public String getSuffix(int arg0) {
        return arg0 + ".xml";
    }
}
```

In Listing 9-75, you implement the ResourceSuffixCreator's only method, getSuffix, and return a suffix of the number provided and an .xml extension. The number provided is the number file that is being created. If you were to re-create the same extension as the default, you would return a dot plus the number provided.

To use CustomerOutputFileSuffixCreator, you configure it as a bean and add it as a dependency to the multiResourceItemWriter bean using the property resourceSuffixCreator. Listing 9-76 shows the added configuration.

Listing 9-76. Configuring CustomerOutputFileSuffixCreator

```
...
@Bean
public MultiResourceItemWriter<Customer> multiCustomerFileWriter(CustomerOutputFileSuffixCre
ator suffixCreator) throws Exception {

        return new MultiResourceItemWriterBuilder<Customer>()
                        .name("multiCustomerFileWriter")
                        .delegate(delegateItemWriter())
                        .itemCountLimitPerResource(25)
                        .resource(new FileSystemResource("Chapter9/target/customer"))
                        .resourceSuffixCreator(suffixCreator)
                        .build();
}
...
```

By running the job again with the additional configuration provided in Listing 9-76, you get a slightly different result, as shown in Listing 9-77.

Listing 9-77. Results Using `ResourceSuffixCreator`

```
michael-minellas-macbook-pro:output mminella$ ls Chapter9/target/customer
customer1.xml      customer2.xml      customer3.xml      customer4.xml
```

You surely agree that the file names in Listing 9-77 are more like what you would expect when generating XML files.

Header and Footer XML Fragments

When creating files, whether a single file for a step/job or multiple files as you saw in the previous example, it's common to need to be able to generate a header or footer on the file. You can use a header to define the format of a flat file (what fields exist in a file or in what order) or include a separate, non-item-related section in an XML file. A footer may include the number of records processed in the file or totals to use as integrity checks after a file has been processed. This section looks at how to generate header and footer records using Spring Batch's callbacks available for them.

When opening or closing a file, Spring Batch provides the ability to add either a header or footer (whichever is appropriate) to your file. Adding a header or footer to a file means different things based on whether it's a flat file or an XML file. For a flat file, adding a header means adding one or more records to the top or bottom of the file. For an XML file, you may want to add an XML segment at either the top or bottom of the file. Because the generation of plain text for a flat file is different from generating an XML segment for an XML file, Spring Batch offers two different interfaces to implement and make this happen. Let's begin by looking at the XML callback interface, `org.springframework.batch.item.xml.StaxWriterCallback`.

The `StaxWriterCallback` interface consists of a single `write(XMLEventWriter writer)` method that is used to add XML to the current XML document. Spring Batch executes a configured callback once at either the header or footer of the file (based on the configuration). To see how this works, in this example you write a `StaxWriterCallback` implementation that adds an XML fragment containing the name of the person who wrote the job (me). Listing 9-78 shows the code for the implementation.

Listing 9-78. `CustomerXmlHeaderCallback`

```
...
@Component
public class CustomerXmlHeaderCallback implements StaxWriterCallback {

    @Override
    public void write(XMLEventWriter writer) throws IOException {
        XMLEventFactory factory = XMLEventFactory.newInstance();

        try {
            writer.add(factory.createStartElement("", "", "identification"));
            writer.add(factory.createStartElement("", "", "author"));
            writer.add(factory.createAttribute("name", "Michael Minella"));
            writer.add(factory.createEndElement("", "", "author"));
            writer.add(factory.createEndElement("", "", "identification"));
```

```
        } catch (XMLStreamException xmlse) {
            System.err.println("An error occured: " + xmlse.getMessage());
            xmlse.printStackTrace(System.err);
        }
    }
}
```

Listing 9-78 shows `CustomerXmlHeaderCallback`. In the callback, you add two tags to the XML file: an identification section and a single author section. The author section contains a single attribute called name with the value `Michael Minella`. To create a tag, you use the `javax.xml.stream.XMLEventFactory`'s `createStartElement` and `createEndElement` methods. Each of these methods takes three parameters: a prefix, a namespace, and the name of the tag. Because you aren't using a prefix or namespace, you pass in empty strings for those. To put this implementation to use, you need to configure `StaxEventItemWriter` to call the callback as the `headerCallback`. Listing 9-79 shows the configuration for this example.

Listing 9-79. XML Configuration for `CustomerXmlHeaderCallback`

```
...
@Bean
public MultiResourceItemWriter<Customer> multiCustomerFileWriter(CustomerOutputFileSuffix
Creator suffixCreator) throws Exception {

        return new MultiResourceItemWriterBuilder<Customer>()
                        .name("multiCustomerFileWriter")
                        .delegate(delegateItemWriter(null))
                        .itemCountLimitPerResource(25)
                        .resource(new FileSystemResource("Chapter9/target/customer"))
                        .resourceSuffixCreator(suffixCreator)
                        .build();
}
...
```

When you execute the multiresource job from the previous example using the header configuration in Listing 9-79, each of the output files begins with the XML fragment as shown in Listing 9-80.

Listing 9-80. XML Header

```
<?xml version="1.0" encoding="UTF-8"?>
<customers>
<identification>
<author name="Michael Minella"/>
</identification>
<customer>
    ...
```

As you can see, adding an XML fragment at either the start or end of an XML file is quite easy. Implement the `StaxWriterCallback` interface and configure the `ItemWriter` to call it as either the header or the footer, and you're done!

Header and Footer Records in a Flat File

Next you can look at adding headers and footers to a flat file. Unlike the XML header and footer generation that use the same interface for either, writing a header in a flat file requires the implementation of a different interface than that of a footer. For the header, you implement the org.springframework.batch.item.file.
FlatFileHeaderCallback interface; and for the footer, you implement the org.springframework.batch.
item.file.FlatFileFooterCallback interface. Both consist of a single method: writeHeader(Writer writer) and writeFooter(Writer writer), respectively. Let's look at how to write a footer that writes the number of records you've processed in the current file.

For this example, you use the MultiResourceItemWriter to write files with 30 formatted records in each file plus a single footer record that states how many records were written in each file. To be able to keep count of the number of items you've written into a file, we'll need to use a couple aspects. The first pointcut will be before any calls to the FlatFileItemWriter.open(ExecutionContext ec) method. We'll use this pointcut to reset a counter each time a new file is opened up. The second pointcut will be before any calls to FlatFileItemWriter.write(List<T> items). Here is where we will increment the counter.

Now you may wonder why not just use an ItemWriteListener.beforeWrite(List<T> items) call instead of jumping through the hoops of an aspect. The reason is the ordering of calls. The beforeWrite(List<T> items) call is called before the call to write(List<T> items). However, the call to open(ExecutionContext ec) is called within that method. Since we need to reset the counter *before* the call to write(List<T> items), we need to use an aspect.

What we will do is create a component that is both an aspect as well as implements the FlatFileFooterCallback. The aspects will manage the state of the callback (how many records have been written in the current file), and the FlatFileFooterCallback.writeFooter(Writer writer) method will write out the results. Listing 9-81 illustrates the code for this implementation.

Listing 9-81. CustomerRecordCountFooterCallback

```
...
@Component
@Aspect
public class CustomerRecordCountFooterCallback implements FlatFileFooterCallback {

        private int itemsWrittenInCurrentFile = 0;

        @Override
        public void writeFooter(Writer writer) throws IOException {
                writer.write("This file contains " +
                                itemsWrittenInCurrentFile + " items");
        }

        @Before("execution(* org.springframework.batch.item.file.FlatFileItemWriter.
        open(..))")
        public void resetCounter() {
                this.itemsWrittenInCurrentFile = 0;
        }

        @Before("execution(* org.springframework.batch.item.file.FlatFileItemWriter.
        write(..))")
```

```
        public void beforeWrite(JoinPoint joinPoint) {
                List<Customer> items = (List<Customer>) joinPoint.getArgs()[0];

                this.itemsWrittenInCurrentFile += items.size();
        }
}
```

As you can see in Listing 9-81, CustomerRecordCountFooterCallback is annotated with the @Component and @Aspect annotations. The first wires it up as a bean for Spring. The second identifies it as an aspect for AspectJ. The class implements the FlatFileFooterCallback and its writeFooter(Writer writer) method. This method writes out the actual footer. The next two methods are the aspect methods. The first, resetCounter() is configured to be called before the open(ExecutionContext ec) method on the FlatFileItemWriter and resets the current count to 0. This will be called once per file. The last method is the beforeWrite(List<T> items) method which increments the count with the number of items passed to the FlatFileItemWriter.write(List<T> items) method.

In order to put our callback to use, we'll need to update the MultiResourceJob in two ways. First, we'll need to use a FlatFileItemWriter as the delegate instead of the XML based ItemWriter we used previously. Second we will need to configure that ItemWriter to use our callback. Listing 9-82 shows the configuration of the new ItemWriter.

Listing 9-82. delegateCustomerItemWriter

```
@Bean
@StepScope
public FlatFileItemWriter<Customer> delegateCustomerItemWriter(CustomerRecordCountFooterCall
back footerCallback) throws Exception {
        BeanWrapperFieldExtractor<Customer> fieldExtractor = new BeanWrapperField
        Extractor<>();
        fieldExtractor.setNames(new String[] {"firstName", "lastName", "address", "city",
        "state", "zip"});
        fieldExtractor.afterPropertiesSet();

        FormatterLineAggregator<Customer> lineAggregator = new FormatterLineAggregator<>();

        lineAggregator.setFormat("%s %s lives at %s %s in %s, %s.");
        lineAggregator.setFieldExtractor(fieldExtractor);

        FlatFileItemWriter<Customer> itemWriter = new FlatFileItemWriter<>();

        itemWriter.setName("delegateCustomerItemWriter");
        itemWriter.setLineAggregator(lineAggregator);
        itemWriter.setAppendAllowed(true);
        itemWriter.setFooterCallback(footerCallback);

        return itemWriter;
}
```

Writing to multiple files based on the number of records per file is made easy using MultiResourceItemWriter. Spring's ability to add a header and/or footer record is also managed in a simple and practical way using the appropriate interfaces and configuration. The next section looks at how to write the same item to multiple writers with the addition of no code.

CompositeItemWriter

Although it may not seem like it, the examples you've reviewed in this chapter up to this point have been simple. A step writes to a single output location. That location may be a database, a file, an e-mail, and so on, but they each have written to one endpoint. However, it's not always that simple. An enterprise may need to write to a database that a web application uses as well as a data warehouse. While items are being processed, various business metrics may need to be recorded. Spring Batch allows you to write to multiple places as you process each item of a step. This section looks at how the CompositeItemWriter lets a step write items to multiple ItemWriters.

Like most things in Spring Batch, the ability to call multiple ItemWriters for each item you process is quite easy. Before you get into the code, however, let's look at the flow of writing to multiple ItemWriters with the same item. Figure 9-11 shows a sequence diagram of the process.

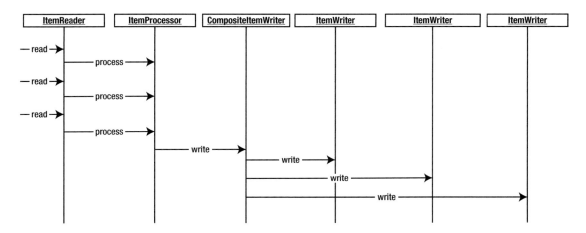

Figure 9-11. *Sequence diagram of writing to multiple ItemWriters*

As Figure 9-11 shows, reading in occurs one at a time, as does processing. However, the diagram also shows that writing occurs in chunks as you would expect, calling each ItemWriter with the items in the current chunk in the order they're configured.

To look at how this feature works, you create a job that reads in the customerWithEmail.csv file from earlier in the chapter. Let's start with the input. Listing 9-83 shows the configuration to read in the customerWithEmail.csv file.

Listing 9-83. Reading in the customerWithEmail.csv File

```
...
@Bean
@StepScope
public FlatFileItemReader<Customer> compositewriterItemReader(
                @Value("#{jobParameters['customerFile']}")Resource inputFile) {

        return new FlatFileItemReaderBuilder<Customer>()
                        .name("compositewriterItemReader")
                        .resource(inputFile)
                        .delimited()
                        .names(new String[] {"firstName",
```

319

```
                                        "middleInitial",
                                        "lastName",
                                        "address",
                                        "city",
                                        "state",
                                        "zip",
                                        "email"})
                        .targetType(Customer.class)
                        .build();
}
...
```

Nothing in Listing 9-83 should be unfamiliar. You're using the same input file you used in the previous examples in this chapter. The configuration consists of the configuration of the FlatFileItemReader using DelimitedLineTokenizer (via the call to .delimited()) and BeanWrapperFieldSetMapper (via the call to .targetType(Customer.class)) to read the file.

On the output side, you need to create three ItemWriters: the XML writer and its dependencies, the JDBC writer and its dependencies, and the CompositeItemWriter that wraps both of the other writers. Listing 9-84 shows the configuration for the output of this step as well as the configuration for the step and job.

Listing 9-84. Output, Step, and Job Configuration

```
...

@Bean
@StepScope
public StaxEventItemWriter<Customer> xmlDelegateItemWriter(
                @Value("#{jobParameters['outputFile']}") Resource outputFile) throws
                Exception {

        Map<String, Class> aliases = new HashMap<>();
        aliases.put("customer", Customer.class);

        XStreamMarshaller marshaller = new XStreamMarshaller();

        marshaller.setAliases(aliases);

        marshaller.afterPropertiesSet();

        return new StaxEventItemWriterBuilder<Customer>()
                        .name("customerItemWriter")
                        .resource(outputFile)
                        .marshaller(marshaller)
                        .rootTagName("customers")
                        .build();
}
```

```java
@Bean
public JdbcBatchItemWriter<Customer> jdbcDelgateItemWriter(DataSource dataSource) {

        return new JdbcBatchItemWriterBuilder<Customer>()
                        .namedParametersJdbcTemplate(new NamedParameterJdbcTemplate
                        (dataSource))
                        .sql("INSERT INTO CUSTOMER (first_name, " +
                                "middle_initial, " +
                                "last_name, " +
                                "address, " +
                                "city, " +
                                "state, " +
                                "zip, " +
                                "email) " +
                                "VALUES(:firstName, " +
                                ":middleInitial, " +
                                ":lastName, " +
                                ":address, " +
                                ":city, " +
                                ":state, " +
                                ":zip, " +
                                ":email)")
                        .beanMapped()
                        .build();
}

@Bean
public CompositeItemWriter<Customer> compositeItemWriter() throws Exception {
        return new CompositeItemWriterBuilder<Customer>()
                        .delegates(Arrays.asList(xmlDelegateItemWriter(null),
                                        jdbcDelgateItemWriter(null)))
                        .build();
}

@Bean
public Step compositeWriterStep() throws Exception {
        return this.stepBuilderFactory.get("compositeWriterStep")
                        .<Customer, Customer>chunk(10)
                        .reader(compositewriterItemReader(null))
                        .writer(compositeItemWriter())
                        .build();
}

@Bean
public Job compositeWriterJob() throws Exception {
        return this.jobBuilderFactory.get("compositeWriterJob")
                        .start(compositeWriterStep())
                        .build();
}
...
```

The configuration for the ItemWriters is about what you would expect. You begin the configuration with the XML writer you're using (xmlDelegateItemWriter) as configured as in the example earlier on in the chapter. The JDBC ItemWriter is next, with the PreparedStatement configured using named parameters and a NamedParameterTemplate injected to be able to handle that. The call to .beanMapped() indicates to Spring Batch that the names of the fields in the item will be used to map them to the names in the SQL statement. Finally you get to the CompositeItemWriter definition (compositeItemWriter). For compositeItemWriter, you configure a list of ItemWriters for the wrapper to call. It's important to note that the ItemWriters are called in the order they're configured with all of the items in a chunk. So if there are ten items in a chunk, the first ItemWriter is called with all ten items followed by the next ItemWriter and so on. It is also important to note that although the execution of the writing is serial (one writer at a time), all of the writes across all of the ItemWriters occur in the same transaction. Because of that, if an item fails to be written at any point in the chunk, the entire chunk is rolled back.

When you run this job as configured via the command java -jar itemWriters-0.0.1-SNAPSHOT.jar customerFile=/input/customerWithEmail.csv outputFile=/output/xmlCustomer.xml, you can see that the output consists of all the records being written to both the database and an XML file. You would think that if the file had 100 customers in it, Spring Batch would consider this to be 200 writes. But if you look at what Spring Batch recorded in the JobRepository, it says that 100 writes were executed.

The reasoning is that Spring Batch is counting the number of items that were written. It doesn't care how many places you write the item to. If the job fails, the restart point depends on how many items you read and processed, not how many you wrote to each location (because those are rolled back anyway).

The CompositeItemWriter makes writing all the items to multiple locations easy. But sometimes you want to write some things to one place and some things to another place. The last ItemWriter you look at in this chapter is ClassifierCompositeItemWriter, which handles just that.

ClassifierCompositeItemWriter

In Chapter 7, you looked at the scenario where you had a single file that contained multiple record types. Handling the ability to map different types of lines to different parsers and mappers so that each would end up in the correct object was no trivial task. But on the writing side, Spring Batch has made life a bit easier. This section looks at how ClassifierCompositeItemWriter allows you to choose where to write items based on a predetermined criterion.

org.springframework.batch.item.support.ClassifierCompositeItemWriter is used to look at items of different types, determine what ItemWriter they should be written to, and forward them accordingly. This functionality is based on two things: ClassifierCompositeItemWriter and an implementation of the org.springframework.batch.classify.Classifier interface. Let's start by looking at the Classifier interface.

The Classifier interface, shown in Listing 9-85, consists of a single method, classify(C item). In the case of what ClassifierCompositeItemWriter uses a Classifier implementation for, the classify(C item) method accepts an item as input and returns the ItemWriter to write the item to. In essence, the Classifier implementation serves as a context, with the ItemWriters as strategy implementations.

Listing 9-85. The Classifier Interface

```
package org.springframework.batch.classify;

public interface Classifier<C, T> {

    T classify(C classifiable);
}
```

ClassifierCompositeItemWriter takes a single dependency, an implementation of the Classifier interface. From there it gets the ItemWriter required for each item as it's processed.

Unlike the regular CompositeItemWriter, which writes all items to all ItemWriters, ClassifierCompositeItemWriter ends up with a different number of items written to each ItemWriter. Let's look at an example where you write all customers who live in a state that starts with the letters *A* through *M* to a flat file and items with a state name starting with the letters *N* through *Z* to the database.

As you've probably gathered, the Classifier implementation is the key to making CompositeItemWriter work, so that is where you start. To implement this Classifier as Listing 9-86 shows, you take a Customer object as the sole parameter to the classify(C item) method. From there, you use a regular expression to determine whether it should be written to a flat file or the database and return the ItemWriter as required.

Listing 9-86. CustomerClassifier

```
...
public class CustomerClassifier implements
            Classifier<Customer, ItemWriter<? super Customer>> {

      private ItemWriter<Customer> fileItemWriter;
      private ItemWriter<Customer> jdbcItemWriter;

      public CustomerClassifier(StaxEventItemWriter<Customer> fileItemWriter,
      JdbcBatchItemWriter<Customer> jdbcItemWriter) {
            this.fileItemWriter = fileItemWriter;
            this.jdbcItemWriter = jdbcItemWriter;
      }

      @Override
      public ItemWriter<Customer> classify(Customer customer) {
            if(customer.getState().matches("^[A-M].*")) {
                  return fileItemWriter;
            } else {
                  return jdbcItemWriter;
            }
      }
}
```

With the CustomerClassifier coded, you can configure the Job and ItemWriters. You reuse the same input and individual ItemWriters you used in the CompositeItemWriter example in the previous section, leaving only ClassifierCompositeItemWriter to configure. The configuration for ClassifierCompositeItemWriter and CustomerClassifier is shown in Listing 9-87.

Listing 9-87. Configuration of the ClassifierCompositeItemWriter and Dependencies

```
      ...

@Bean
public ClassifierCompositeItemWriter<Customer> classifierCompositeItemWriter() throws
Exception {
      Classifier<Customer, ItemWriter<? super Customer>> classifier = new CustomerClassifier
      (xmlDelegate(null), jdbcDelgate(null));
```

```
            return new ClassifierCompositeItemWriterBuilder<Customer>()
                    .classifier(classifier)
                    .build();
}

@Bean
public Step classifierCompositeWriterStep() throws Exception {
    return this.stepBuilderFactory.get("classifierCompositeWriterStep")
                    .<Customer, Customer>chunk(10)
                    .reader(classifierCompositeWriterItemReader(null))
                    .writer(classifierCompositeItemWriter())
                    .build();
}

@Bean
public Job classifierCompositeWriterJob() throws Exception {
    return this.jobBuilderFactory.get("classifierCompositeWriterJob")
                    .start(classifierCompositeWriterStep())
                    .build();
}
...
```

When you build and run classifierFormatJob via the statement java -jar itemWriters-0.0.1-SNAPSHOT.jar jobs/formatJob.xml formatJob customerFile=/input/customerWithEmail.csv outputFile=/output/xmlCustomer.xml, you're met with a bit of a surprise. It doesn't work. Instead of the normal output of Spring telling you the job completed as expected, you're met with an exception, as shown in Listing 9-88.

Listing 9-88. Results of classifierFormatJob

```
2018-05-10 22:51:23.691  INFO 11102 --- [          main] o.s.b.c.l.support.SimpleJob
Launcher     : Job: [SimpleJob: [name=classifierCompositeWriterJob]] launched with the
following parameters: [{customerFile=/data/customerWithEmail.csv, outputFile=file:/
Users/mminella/Documents/IntelliJWorkspace/def-guide-spring-batch/Chapter9/target/
formattedCustomers.xml}]
2018-05-10 22:51:23.701  INFO 11102 --- [          main] o.s.batch.core.job.SimpleStep
Handler      : Executing step: [classifierCompositeWriterStep]
2018-05-10 22:51:23.900 ERROR 11102 --- [          main] o.s.batch.core.step.Abstract
Step         : Encountered an error executing step classifierCompositeWriterStep in job
classifierCompositeWriterJob

org.springframework.batch.item.WriterNotOpenException: Writer must be open before it can be
written to
    at org.springframework.batch.item.xml.StaxEventItemWriter.write(StaxEventItemWriter.
    java:761) ~[spring-batch-infrastructure-4.0.1.RELEASE.jar:4.0.1.RELEASE]
    at org.springframework.batch.item.xml.StaxEventItemWriter$$FastClassBySpringCGLIB$$d105
    dd1.invoke(<generated>) ~[spring-batch-infrastructure-4.0.1.RELEASE.jar:4.0.1.RELEASE]
    at org.springframework.cglib.proxy.MethodProxy.invoke(MethodProxy.java:204) ~[spring-
    core-5.0.5.RELEASE.jar:5.0.5.RELEASE]
    at org.springframework.aop.framework.CglibAopProxy$CglibMethodInvocation.
    invokeJoinpoint(CglibAopProxy.java:747) ~[spring-aop-5.0.5.RELEASE.jar:5.0.5.RELEASE]
    at org.springframework.aop.framework.ReflectiveMethodInvocation.proceed(ReflectiveMethodI
    nvocation.java:163) [spring-aop-5.0.5.RELEASE.jar:5.0.5.RELEASE]
```

What went wrong? All you really did was swap out the CompositeItemWriter you used in the previous section with the new ClassifierCompositeItemWriter. The issue centers around the ItemStream interface.

The ItemStream Interface

The ItemStream interface serves as the contract to be able to periodically store and restore state. Consisting of three methods, open(ExecutionContext ec), update(ExecutionContext ec), and close(), the ItemStream interface is implemented by any stateful component in Spring Batch. In cases, for example, where a file is involved in the input or output, the open(ExecutionContext ec) method opens the required file, and the close() method closes the required file. The update(ExecutionContext ec) method records the current state (number of records written, and so on) as each chunk is completed.

The reason for the difference between CompositeItemWriter and ClassifierCompositeItemWriter is that CompositeItemWriter implements the org.springframework.batch.item.ItemStream interface. In CompositeItemWriter, the open(ExecutionContext ec) method loops through the delegate ItemWriters and calls the open(ExecutionContext ec) method on each of them as required. The close() and update(ExecutionContext ec) methods work the same way. However, ClassifierCompositeItemWriter doesn't implement the ItemStream method. Because of this, the XML file is never opened or XMLEventFactory (or the underlying XML writing) created, throwing the exception shown in Listing 9-88.

How do you fix this error? Spring Batch provides the ability to register ItemStreams to be handled in a step manually. If an ItemReader or ItemWriter implements ItemStream, the methods are handled for you. If they don't (as in the case of ClassifierCompositeItemWriter), you're required to register the ItemReader or ItemWriter as a stream to be able to work with it if it maintains state. Listing 9-89 shows the updated configuration for the job, registering the xmlOutputWriter as an ItemStream.[6]

Listing 9-89. Updated Configuration Registering the Appropriate ItemStream for Processing

```
    ...
@Bean
public Step classifierCompositeWriterStep() throws Exception {
        return this.stepBuilderFactory.get("classifierCompositeWriterStep")
                        .<Customer, Customer>chunk(10)
                        .reader(classifierCompositeWriterItemReader(null))
                        .writer(classifierCompositeItemWriter())
                        .stream(xmlDelegate(null))
                        .build();
}

@Bean
public Job classifierCompositeWriterJob() throws Exception {
        return this.jobBuilderFactory.get("classifierCompositeWriterJob")
                        .start(classifierCompositeWriterStep())
                        .build();
}

    ...
```

---

[6]You only need to register the xmlDelegate as a stream. JdbcBatchItemWriter doesn't implement the ItemStream interface because it doesn't maintain any state.

If you rebuild and rerun the job with the updated configuration, you see that all the records are processed as expected.

Summary

Spring Batch's ItemWriter implementations provide a wide range of output options. From writing to a simple flat file to choosing which items get written to which ItemWriters on the fly, there aren't many scenarios that aren't covered by the components Spring Batch provides out of the box.

This chapter has covered the majority of the ItemWriters available in Spring Batch. You also looked at how to use the ItemWriters provided by the framework to complete the sample application. In the next chapter, you look at how to use the scalability features of the framework to allow the jobs to scale and perform as required.

CHAPTER 10

Sample Application

Tutorials you find on the Internet in technology can be funny. Most of them rarely extend past a "Hello, World!" level of complexity for any new concept. And although that may be great for a basic understanding of a technology, you know that life is never as simple as a tutorial makes it out to be. Because of this, in this chapter you look at a more real-world example of a Spring Batch job.

This chapter covers the following:

- *Reviewing the statement job*: Before developing any new functionality, you review the goals of the job to be developed, as outlined in Chapter 3.

- *Project setup*: You create a brand-new Spring Batch project from Spring Initializr.

- *Job development:* You walk through the entire development process for the statement job outlined in Chapter 3.

Let's get started by reviewing what the statement job you develop is required to do.

Reviewing the Statement Job

The job you develop in this chapter is for a mythical bank called Apress Banking. Apress Banking has a large number of clients that have multiple traditional bank accounts. At the end of each month, the clients receive a composite statement that lists all their accounts, all of the transactions that occurred over the past month, the total amount that was credited to their account, total amount that was debited from their account, and their current balance.

To accomplish these requirements, a four step job is used as outlined in Figure 10-1.

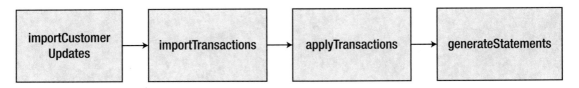

Figure 10-1. *The flow for the statement job*

© Michael T. Minella 2019
M. T. Minella, *The Definitive Guide to Spring Batch*, https://doi.org/10.1007/978-1-4842-3724-3_10

The job starts in step 1 with importing customer data. This step imports a single CSV file that contains a number of different record formats, each one for a different type of update to a customer. We apply these updates to the customer records in our database before importing transactions in the second step.

The transactions that have occurred are provided via an XML document that we'll read into the database as new records. Once all the transaction records have been successfully imported into the database, we'll need to apply them to the current balance, adding all credits and deducting all debits from the running total. Step 3 of the job is where this occurs.

The final step of the job is to generate the statement files themselves. For each customer a single file is created that has a header with the customer's address in it, and for each account the customer has (they can have more than one account), a header for the account, a list of all transactions, a total for all credits, a total for all debits, and the current balance are all printed. In this chapter, we'll implement these steps and go into detail as to why the job is designed the way it is.

While we've done this many times throughout this book, we'll begin this work by creating a new shell project from Spring Initializr in the next section.

Setting Up a New Project

To begin any Spring Boot based project, the best place to begin is `https://start.spring.io` (It's Spring Developer Advocate Josh Long's second favorite place on the Internet). Using either an IDE like Spring Tool Suite or IntelliJ IDEA, we can do this directly from our IDE. I'm an IDEA user so I'll walk you through those steps.

Begin by going to File ➤ New Project. Along the left hand side, you'll be able to select Spring Initializr. Once that is selected, you will be able to select the Project SDK and the Service URL. For this project, we'll use Java 8 since that's the default for Spring Boot 2. We'll also use the default Service URL.[1] Figure 10-2 shows the selections made.

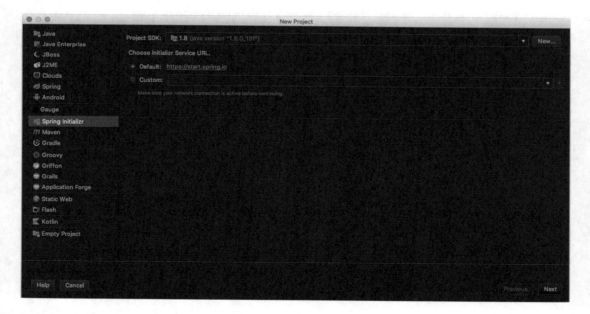

Figure 10-2. *Spring Initializr being used within IntelliJ IDEA*

---

[1]Some enterprises choose to run their own instance of Spring Initializr so that they can customize it or not have developer machines need to hit the Internet. If that is the case for you, you'll want to enter your custom URL in your IDE.

With those selections made we can click Next and enter our project metadata. We've used Apache Maven throughout this book, so we won't change that here. We can fill in the group id, artifact id, select Maven Project (this will give us the full shell of a Maven based project), and Java for the language. We'll select JAR for the packaging so that the POM that is generated is configured with the Spring Boot plug-in to generate an über jar for us. We'll select 8 for the Java version. Why do we need to select the Java version twice? The first one was what your IDE will use to build and run the project in. The second one is what the Maven POM will be configured to compile to. We finish up by configuring a version, name, description, and default package. In our case, the values for all of these are as follows:

- Group ID: com.apress.batch

- Artifact ID: chapter10

- Version: 0.0.1-SNAPSHOT

- Name: Statement Batch Job

- Description: Apress Banking statement generation batch job

- Default Package: com.apress.batch.chapter10

When we click Next, we're taken to where we can select our dependencies. These dependencies are Spring Boot starters that we'll select to be included within our project. For this project, we'll need to select Batch, JDBC, and HSQLDB. Each of these can be added by simply typing them in the search box at the top of the window and pressing enter. We'll have a few other dependencies we need to add that don't have Spring Boot starters for them once our project is loaded. Figure 10-3 illustrates our selections made.

Figure 10-3. *Selecting dependencies for our project in IDEA*

The final screen asks us to name our project and select the directory we want the project to be downloaded into. Enter the values that work best for your environment and click Finish. A new IDEA window will open with your project configured.

The project should look like any other Maven project. Spring Initializr provides a basic class in the root of the default package with the required main method for bootstrapping Spring Boot. It also provides a single test that does nothing but boostraps the ApplicationContext. Since Spring Boot launches any Spring Batch jobs it finds by default, that test isn't all that useful, so we can delete it.

And we're set. We can run a Maven build from the command line in the root of our project using the command ./mvnw clean install and it should build successfully. With that, we can begin to build our batch job. We need to add the @EnableBatchProcessing annotation to our main class. Listing 10-1 shows our main class updated with the annotation applied.

Listing 10-1. Chapter10Application

```
...
@EnableBatchProcessing
@SpringBootApplication
public class Chapter10Application {

        public static void main(String[] args) {
                SpringApplication.run(Chapter10Application.class, realArgs);
        }
}
```

With our main class configured, we can begin working on the job and first step. As our diagram illustrated in Figure 10-1, updating the customer information is the first step. The next section will walk through how we accomplish the functionality required for that step.

Importing Customer Updates

The first step in our job, as mentioned earlier is to import customer updates. We receive a CSV file that contains three different record formats. This section will look into how to parse those records and apply the related updates to date within our database.

Before we dig into the batch code, however, let's review the data model for this job. The data model for this job is pretty simple compared to most enterprise data models, but it contains what we'd expect. The model begins with the CUSTOMER table. This table contains name, address, and various contact information including email address and different types of phone numbers. It also has a field for the user to indicate their preference for what type of communication should be used to contact them. The ACCOUNT table is next. The CUSTOMER table has a many to many relationship with the ACCOUNT table (a Customer can have many Accounts and an Account can have many Customers). The ACCOUNT table is very simple, containing only an id, the current balance, and the date of the last statement that was issued. The last table with business data in it is the TRANSACTION table. This table contains the details of each transaction that has occurred within an account. As you'd expect, the ACCOUNT table has a one to many relationships with the TRANSACTION table. With all of the business tables defined, the fourth and final table in our data model is the CUSTOMER_ACCOUNT table which serves as a join table between the CUSTOMER table and the ACCOUNT table. Figure 10-4 illustrates the data model for this job.

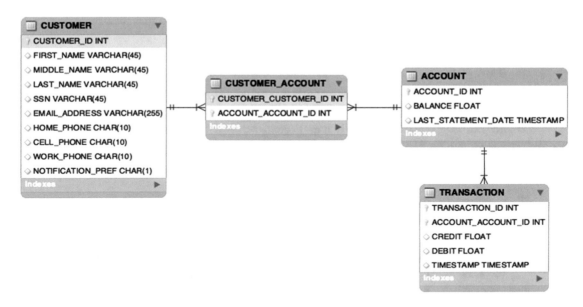

Figure 10-4. *Data model for the statement job*

In order to import the data from the customer file, we need to understand the file format we'll receive it in. The file consists of three record formats: one for updates to customer name fields, one for updates to customer address related fields, and one for updates to customer communication mechanisms. Listing 10-2 shows a sample of the file format.

Listing 10-2. Sample of the `customer_update.csv` File

```
2,2,,,Montgomery,Alabama,36134
2,2,,,Montgomery,Alabama,36134
3,441,,,,316-510-9138,2
3,174,trothchild3o@pinterest.com,,785-790-7373,467-631-6632,5
2,287,,,Rochester,New York,14646
2,287,,,Rochester,New York,14646
1,168,Rozelle,Heda,Farnill
2,204,2 Warner Junction,,Akron,Ohio,44305
```

Each record type is indicated via the first field in the record. We'll be able to use that to our advantage when we parse the file. For now, let's look at each record type. Record type 1 (which begins with the number 1) is to update the customer record fields. This record type has five fields:

1. *Record type:* This will always be 1 for a record type 1.

2. *Customer ID:* The id of the customer record to be updated.

3. *First name:* The first name the customer record should be updated with. If blank, no update to first name should be done.

4. *Middle Name:* The middle name the customer record should be updated with. If blank, no update to the middle name should be done.

5. *Last Name:* The last name of the customer record should be updated with. If blank, no update to the last name should be done.

It's important to note that it is expected that the customer exist in the database prior to the update record coming through. We'll validate that in this import step.

The next record format is record type 2. Record type 2 has seven fields:

1. *Record type:* This will always be 2 for a record type 2.

2. *Customer ID:* The id of the customer record to be updated.

3. *Address 1:* The first line of the address to be updated. If blank, no update should be executed.

4. *Address 2:* The second (optional) line of the address to be updated. If blank, no update should be executed.

5. *City:* The city of the customer. If blank, no update should be executed.

6. *State:* The state of the customer. If blank, no update should be executed.

7. *Postal Code:* The postal code of the customer. If blank, no update should be executed.

The final record format in the customer update file is record type 3. This record type is used to update customer contact information. The record has seven fields:

1. *Record type:* This will always be 3 for a record type 3.

2. *Customer ID:* The id of the customer record to be updated.

3. *Email address:* The email address to update the customer record to. If blank, no update should be executed.

4. *Home phone:* The home phone number to update the customer record with. If blank, no update should be executed.

5. *Cell phone:* The cell phone number to update the customer record with. If blank, no update should be executed.

6. *Work phone:* The work phone number to update the customer record with. If blank, no update should be executed.

7. *Notification preference:* The indicator of what mechanism to use to contact a customer. If blank, no update should be executed.

To process this file, we'll begin by defining the job and its first step. The configuration for our job will live in the class `com.apress.batch.chapter10.configuration.ImportJobConfiguration`. Listing 10-3 illustrates the configuration of the job and first step.

Listing 10-3. Definition of Import Job

```
...
@Configuration
public class ImportJobConfiguration {

        @Autowired
        private JobBuilderFactory jobBuilderFactory;

        @Autowired
        private StepBuilderFactory stepBuilderFactory;

        @Bean
        public Job job() throws Exception {
                return this.jobBuilderFactory.get("importJob")
                                .start(importCustomerUpdates())
                                .build();
        }

        @Bean
        public Step importCustomerUpdates() throws Exception {
                return this.stepBuilderFactory.get("importCustomerUpdates")
                                .<CustomerUpdate, CustomerUpdate>chunk(100)
                                .reader(customerUpdateItemReader(null))
                                .processor(customerValidatingItemProcessor(null))
                                .writer(customerUpdateItemWriter())
                                .build();
        }
...
```

This listing is simple enough. We begin by defining a Spring Configuration class (via the @Configuration) annotation. From there, we can autowire in the JobBuilderFactory and StepBuilderFactory provided by @EnableBatchProcessing (which lives on our main class). With the builders wired in, we can define our Job and Step. The Job is defined using the jobBuilderFactory, starting with the step importCustomerUpdates, and calling build() to construct the Job.

The importCustomerUpdates Step is defined using the StepBuilderFactor to get a StepBuilder which is configured for chunk-based processing. Each chunk consists of 100 items. This step will use an ItemReader called customerUpdateItemReader, an ItemProcessor called customerValidatingItemProcessor, and an ItemWriter called customerUpdateItemWriter. All of which will read, process, and write CustomerUpdate objects. That all sounds nice, but what do those ItemReader, ItemProcessor, and ItemWriter look like? Let's start with the ItemReader.

The ItemReader in this step is a FlatFileItemReader. Configuring it is actually very straightforward. We use the FlatFileItemReaderBuilder to configure a name (to support restartability), a Resource (the file we'll be reading from), a LineTokenizer that knows how to parse the records in the file, and a FieldSetMapper that knows how to map the parsed tokens to a domain object. Listing 10-4 lists the code for this ItemReader.

Listing 10-4. Reading the Customer Update File

```
...
@Bean
@StepScope
public FlatFileItemReader<CustomerUpdate> customerUpdateItemReader(
@Value("#{jobParameters['customerUpdateFile']}") Resource inputFile) throws Exception {

        return new FlatFileItemReaderBuilder<CustomerUpdate>()
                        .name("customerUpdateItemReader")
                        .resource(inputFile)
                        .lineTokenizer(customerUpdatesLineTokenizer())
                        .fieldSetMapper(customerUpdateFieldSetMapper())
                        .build();
}
...
```

You'll notice that the customerUpdateItemReader is step scoped and we're using a job parameter to specify where the file is to be read.

While we have the ItemReader defined, we need to define the LineTokenizer and FieldSetMapper. We've already discussed that there are three record formats in this file, so how do we accomplish the tokenizing in a single LineTokenizer? Simple, we don't. We use composition to create a composite that delegates to the correct LineTokenizer based on a pattern for each file. Spring Batch provides the PatternMatchingCompositeLineTokenizer for this very use case. It requires that you create a Map<String, LineTokenizer>. The String in each Map entry defines a pattern the record must match in order to use that LineTokenzier. So what we really will do is define three LineTokenizer implementations and define a pattern to identify when to use each. Listing 10-5 illustrates that configuration.

Listing 10-5. LineTokenizer Configurations for the Customer Update File

```
...
@Bean
public LineTokenizer customerUpdatesLineTokenizer() throws Exception {
        DelimitedLineTokenizer recordType1 = new DelimitedLineTokenizer();

        recordType1.setNames("recordId", "customerId", "firstName",
        "middleName", "lastName");

        recordType1.afterPropertiesSet();

        DelimitedLineTokenizer recordType2 = new DelimitedLineTokenizer();

        recordType2.setNames("recordId", "customerId", "address1",
        "address2", "city", "state", "postalCode");

        recordType2.afterPropertiesSet();

        DelimitedLineTokenizer recordType3 = new DelimitedLineTokenizer();

        recordType3.setNames("recordId", "customerId", "emailAddress",
                "homePhone", "cellPhone", "workPhone", "notificationPreference");
```

```
        recordType3.afterPropertiesSet();

        Map<String, LineTokenizer> tokenizers = new HashMap<>(3);
        tokenizers.put("1*", recordType1);
        tokenizers.put("2*", recordType2);
        tokenizers.put("3*", recordType3);

        PatternMatchingCompositeLineTokenizer lineTokenizer =
                    new PatternMatchingCompositeLineTokenizer();

        lineTokenizer.setTokenizers(tokenizers);

        return lineTokenizer;
    }
...
```

As you can see in Listing 10-5, we configure three DelimitedLineTokenizer instances, each defining the fields in each record type. We then map each one to the pattern for the prefix identifying each record type. The PatternMatchingCompositeLineTokenizer will take care of delegating to the correct LineTokenizer based on those patterns.

The last piece to reading this file is mapping them to a domain object. Now we could go with an über domain object that has all the fields for each of the three record types in it and have a simple FieldSetMapper that handles populating the fields it has. However, that's a rather messy way to deal with the data. So we have three distinct domain objects, one for each record type. Record type 1 uses a CustomerNameUpdate object, record type 2 uses a CustomerAddressUpdate object, and record type 3 uses a CustomerContactUpdate object. Each of these extends a common class, CustomerUpdate. CustomerUpdate contains the customerId field. This super class serves two purposes. The first is a common place to hold the customerId. However, the more important purpose is to allow us to use CustomerUpdate in the generics for Step configuration (take another look at Listing 10-3 to see it in action). Listing 10-6 shows the domain objects we're using for this step.

Listing 10-6. Domain Objects for the Customer Update Step

```
...
public class CustomerUpdate {
        protected final long customerId;

        public CustomerUpdate(long customerId) {
                this.customerId = customerId;
        }
// accessors removed
}

...
public class CustomerNameUpdate extends CustomerUpdate {

        private final String firstName;

        private final String middleName;

        private final String lastName;
```

```java
        public CustomerNameUpdate(long customerId, String firstName,
                        String middleName, String lastName) {

                super(customerId);
                this.firstName = StringUtils.hasText(firstName) ? firstName : null;
                this.middleName = StringUtils.hasText(middleName) ? middleName : null;
                this.lastName = StringUtils.hasText(lastName) ? lastName : null;
        }
// accessors removed
}


...
public class CustomerAddressUpdate extends CustomerUpdate {

        private final String address1;

        private final String address2;

        private final String city;

        private final String state;

        private final String postalCode;

        public CustomerAddressUpdate(long customerId, String address1,
                        String address2, String city, String state, String postalCode) {

                super(customerId);
                this.address1 = StringUtils.hasText(address1) ? address1 : null;
                this.address2 = StringUtils.hasText(address2) ? address2 : null;
                this.city = StringUtils.hasText(city) ? city : null;
                this.state = StringUtils.hasText(state) ? state : null;
                this.postalCode = StringUtils.hasText(postalCode) ? postalCode : null;
        }
// accessors removed
}


...

public class CustomerContactUpdate extends CustomerUpdate {

        private final String emailAddress;

        private final String homePhone;

        private final String cellPhone;

        private final String workPhone;

        private final Integer notificationPreferences;
```

```
        public CustomerContactUpdate(long customerId, String emailAddress, String homePhone,
        String cellPhone, String workPhone, Integer notificationPreferences) {
                super(customerId);
                this.emailAddress = StringUtils.hasText(emailAddress) ? emailAddress : null;
                this.homePhone = StringUtils.hasText(homePhone) ? homePhone : null;
                this.cellPhone = StringUtils.hasText(cellPhone) ? cellPhone : null;
                this.workPhone = StringUtils.hasText(workPhone) ? workPhone : null;
                this.notificationPreferences = notificationPreferences;
        }
// accessors removed
}
```

In order to determine which object to return, we need a FieldSetMapper that will create and return the right domain object based on the record type. Listing 10-7 shows how we can keep things simple and use a lambda expression to create a FieldSetMapper that handles this mapping.

Listing 10-7. FieldSetMapper Configurations for the Customer Update File

```
...
@Bean
public FieldSetMapper<CustomerUpdate> customerUpdateFieldSetMapper() {
        return fieldSet -> {
                switch (fieldSet.readInt("recordId")) {
                        case 1: return new CustomerNameUpdate(
                                        fieldSet.readLong("customerId"),
                                        fieldSet.readString("firstName"),
                                        fieldSet.readString("middleName"),
                                        fieldSet.readString("lastName"));
                        case 2: return new CustomerAddressUpdate(
                                        fieldSet.readLong("customerId"),
                                        fieldSet.readString("address1"),
                                        fieldSet.readString("address2"),
                                        fieldSet.readString("city"),
                                        fieldSet.readString("state"),
                                        fieldSet.readString("postalCode"));
                        case 3:
                                String rawPreference =
                                        fieldSet.readString("notificationPreference");

                                Integer notificationPreference = null;

                                if(StringUtils.hasText(rawPreference)) {
                                        notificationPreference = Integer.
                                                        parseInt(rawPreference);
                                }

                                return new CustomerContactUpdate(fieldSet.
                                readLong("customerId"),
                                        fieldSet.readString("emailAddress"),
                                        fieldSet.readString("homePhone"),
                                        fieldSet.readString("cellPhone"),
                                        fieldSet.readString("workPhone"),
                                                notificationPreference);
```

```
                            default: throw new IllegalArgumentException(
                                    "Invalid record type was found:" +
                                            fieldSet.readInt("recordId"));
                }
        };
}
...
```

The lambda illustrated in Listing 10-7 looks at the record type field in each record and creates a new instance of the appropriate domain object for each record. If no record can be found, an exception is thrown indicating that the record type was invalid.

Reading is only the first part of the process of applying the customer updates. The goal of this step is to get the data into your database. To do that, we need to first validate that each record actually has a valid customer id. This next section will look at how to do that with Spring Batch.

Validating Customer ID

In our step definition we defined an `ItemProcessor` called `customerValidatingItemProcessor`. The intent of this component will be to look up the customer id in the `CustomerUpdate` object it receives. If it exists in the database, we'll let the record through. If it does not exist, we'll filter those records out. In a real-world scenario we may want to write these items to a new file for future debugging; for the case of this job, filtering them out is good enough.

To do this, we can use Spring Batch's `ValidatingItemProcessor`. This `ItemProcessor` takes an implementation of a `org.springframework.batch.item.validator.Validator` (this is different from the Spring Framework `Validator` interface). In our case, we'll create a custom `Validator` implementation that looks up the customer id and throws a `ValidationException` if none are found. Listing 10-8 shows the code for our `CustomerItemValidator`.

Listing 10-8. `CustomerItemValidator`

```
...
@Component
public class CustomerItemValidator implements Validator<CustomerUpdate> {

        private final NamedParameterJdbcTemplate jdbcTemplate;

        private static final String FIND_CUSTOMER =
                "SELECT COUNT(*) FROM CUSTOMER WHERE customer_id = :id";

        public CustomerItemValidator(DataSource dataSource) {
                this.jdbcTemplate = new NamedParameterJdbcTemplate(dataSource);
        }

        @Override
        public void validate(CustomerUpdate customer) throws ValidationException {
                Map<String, Long> parameterMap =
                        Collections.singletonMap("id", customer.getCustomerId());

                Long count = jdbcTemplate.queryForObject(FIND_CUSTOMER, parameterMap,
                                                Long.class);
```

```
                if(count == 0) {
                        throw new ValidationException(
                                String.format("Customer id %s was not able to be found",
                                        customer.getCustomerId() ));
                }
        }
}
```

With our `Validator` defined, we can configure our `ItemProcessor`. Listing 10-9 provides the configuration for the `ValidatingItemProcessor`.

Listing 10-9. `customerValidatingItemProcessor`

```
...
@Bean
public ValidatingItemProcessor<CustomerUpdate> customerValidatingItemProcessor(CustomerItem
Validator validator) {

        ValidatingItemProcessor<CustomerUpdate> customerValidatingItemProcessor =
                        new ValidatingItemProcessor<>(validator);

        customerValidatingItemProcessor.setFilter(true);

        return customerValidatingItemProcessor;
}
...
```

With the `ItemProcessor` configured, all that is left is to configure the `ItemWriter` side of this step. However, if we have three items to write, how do we manage that on the write side of the step? The next section will explain how.

Writing Customer Updates

The final piece for our first step is the applying of the updates to the CUSTOMER table. Since we have three record types, meaning three different update types, we'll want three different `ItemWriters` to be able to delegate between. We can use Spring Batch's `ClassifierCompositeItemWriter` to delegate between our `ItemWriter` implementations based on a `Classifier` we implement.

We'll begin by looking at the three `ItemWriter` implementations we'll need to configure. They are all the same except for the SQL they are using for the update. Using the `JdbcBatchItemWriterBuilder`, we can configure our SQL, the `DataSource`, and tell Spring to map our statement's parameters using bean names with just a few lines of code. Listing 10-10 shows the configuration for each of the three `JdbcBatchItemWriters` we'll configure.

Listing 10-10. `customerValidatingItemProcessor`

```
...
@Bean
public JdbcBatchItemWriter<CustomerUpdate> customerNameUpdateItemWriter(DataSource
dataSource) {
        return new JdbcBatchItemWriterBuilder<CustomerUpdate>()
                        .beanMapped()
```

```
                      .sql("UPDATE CUSTOMER " +
                              "SET FIRST_NAME = COALESCE(:firstName, FIRST_NAME), " +
                              "MIDDLE_NAME = COALESCE(:middleName, MIDDLE_NAME), " +
                              "LAST_NAME = COALESCE(:lastName, LAST_NAME) " +
                              "WHERE CUSTOMER_ID = :customerId")
                      .dataSource(dataSource)
                      .build();
}

@Bean
public JdbcBatchItemWriter<CustomerUpdate> customerAddressUpdateItemWriter(DataSource
dataSource) {
        return new JdbcBatchItemWriterBuilder<CustomerUpdate>()
                      .beanMapped()
                      .sql("UPDATE CUSTOMER SET " +
                              "ADDRESS1 = COALESCE(:address1, ADDRESS1), " +
                              "ADDRESS2 = COALESCE(:address2, ADDRESS2), " +
                              "CITY = COALESCE(:city, CITY), " +
                              "STATE = COALESCE(:state, STATE), " +
                              "POSTAL_CODE = COALESCE(:postalCode, POSTAL_CODE) " +
                              "WHERE CUSTOMER_ID = :customerId")
                      .dataSource(dataSource)
                      .build();
}

@Bean
public JdbcBatchItemWriter<CustomerUpdate> customerContactUpdateItemWriter(DataSource
dataSource) {
        return new JdbcBatchItemWriterBuilder<CustomerUpdate>()
                      .beanMapped()
                      .sql("UPDATE CUSTOMER SET " +
                          "EMAIL_ADDRESS = COALESCE(:emailAddress, EMAIL_ADDRESS), " +
                          "HOME_PHONE = COALESCE(:homePhone, HOME_PHONE), " +
                          "CELL_PHONE = COALESCE(:cellPhone, CELL_PHONE), " +
                          "WORK_PHONE = COALESCE(:workPhone, WORK_PHONE), " +
                          "NOTIFICATION_PREF = COALESCE(:notificationPreferences,
                          NOTIFICATION_PREF) " +
                              "WHERE CUSTOMER_ID = :customerId")
                      .dataSource(dataSource)
                      .build();
}
...
```

Each of the ItemWriter configurations in Listing 10-10 do the same thing, just setting different columns with the appropriate values. The reason we use COALESCE for each of the values in the SQL statement is because we only want to update the values that the input file provided. If the input file provided null as the value, we shouldn't update it.

Now that we have those three ItemWriters configured, we need to be able to choose the correct one based on the type of item that we receive (since the type of item is based on the record type from the input file). To make this choice, we'll implement a org.springframework.classify.Classifier that evaluates the item it is given and return the appropriate ItemWriter. Listing 10-11 illustrates the simple Classifier implementation we'll use.

Listing 10-11. CustomerUpdateClassifier

```
...
public class CustomerUpdateClassifier implements
Classifier<CustomerUpdate, ItemWriter<? super CustomerUpdate>> {

        private final JdbcBatchItemWriter<CustomerUpdate> recordType1ItemWriter;
        private final JdbcBatchItemWriter<CustomerUpdate> recordType2ItemWriter;
        private final JdbcBatchItemWriter<CustomerUpdate> recordType3ItemWriter;

        public CustomerUpdateClassifier(
                JdbcBatchItemWriter<CustomerUpdate> recordType1ItemWriter,
                JdbcBatchItemWriter<CustomerUpdate> recordType2ItemWriter,
                JdbcBatchItemWriter<CustomerUpdate> recordType3ItemWriter) {

                this.recordType1ItemWriter = recordType1ItemWriter;
                this.recordType2ItemWriter = recordType2ItemWriter;
                this.recordType3ItemWriter = recordType3ItemWriter;
        }

        @Override
        public ItemWriter<? super CustomerUpdate> classify(CustomerUpdate classifiable) {

                if(classifiable instanceof CustomerNameUpdate) {
                        return recordType1ItemWriter;
                }
                else if(classifiable instanceof CustomerAddressUpdate) {
                        return recordType2ItemWriter;
                }
                else if(classifiable instanceof CustomerContactUpdate) {
                        return recordType3ItemWriter;
                }
                else {
                        throw new IllegalArgumentException("Invalid type: " +
                                classifiable.getClass().getCanonicalName());
                }
        }
}
```

As you can see, the Classifier takes in each of the ItemWriter instances as constructor parameters. Then, based on the type of item passed to it, the correct ItemWriter is returned. The final piece of the first step in our statement job is to configure the ClassifierCompositeItemWriter. This ItemWriter is pretty simple to configure since all the work is done in the Classifier and the delegate ItemWriter instances. Listing 10-12 shows how to configure our customerUpdateItemWriter.

Listing 10-12. `customerUpdateItemWriter`

```
...
@Bean
public ClassifierCompositeItemWriter<CustomerUpdate> customerUpdateItemWriter() {

        CustomerUpdateClassifier classifier =
                        new CustomerUpdateClassifier(customerNameUpdateItemWriter(null),
                                    customerAddressUpdateItemWriter(null),
                                    customerContactUpdateItemWriter(null));

        ClassifierCompositeItemWriter<CustomerUpdate> compositeItemWriter =
                        new ClassifierCompositeItemWriter<>();

        compositeItemWriter.setClassifier(classifier);

        return compositeItemWriter;
}
...
```

With all of the components for the first Step written and configured, you can run your batch job to see the first step run. You'll want to configure a "real database" using the Spring Boot properties in Listing 10-13 to be able to view the results.

Listing 10-13. application.properties

```
spring.datasource.driverClassName=com.mysql.jdbc.Driver
spring.datasource.url=jdbc:mysql://localhost:3306/statement
spring.datasource.username=<USERNAME>
spring.datasource.password=<PASSWORD>
spring.datasource.schema=schema-mysql.sql
spring.datasource.initialization-mode=always
spring.batch.initialize-schema=always
```

The last piece you'll need to configure to run your job and test the first step is add the driver for MySQL (this example is using MySQL. Replace configuration values and driver as needed for other database options). Listing 10-14 has the Maven dependency for MySQL (the version is provided by Spring Boot).

Listing 10-14. MySQL Dependency

```
<dependency>
        <groupId>mysql</groupId>
        <artifactId>mysql-connector-java</artifactId>
</dependency>
```

With all of those values configured, we can build our project from the command line via `./mvnw clean install`. Once that is complete, from the target directory of our project, we can run it via the command `java -jar chapter10-0.0.1-SNAPSHOT.jar customerUpdateFile=<PATH_TO_CUSTOMER_FILE>`. With that, you should be able to validate the data has been correctly applied to the CUSTOMER table.

In the next section, we'll work on the second step of the job, importing the transaction file.

Importing Transactions

With the customers updated, we can now import the transactions from the transaction file. While the customer input file was complex since it had multiple record types to deal with, the transaction file is actually very simple. It is a simple XML file that we'll be importing directly into the transaction table in our database. Listing 10-15 shows an example of the transaction file we'll be importing.

Listing 10-15. Transaction File

```xml
<?xml version='1.0' encoding='UTF-8'?>
<transactions>
        <transaction>
                <transactionId>2462744</transactionId>
                <accountId>405</accountId>
                <description>Skinix</description>
                <credit/>
                <debit>-438</debit>
                <timestamp>2018-06-01 19:39:53</timestamp>
        </transaction>
        <transaction>
                <transactionId>4243424</transactionId>
                <accountId>584</accountId>
                <description>Yakidoo</description>
                <credit>8681.98</credit>
                <debit/>
                <timestamp>2018-06-12 18:39:09</timestamp>
        </transaction>
...
</transactions>
```

The transaction file begins with a `transactions` element that wraps all of the individual `transaction` elements. Each `transaction` chunk represents a single bank transaction and will result in an item in our batch job. These blocks map to a `Transaction` domain object that is listed in Listing 10-16.

Listing 10-16. Transaction Domain Object

```java
...
@XmlRootElement(name = "transaction")
public class Transaction {

        private long transactionId;

        private long accountId;

        private String description;

        private BigDecimal credit;

        private BigDecimal debit;

        private Date timestamp;
```

```java
    public Transaction() {
    }

    public Transaction(long transactionId,
                       long accountId,
                       String description,
                       BigDecimal credit,
                       BigDecimal debit,
                       Date timestamp) {

            this.transactionId = transactionId;
            this.accountId = accountId;
            this.description = description;
            this.credit = credit;
            this.debit = debit;
            this.timestamp = timestamp;
    }

    // accessors removed for brevity

    @XmlJavaTypeAdapter(JaxbDateSerializer.class)
    public void setTimestamp(Date timestamp) {
            this.timestamp = timestamp;
    }

    public BigDecimal getTransactionAmount() {
            if(credit != null) {
                    if(debit != null) {
                            return credit.add(debit);
                    }
                    else {
                            return credit;
                    }
            }
            else if(debit != null) {
                    return debit;
            }
            else {
                    return new BigDecimal(0);
            }
    }
}
```

As you can see, the Transaction domain object has fields that map directly to the XML chunks in the input file. The three notable items on this class are the @XmlRootElement annotation at the class level, the @XmlJavaTypeAdapter on the setter for the timestamp field, and the additional method getTransactionAmount(). The @XmlRootElement is a JAX-B annotation that defines what the root tag is for that domain object. In our case, it is the tag transaction. The @XmlJavaTypeAdapter is used on the setter for the timestamp field because JAX-B doesn't have a nice and simple way to handle the conversion from a String to a java.util.Date. Because of that, we need to provide a bit of code that JAX-B will use to do that conversion which is the JaxbDateSerializer. Listing 10-17 illustrates the JaxbDateSerializer.

Listing 10-17. JaxbDateSerializer

```
...
public class JaxbDateSerializer extends XmlAdapter<String, Date> {

        private SimpleDateFormat dateFormat = new SimpleDateFormat("yyyy-MM-dd hh:mm:ss");

        @Override
        public String marshal(Date date) throws Exception {
                return dateFormat.format(date);
        }

        @Override
        public Date unmarshal(String date) throws Exception {
                return dateFormat.parse(date);
        }
}
```

The JaxbDateSerializer extends XmlAdapter and is used by JAX-B for the type conversion from a String to a java.util.Date in our case. The final addition to the Transaction class is the getTransactionAmount() method. A Transaction contains either a credit or a debit. However, when we need to do math, we don't care about whether the value is a credit or debit, we just care how much the value is. So this method returns the actual value for the transaction.

Once we have the domain object defined, we can configure our second step, importTransactions, and begin to look at its components. We'll do that in the next section.

Reading Transactions

Let's start the importing of our bank transactions by configuring the Step and adding it to our Job. Listing 10-18 has the configuration of our second Step, importTransactions.

Listing 10-18. importTransactions

```
...
@Bean
public Job job() throws Exception {
        return this.jobBuilderFactory.get("importJob")
                        .start(importCustomerUpdates())
                        .next(importTransactions())
                        .build();
}

@Bean
public Step importTransactions() {
        return this.stepBuilderFactory.get("importTransactions")
                        .<Transaction, Transaction>chunk(100)
                        .reader(transactionItemReader(null))
                        .writer(transactionItemWriter(null))
                        .build();
}
...
```

The importTransactions step is a simple one. We define a reader named transactionItemReader and a writer named transactionItemWriter. Let's take a look at the configuration for the ItemReader. Since we'll be reading XML, we'll be using the StaxEventItemReader. As for how we plan on unmarshalling the XML, the JAX-B annotations on our domain object should have been a bit of a giveaway, but we'll be using JAX-B to handle that aspect. By taking this approach, the configuration of our reader becomes very simple. Listing 10-19 shows the 16 lines required to configure it.

Listing 10-19. transactionItemReader

```
...
@Bean
@StepScope
public StaxEventItemReader<Transaction> transactionItemReader(
        @Value("#{jobParameters['transactionFile']}") Resource transactionFile) {

    Jaxb2Marshaller unmarshaller = new Jaxb2Marshaller();
    unmarshaller.setClassesToBeBound(Transaction.class);

    return new StaxEventItemReaderBuilder<Transaction>()
                    .name("fooReader")
                    .resource(transactionFile)
                    .addFragmentRootElements("transaction")
                    .unmarshaller(unmarshaller)
                    .build();
}
...
```

The step scoped transactionItemReader takes the location of the input file as a job parameter named transactionFile. In the method, we create a new Jaxb2Marshaller and bind it to the Transaction domain object we defined in Listing 10-16. Finally, we use the StaxEventItemReaderBuilder to configure our ItemReader. We pass it a name (for restartability), the resource injected via the job parameter, we define the root element for each XML snip to be parsed (transaction), and we pass the StaxEventItemReaderBuilder the Jaxb2Marshaller to use when parsing the XML. Calling build gives us our StaxEventItemReader.

Once we have the ItemReader, we need an ItemWriter. This also will be familiar since it's the same type of configuration from the previous step (just simplified). The next section will review how we configure the transactionItemWriter.

Writing Transactions

The transactionItemWriter is going to be responsible for writing the transactions to the transaction table in the database. To accomplish this, we'll again us the JdbcBatchItemWriter. It's configuration is found in Listing 10-20.

Listing 10-20. transactionItemWriter

```
...
@Bean
public JdbcBatchItemWriter<Transaction> transactionItemWriter(DataSource dataSource) {
    return new JdbcBatchItemWriterBuilder<Transaction>()
                    .dataSource(dataSource)
```

```
                        .sql("INSERT INTO TRANSACTION (TRANSACTION_ID, " +
                                    "ACCOUNT_ACCOUNT_ID, " +
                                    "DESCRIPTION, " +
                                    "CREDIT, " +
                                    "DEBIT, " +
                                    "TIMESTAMP) VALUES (:transactionId, " +
                                    ":accountId, " +
                                    ":description, " +
                                    ":credit, " +
                                    ":debit, " +
                                    ":timestamp)")
                .beanMapped()
                .build();
}
...
```

The `JdbcBatchItemWriterBuilder` is used to configure our `JdbcBatchItemWriter` by taking a DataSource, SQL statement, and telling the `ItemWriter` to use the item's property names as keys to set the SQL statement's values.

That is all we need for our second step. With the `ItemReader` and `ItemWriter` configured, we can build our job via `./mvnw clean install` and run it with the same command as last time plus our new input file parameter: `java -jar chapter10-0.0.1-SNAPSHOT.jar customerUpdateFile=<PATH_TO_CUSTOMER_FILE> transactionFile=<PATH_TO_TRANSACTION_FILE>`. After our job's successful run, we can validate that the values in the transaction XML file got into the transaction table in our database.

With the transactions imported, we now need to apply them to the balance value in the account table. The next section will cover how to accomplish that.

Applying Transactions to Current Balance

The next step in our job is to apply the transactions we just imported to the account balance. This is actually our easiest step to configure. It, like the import transaction step, has a simple `ItemReader` and a simple `ItemWriter`. We'll use the `JdbcBatchItemWriter` like we have in the previous two steps; however, our input will also come from the database in the form of the transactions we just loaded. Let's start by looking at the configuration for the step itself. Listing 10-21 shows the `applyTransactions` step's configuration.

Listing 10-21. applyTransactions Step

```
...
@Bean
public Job job() throws Exception {
        return this.jobBuilderFactory.get("importJob")
                        .start(importCustomerUpdates())
                        .next(importTransactions())
                        .next(applyTransactions())
                        .build();
}

...
```

```
@Bean
public Step applyTransactions() {
        return this.stepBuilderFactory.get("applyTransactions")
                        .<Transaction, Transaction>chunk(100)
                        .reader(applyTransactionReader(null))
                        .writer(applyTransactionWriter(null))
                        .build();
}
...
```

Listing 10-21 adds the new step to our job's configuration then defines the Step bean. The Step is created using the builders configured to read and write Transaction domain objects with a chunk size of 100. The reader will be the applyTransactionReader with the factory method taking a DataSource and the writer will be the applyTransactionWriter also taking a DataSource. In the next section, we'll look at how we define the ItemReader for our job.

Reading the Transaction Data

Reading the transaction data from the database table we just imported it into is simple thanks to the JdbcCursorItemReader. To use that reader, all we need to configure is the name (for restartability), a DataSource, a SQL statement, and in our case a RowMapper implementation. We'll use a lambda expression for that. Listing 10-22 shows the code required to configure this ItemReader.

Listing 10-22. applyTransactionsReader

```
...
@Bean
public JdbcCursorItemReader<Transaction> applyTransactionReader(DataSource dataSource) {
        return new JdbcCursorItemReaderBuilder<Transaction>()
                        .name("applyTransactionReader")
                        .dataSource(dataSource)
                        .sql("select transaction_id, " +
                                        "account_account_id, " +
                                        "description, " +
                                        "credit, " +
                                        "debit, " +
                                        "timestamp " +
                                        "from TRANSACTION " +
                                        "order by timestamp")
                        .rowMapper((resultSet, i) ->
                                        new Transaction(
                                                resultSet.getLong("transaction_id"),
                                                resultSet.getLong("account_account_id"),
                                                resultSet.getString("description"),
                                                resultSet.getBigDecimal("credit"),
                                                resultSet.getBigDecimal("debit"),
                                                resultSet.getTimestamp("timestamp")))
                        .build();
}
...
```

That really is all that's needed to read the Transaction data. In the next section, we'll look at how to apply the transactions to the account balance using a `JdbcBatchItemWriter`.

Updating the Account Balance

With each item read, we can apply the result of that bank transaction to the appropriate account. The nice thing is that we only need a SQL query to do this. I'll take a minute to acknowledge that there are much more performant ways of doing this. Querying the items for a sum and applying that in a single item per account would be just as effective and more performant. However, business rules may make this approach more realistic (tracking when a balance goes negative or below a given threshold may be required, for example). However, in this example, I'm looking to keep things simple and reuse as much as possible. Listing 10-23 illustrates the applying of each transaction to the account balance.

Listing 10-23. `applyTransactionsWriter`

```
...
@Bean
public JdbcBatchItemWriter<Transaction> applyTransactionWriter(DataSource dataSource) {
        return new JdbcBatchItemWriterBuilder<Transaction>()
                        .dataSource(dataSource)
                        .sql("UPDATE ACCOUNT SET " +
                                    "BALANCE = BALANCE + :transactionAmount " +
                                    "WHERE ACCOUNT_ID = :accountId")
                        .beanMapped()
                        .assertUpdates(false)
                        .build();
}
...
```

In Listing 10-23, we configure the `JdbcBatchItemWriter` with the `DataSource`, a SQL statement that adds the amount of the item's transaction to the current balance and identifies that the parameters in our SQL statement can be populated via calling bean properties.

With our reader and writer built we can now build and run our job. Using the same commands we did after we built the `importTransactions` Step (`./mvnw clean install` to build the project and `java -jar chapter10-0.0.1-SNAPSHOT.jar customerUpdateFile=<PATH_TO_CUSTOMER_FILE> transactionFile=<PATH_TO_TRANSACTION_FILE>` to execute the job) we can validate that our step has worked and the transactions are applied correctly.

The next section brings us to the last step in our job, actually generating the statements. While this step is simple on the surface, there is a bit more code involved. The next section will take a look.

Generating Monthly Statement

The end goal of this batch job is to generate a statement for each customer with a summary of their account. All the processing up to this point has been about updating and preparing to write the statement. Step 4 is where you do that work. This section looks at the processing involved writing the statements.

Reading the Statement Data

When you look at the expected output of this last step, you quickly realize that a large amount of data needs to be pulled in order to generate the statement. Before you get into how to pull that data, let's look at the domain object you use to represent the data: the Statement object (see Listing 10-24).

Listing 10-24. Statement.java

```
...
public class Statement {

        private final Customer customer;
        private List<Account> accounts = new ArrayList<>();

        public Statement(Customer customer, List<Account> accounts) {
                this.customer = customer;
                this.accounts.addAll(accounts);
        }

        // accessors removed for brevity
...
}
```

The Statement object consists of a Customer instance (for who the statement is being generated for) and a list of Account objects representing each of the accounts the customer has. Each Customer object contains all of the data in the CUSTOMER table in our database. The Account object, as you'd expect also maps directly to the ACCOUNT table in the database. Listing 10-25 shows the code for both of these domain objects.

Listing 10-25. Customer.java and Account.java

```
...
public class Customer {

        private final long id;
        private final String firstName;
        private final String middleName;
        private final String lastName;
        private final String address1;
        private final String address2;
        private final String city;
        private final String state;
        private final String postalCode;
        private final String ssn;
        private final String emailAddress;
        private final String homePhone;
        private final String cellPhone;
        private final String workPhone;
        private final int notificationPreferences;

        public Customer(long id, String firstName, String middleName, String lastName,
        String address1, String address2, String city, String state, String postalCode,
        String ssn, String emailAddress, String homePhone, String cellPhone, String
        workPhone, int notificationPreferences) {
```

```
                this.id = id;
                this.firstName = firstName;
                this.middleName = middleName;
                this.lastName = lastName;
                this.address1 = address1;
                this.address2 = address2;
                this.city = city;
                this.state = state;
                this.postalCode = postalCode;
                this.ssn = ssn;
                this.emailAddress = emailAddress;
                this.homePhone = homePhone;
                this.cellPhone = cellPhone;
                this.workPhone = workPhone;
                this.notificationPreferences = notificationPreferences;
        }

        // accessors removed

        ...
}

...
public class Account {

        private final long id;
        private final BigDecimal balance;
        private final Date lastStatementDate;
        private final List<Transaction> transactions = new ArrayList<>();

        public Account(long id, BigDecimal balance, Date lastStatementDate) {
                this.id = id;
                this.balance = balance;
                this.lastStatementDate = lastStatementDate;
        }

        // accessors removed

        ...
}
```

While our domain object consists of all of the components for our statement, our reader won't populate them all. For this step, we will use what's called the driving query pattern. This means our ItemReader will read just the basics (the Customer in this case). The ItemProcessor will enrich the Statement object with the account information before going to the ItemWriter for the final generation of the statement. Let's begin by taking a look at the configuration for the step itself. Listing 10-26 shows the configuration of our final Step and its addition to the Job.

Listing 10-26. generateStatements Step

```
...
@Bean
public Job job() throws Exception {
        return this.jobBuilderFactory.get("importJob")
                        .start(importCustomerUpdates())
                        .next(importTransactions())
                        .next(applyTransactions())
                        .next(generateStatements(null))
                        .build();
}
...
@Bean
public Step generateStatements(AccountItemProcessor itemProcessor) {
        return this.stepBuilderFactory.get("generateStatements")
                        .<Statement, Statement>chunk(1)
                        .reader(statementItemReader(null))
                        .processor(itemProcessor)
                        .writer(statementItemWriter(null))
                        .build();
}
...
```

The last step in our job, generateStatements, consists of a simple ItemReader, an ItemProcessor, and an ItemWriter. You'll notice that the chunk size is one for our final step. The reason for this is that we want a single file per statement. To do that, we'll use the MultiResourceItemWriter. This, however, only rotates files once per chunk. If we want one file per item, our chunk size then needs to be 1.

With our step configured we can configure the ItemReader for our step. The ItemReader for the generateStatements step is a simple JdbcCursorItemReader. We configure the JdbcCursorItemReader with a name (for restartability), a DataSource, the SQL statement we want to run, and a RowMapper (in our case, we'll just us a lambda expression). Listing 10-27 shows the configuration of our ItemReader.

Listing 10-27. statementItemReader

```
...
@Bean
public JdbcCursorItemReader<Statement> statementItemReader(DataSource dataSource) {
        return new JdbcCursorItemReaderBuilder<Statement>()
                        .name("statementItemReader")
                        .dataSource(dataSource)
                        .sql("SELECT * FROM CUSTOMER")
                        .rowMapper((resultSet, i) -> {
                                Customer customer =
                                        new Customer(resultSet.getLong("customer_id"),
                                                resultSet.getString("first_name"),
                                                resultSet.getString("middle_name"),
                                                resultSet.getString("last_name"),
                                                resultSet.getString("address1"),
                                                resultSet.getString("address2"),
                                                resultSet.getString("city"),
                                                resultSet.getString("state"),
```

```
                            resultSet.getString("postal_code"),
                            resultSet.getString("ssn"),
                            resultSet.getString("email_address"),
                            resultSet.getString("home_phone"),
                            resultSet.getString("cell_phone"),
                            resultSet.getString("work_phone"),
                            resultSet.getInt("notification_pref"));

                return new Statement(customer);
        }).build();
}
...
```

With our reader configured, we will need to write our `ItemProcessor` to enrich the `Statement` object that was just returned via the `ItemReader` with the `Accounts` associated with the customer. The next section will cover this in detail.

Enrich the Statement with Accounts

Once we have each customer read, we can read the `Account` and `Transactions` needed to generate the statement. We'll use Spring's `JdbcTemplate` to do this. However, since the query will result in a parent child relationship (one parent `Account` to multiple `Transaction` children), we won't be able to use a `RowMapper`. Instead we'll use a `ResultSetExtractor`. Unlike the `RowMapper` interface which is intended to map a single row to an object, the `ResultSetExtractor` looks at a `ResultSet` as a whole (when using a `RowMapper`, if you advance the `ResultSet` yourself, an exception is thrown). We'll use a `ResultSetExtractor` because the query we'll be running results in a parent child relationship with one account having many transactions. We'll need to read multiple rows from the `ResultSet` to create each `Account`. Let's start digging into this code by looking at the code for the `AccountItemProcessor` in Listing 10-28 where we're executing our query and enriching our `Statement` object.

Listing 10-28. `AccountItemProcessor`

```
...
@Component
public class AccountItemProcessor implements ItemProcessor<Statement, Statement> {

        @Autowired
        private final JdbcTemplate jdbcTemplate;

        public AccountItemProcessor(JdbcTemplate jdbcTemplate) {
                this.jdbcTemplate = jdbcTemplate;
        }

        @Override
        public Statement process(Statement item) throws Exception {

                item.setAccounts(this.jdbcTemplate.query("select a.account_id," +
                        "       a.balance," +
                        "       a.last_statement_date," +
                        "       t.transaction_id," +
                        "       t.description," +
```

```
"          t.credit," +
"          t.debit," +
"          t.timestamp " +
"from account a left join " +  //HSQLDB
"     transaction t on a.account_id = t.account_account_id "+
"where a.account_id in " +
"         (select account_account_id " +
"          from customer_account " +
"          where customer_customer_id = ?) " +
"order by t.timestamp",
new Object[] {item.getCustomer().getId()},
          new AccountResultSetExtractor()));

        return item;
    }
}
```

This ItemProcessor runs a query to find all the accounts and their transactions for a specified customer. The code for that is pretty simple (beyond the SQL query itself). The real "work" is done in the AccountResultSetExtractor which we'll look at in Listing 10-29.

Listing 10-29. AccountResultSetExtractor

```
...
public class AccountResultSetExtractor implements ResultSetExtractor<List<Account>> {

    private List<Account> accounts = new ArrayList<>();
    private Account curAccount;

    @Nullable
    @Override
    public List<Account> extractData(ResultSet rs) throws SQLException, DataAccessException {

        while (rs.next()) {

            if(curAccount == null) {
                curAccount = new Account(
                            rs.getLong("account_id"),
                            rs.getBigDecimal("balance"),
                            rs.getDate("last_statement_date"));
            }
            else if (rs.getLong("account_id") != curAccount.getId()) {
                accounts.add(curAccount);

                curAccount = new Account(rs.getLong("account_id"),
                            rs.getBigDecimal("balance"),
                            rs.getDate("last_statement_date"));
            }

            if(StringUtils.hasText(rs.getString("description"))) {
                curAccount.addTransaction(
                        new Transaction(rs.getLong("transaction_id"),
                                    rs.getLong("account_id"),
```

```
                                        rs.getString("description"),
                                        rs.getBigDecimal("credit"),
                                        rs.getBigDecimal("debit"),
                        new Date(rs.getTimestamp("timestamp").getTime()))));
                }
        }

        if(curAccount != null) {
                accounts.add(curAccount);
        }

        return accounts;
    }
}
```

As Listing 10-29 illustrates, we iterate over the ResultSet building an Account object. If the current Account is null or the account id does not equal the current one, we'll create a new Account object. Once we have the Account object, for each record that has a transaction, we add a Transaction object. This allows us to build up a list of Account objects that we return to the ItemProcessor which adds them to the Statement item to be written. The last piece of this puzzle is the configuration of the ItemWriter. Saving the best for last, let's dig into the ItemWriter used for writing the statement files.

Writing Statements

It is common to see there be a lot of pre-processing before the final thing is generated. The job we have been working on is no different. However, this is where things arrive. We are going to write our statements, one statement per file. For this ItemWriter we'll need a MultiResourceItemWriter to write one statement per file. That will delegate to a FlatFileItemWriter. For each file, we'll need to generate a header with the customer information as well as the information about each account. Let's start by looking at the detail pieces then put them together. Starting with the custom LineAggregator we need to create for outputting each statement's accounts. Listing 10-30 has the code for the StatementLineAggregator.

Listing 10-30. StatementLineAggregator

```
public class StatementLineAggregator implements LineAggregator<Statement> {

        private static final String ADDRESS_LINE_ONE =
                        String.format("%121s\n", "Apress Banking");
        private static final String ADDRESS_LINE_TWO =
                        String.format("%120s\n", "1060 West Addison St.");
        private static final String ADDRESS_LINE_THREE =
                        String.format("%120s\n\n", "Chicago, IL 60613");
        private static final String STATEMENT_DATE_LINE =
                        String.format("Your Account Summary %78s ", "Statement Period") +
                                "%tD to %tD\n\n";

        public String aggregate(Statement statement) {
                StringBuilder output = new StringBuilder();
```

```
                formatHeader(statement, output);
                formatAccount(statement, output);

                return output.toString();
        }

        private void formatAccount(Statement statement, StringBuilder output) {
                if(!CollectionUtils.isEmpty(statement.getAccounts())) {

                        for (Account account : statement.getAccounts()) {

                                output.append(
                                                String.format(STATEMENT_DATE_LINE,
                                                        account.getLastStatementDate(),
                                                                new Date())));

                                BigDecimal creditAmount = new BigDecimal(0);
                                BigDecimal debitAmount = new BigDecimal(0);
                                for (Transaction transaction : account.getTransactions()) {
                                        if(transaction.getCredit() != null) {
                                                creditAmount =
                                                creditAmount.add(transaction.getCredit());
                                        }

                                        if(transaction.getDebit() != null) {
                                                debitAmount =
                                                        debitAmount.add(transaction.getDebit());
                                        }

                                        output.append(
                        String.format("                  %tD         %-50s    %8.2f\n",
                                                        transaction.getTimestamp(),
                                                        transaction.getDescription(),
                                                transaction.getTransactionAmount())));
                                }

                                output.append(
                        String.format("%80s %14.2f\n", "Total Debit:" , debitAmount));
                                output.append(
                        String.format("%81s %13.2f\n", "Total Credit:", creditAmount));
                                output.append(
                        String.format("%76s %18.2f\n\n", "Balance:", account.getBalance()));
                        }
                }
        }

        private void formatHeader(Statement statement, StringBuilder output) {
                Customer customer = statement.getCustomer();
```

```
            String customerName =
                    String.format("\n%s %s",
                                    customer.getFirstName(),
                                    customer.getLastName());
            output.append(customerName +
                    ADDRESS_LINE_ONE.substring(customerName.length()));

            output.append(customer.getAddress1() +
                    ADDRESS_LINE_TWO.substring(customer.getAddress1().length()));

            String addressString =
                    String.format("%s, %s %s",
                                    customer.getCity(),
                                    customer.getState(),
                                    customer.getPostalCode());
            output.append(addressString +
                    ADDRESS_LINE_THREE.substring(addressString.length()));
        }
}
```

That's a lot of code; however, the majority of it is `String.format` calls with well-defined expressions. The `formatHeader(Statement statement, StringBuilder output)` method is responsible for formatting and appending the strings to the output. The `formatAccount(Statement statement, StringBuilder output)` method does essentially the same thing, only doing it for each account and the transactions within the account.

The next component for the `ItemWriter` is the `HeaderCallback`. This will provide the generic elements of each statement. Listing 10-31 shows the code for the `HeaderCallback`.

Listing 10-31. StatementHeaderCallback

```
...
public class StatementHeaderCallback implements FlatFileHeaderCallback {

        public void writeHeader(Writer writer) throws IOException {
                writer.write(String.format("%120s\n", "Customer Service Number"));
                writer.write(String.format("%120s\n", "(800) 867-5309"));
                writer.write(String.format("%120s\n", "Available 24/7"));
                writer.write("\n");
        }

}
```

This class does essentially the same thing as the `StatementLineAggregator` in that all it's doing is formatting strings and appending them to the current stream. However, this data is static so there is nothing that changes.

Those are the components required for the FlatFileItemWriter used to generate the statements. To configure the actual FlatFileItemWriter, we just need to pass the previous two instances to the builder. Listing 10-32 demonstrates that.

Listing 10-32. individualStatementItemWriter

```
...
@Bean
public FlatFileItemWriter<Statement> individualStatementItemWriter() {
        FlatFileItemWriter<Statement> itemWriter = new FlatFileItemWriter<>();

        itemWriter.setName("individualStatementItemWriter");
        itemWriter.setHeaderCallback(new StatementHeaderCallback());
        itemWriter.setLineAggregator(new StatementLineAggregator());

        return itemWriter;
}
...
```

The FlatFileItemWriter is configured with the name, HeaderCallback, and the LineAggregator we've just gone through. With those configured, the last piece to configure is the MultiResourceItemWriter. This component is used to generate a file per customer. Listing 10-33 illustrates its configuration.

Listing 10-33. statementItemWriter

```
...
@Bean
@StepScope
public MultiResourceItemWriter<Statement> statementItemWriter(@Value("#{jobParameters['outpu
tDirectory']}") Resource outputDir) {
        return new MultiResourceItemWriterBuilder<Statement>()
                        .name("statementItemWriter")
                        .resource(outputDir)
                        .itemCountLimitPerResource(1)
                        .delegate(individualStatementItemWriter())
                        .build();
}
```

The last piece of this puzzle is the MultiResourceItemWriter. We configure that with a name, resource representing the directory to write into, the item count per resource (in our case it is 1), and finally the delegate.

That's it. With the components for our job configured, we can build the job via the ./mvnw clean install command and run it via the java -jar chapter10-0.0.1-SNAPSHOT.jar customerUpdateFile=<PATH_TO_CUSTOMER_FILE> transactionFile=<PATH_TO_TRANSACTION_FILE> outputDirectory=L<PATH_TO_OUTPUT_DIR>. The output for our job is the full statement as shown in Listing 10-34.

Listing 10-34. A Sample Statement

```
                                           Customer Service Number
                                                  (800) 867-5309
                                                  Available 24/7
```

```
Elliot Winslade                                            Apress Banking
3 Clyde Gallagher Parkway                              1060 West Addison St.
San Antonio, Texas 78250                                   Chicago, IL 60613


Your Account Summary                    Statement Period 05/08/18 to 06/20/18

                                        Total Debit:            0.00
                                        Total Credit:           0.00
                                        Balance:            24082.61


Your Account Summary                    Statement Period 05/06/18 to 06/20/18

        06/05/18        Quinu                              10733.88
        06/15/18        Jabbercube                         -1061.00
                                        Total Debit:        -1061.00
                                        Total Credit:       10733.88
                                        Balance:            11413.68
```

Summary

Learning how to do something without context makes it hard to take what you've learned and apply it to the real world. This chapter has taken commonly used elements of the Spring Batch framework and put them together into a realistic example of a batch job.

With the basics covered, we will dive deeper into the more advanced topics of Spring Batch in the upcoming chapters. In Chapter 11, you will look at how to scale batch jobs beyond a single threaded execution like you have used up to this point.

CHAPTER 11

■ ■ ■

Scaling and Tuning

The IRS processed over 115 million individual tax returns in 2018. Atlanta's Hartsfield-Jackson airport handled nearly 104 million passengers in 2017. Facebook has more than 300 million photos uploaded a day. Apple sold more than 216 million iPhones in 2017. The amount of data the world generates every day is staggering. It used to be that as the data increased, so did the processors to process it. If your app wasn't fast enough, you could wait a year and buy a new server, and all was fine.

But that isn't the case anymore. CPUs are not getting faster anymore. However, the overall cost of compute is dropping. Instead of getting faster processing, you get more compute power through either more cores on a single chip or more chips via distributed systems. The developers behind Spring Batch understand this and made parallel processing one of the primary focuses of the framework. This chapter looks at the following:

- *Profiling batch jobs:* You see a process for profiling your batch jobs so that the optimization decisions you make positively impact your performance and not the other way around.

- *Evaluating each of the scalability options in Spring Batch:* Spring Batch provides a number of different scalability options, each of which is reviewed in detail.

Profiling Your Batch Process

Michael A. Jackson put forth the best two rules of optimization in his 1975 book *Principals of Program Design*:

- *Rule 1. Don't do it.*

- *Rule 2. (for experts only) Don't do it yet.*

The idea behind this is simple. Software changes over the course of its development. Because of this, it's virtually impossible to make accurate decisions about how to design a system until the system has been developed. After the system has been developed, you can test it for performance bottlenecks and address those as required. By not taking this approach, you risk being described by my second most favorite quote on optimization, this one by W. A. Wulf:

> *More computing sins are committed in the name of efficiency (without necessarily achieving it) than for any other single reason—including blind stupidity.*

To profile any Java application there are many options, ranging from free to very expensive. However, one of the best free options is available on Github: VisualVM. This is the tool you can use to profile batch jobs. Before you begin profiling your jobs, let's take a quick tour of the VisualVM tool.

M. T. Minella, *The Definitive Guide to Spring Batch*, https://doi.org/10.1007/978-1-4842-3724-3_11

A Tour of VisualVM

Oracle's VisualVM is a tool that gives you insights into what is going on in your JVM. As JConsole's big brother, VisualVM provides not only JMX administration like JConsole but also information about CPU and memory usage, method execution times, as well as thread management and garbage collection. This section looks at the capabilities of the VisualVM tool.

Before you can try VisualVM, you have to install it. Before Java 9, VisualVM was provided with your JVM. However, that changed with Java 9 and is now only available via Github. You can obtain the latest version of VisualVM and the installation instructions at: `https://visualvm.github.io/index.html`.

With VisualVM installed, you can launch it. VisualVM greets you with a menu on the left and a Start Page on the right, as shown in Figure 11-1.

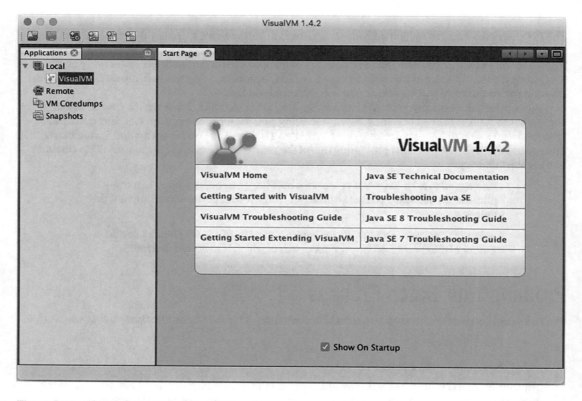

Figure 11-1. *The start screen for VisualVM*

The menu on the left is broken into four sections: Local and Remote are where you find applications that you can connect to, to profile. When you start VisualVM, because it's itself a Java application, it appears in the Local section. Below the Local and Remote sections is where you can load either Java VM coredumps that you've collected previously that you want to analyze, or snapshots, which are the state of a VM at a certain point in time that you can capture using VisualVM. To see some of the capabilities of the VisualVM tool, let's connect VisualVM to an instance of Eclipse.

When you first connect to a running JVM, VisualVM displays the screen shown in Figure 11-2.

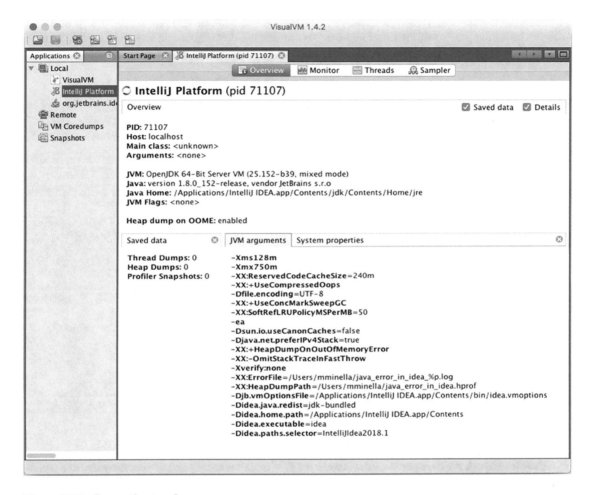

Figure 11-2. *Connecting to a Java process*

Along the top of the screen are four tabs:

- *Overview:* Provides an overview of the Java application running, including the main class, application name, process id, and arguments passed into the JVM on startup.

- *Monitor:* Displays charts showing CPU utilization, memory utilization (both heap and PermGen), the number of classes loaded, and the number of live and daemon threads. The Monitor tab also lets you perform garbage collection as well as generate a heap dump for later analysis.

- *Threads:* Displays information about all threads the application has launched and what they're doing (running, sleeping, waiting, or monitoring). This data is shown in either timeline, table, or detail form.

- *Sampler:* Allows you to take a sample of the CPU utilization and memory allocation for your application as well as take snapshots. CPU shows what methods are taking how long to run. Memory utilization shows what classes are taking how much memory.

In addition to the tabs, Overview shows you information about the current Java process that is being analyzed including process id, the host the process is running on, JVM arguments, as well as the full list of system properties the JVM knows.

The second tab is the Monitor tab, as shown in Figure 11-3.

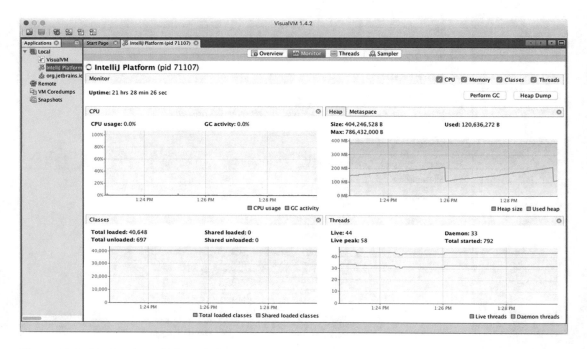

Figure 11-3. *The Monitor tab for an Eclipse instance*

The Monitor tab is where you view the state of the JVM from a memory and CPU perspective as a whole. The other tabs are more useful when you're determining the cause of a problem identified in the Monitor tab (if you keep running out of memory, or the CPU is pegged for some reason). All the charts on the Monitor tab are resizable, and they can be hidden as required.

The next tab available in VisualVM is the Threads tab, displayed in Figure 11-4.

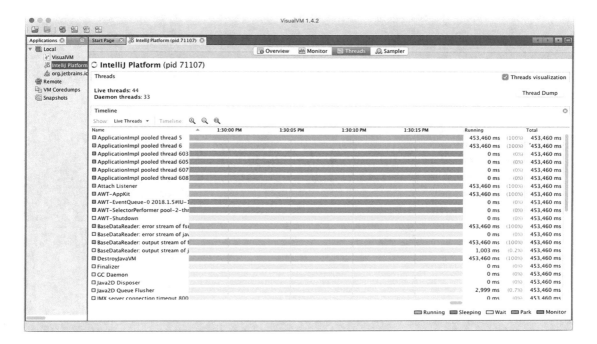

Figure 11-4. *The Threads tab in VisualVM*

All Java applications are multithreaded. At the least, you have the main execution thread and an additional thread for garbage collection. However, most Java applications spawn many additional threads for various reasons. This tab allows you to see information about the various threads your application has spawned and what they're doing. Figure 11-4 shows the data as a timeline, but the data is also available as a table and as detailed graphs for each thread.

The last tab, as shown in Figure 11-5, is the Sampler tab.

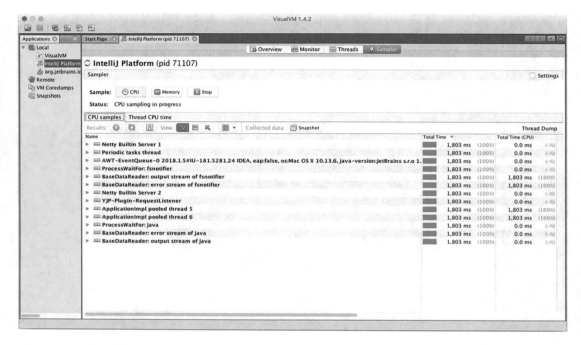

Figure 11-5. *VisualVM's Sampler tab*

In this tab, you're presented with a screen which includes CPU and Memory buttons as well as a Stop button. To begin sampling either CPU execution by method or memory footprint by class, click the appropriate button. The tables update periodically with the current state of the VM VisualVM is studying.

VisualVM is a powerful and extendable tool. Many plug-ins are available to extend the feature set provided out of the box. You can add things like the ability to view the stack trace of currently executing threads with the Thread Inspector plug-in, visual garbage collection with the Visual GC plug-in, and access to MBeans via the MBean browser, to extend VisualVM's already powerful suite of tools.

Now that you have an idea of what Oracle's VisualVM can do, let's see how you can use it to profile Spring Batch applications.

Profiling Spring Batch Applications

When you profile your applications, you're typically looking at one of two things: how hard the CPU is working and where, and how much memory is being used and on what. The first questions, how hard the CPU is working and where, relate to what your CPU is working on. Is your job computationally difficult? Is your CPU using a lot of its effort in places other than your business logic—for example, is it spending more time working on parsing files than actually doing the calculations you want it to? The second set of questions revolves around memory. Are you using most if not all of the available memory? If so, what is taking up all the memory? Do you have a Hibernate object that isn't lazily loading a collection, which is causing memory pressure? This section looks at how to see where resources are being used in your Spring Batch applications.

CPU Profiling

It would be nice to have a straightforward checklist of things to check when you're profiling applications. But it just isn't that easy. Profiling an application can, at times, feel more like an art than a science. This section walks through how to obtain data related to the performance of your applications and their utilization of the CPU.

When you look at how a CPU is performing within your application, you typically use the measure of time to determine the hot spots (the areas that aren't performing to your expectations). What areas is the CPU working in the most? For example, if you have an infinite loop somewhere in your code, the CPU will spend a large amount of time there after it's triggered. However, if everything is running fine, you can expect to see either no bottlenecks or at bottlenecks that you would expect (I/O is typically the bottleneck of most modern systems).

To view the CPU profiling functionality at work, let's use the statement job that you completed in the last chapter. This job consists of four steps and interacts with both files and a database. Figure 11-6 shows from a high level what the job does as it's currently configured.

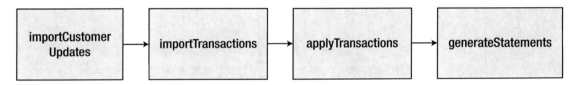

Figure 11-6. *Statement job*

To execute the job, you use the command `java -jar chapter10-0.0.1-SNAPSHOT.jar customerUpdateFile=<PATH_TO_CUSTOMER_FILE> transactionFile=<PATH_TO_TRANSACTION_FILE> outputDirectory=<PATH_TO_OUTPUT_DIR>`. After you've launched the job, it appears in the VisualVM menu on the left under Local. To connect to it, all you need to do is double-click it.

Now that you've connected to the running statement job, you can begin to look at how things operate within it. Let's first look at the Monitor tab to see how busy the CPU is in the first place. After running the statement job with a customer transaction file containing 100 customers and more than 20,000 transactions, you can see that the CPU utilization for this job is minimal. Figure 11-7 shows the charts from the Monitor tab after a run of the job.

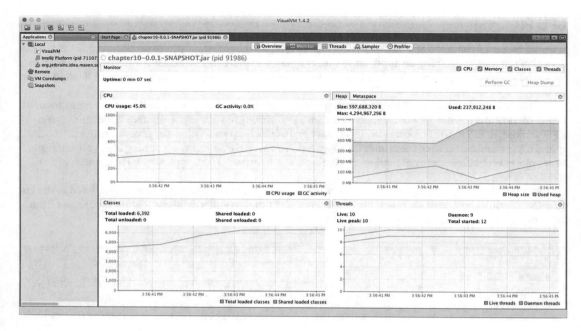

Figure 11-7. *Resource utilization for the statement job*

As Figure 11-7 shows, the statement job isn't a CPU-intensive process. In fact, if you look at the memory profile, the job isn't very memory intensive either. However, you can easily change that. If you add a small loop into the ItemProcessor used in step 4 (AccountItemProcessor) you can quickly make your CPU busy. Listing 11-1 shows the loop you add.

Listing 11-1. Using PricingTiersItemProcessor to Calculate Prime Numbers

```
...
@Component
public class AccountItemProcessor implements ItemProcessor<Statement, Statement> {

        @Autowired
        private final JdbcTemplate jdbcTemplate;

        public AccountItemProcessor(JdbcTemplate jdbcTemplate) {
                this.jdbcTemplate = jdbcTemplate;
        }

        @Override
        public Statement process(Statement item) throws Exception {

                int threadCount = 10;
                CountDownLatch doneSignal = new CountDownLatch(threadCount);
```

```
            for(int i = 0; i < threadCount; i++) {
                    Thread thread = new Thread(() -> {
                            for (int j = 0; j < 1000000; j++) {
                                    new BigInteger(String.valueOf(j))
.isProbablePrime(0);
                            }
                            doneSignal.countDown();
                    });
                    thread.start();
            }

            doneSignal.await();

            item.setAccounts(this.jdbcTemplate.query("select a.account_id," +
                            "       a.balance," +
                            "       a.last_statement_date," +
                            "       t.transaction_id," +
                            "       t.description," +
                            "       t.credit," +
                            "       t.debit," +
                            "       t.timestamp " +
                            "from account a left join " +  //HSQLDB
                            "    transaction t on a.account_id = t.account_account_id " +
//                           "from account a left join " +  //MYSQL
//                           "    transaction t on a.account_id = t.account_account_id " +
                            "where a.account_id in " +
                            "       (select account_account_id " +
                            "       from customer_account " +
                            "       where customer_customer_id = ?) " +
                            "order by t.timestamp",
                            new Object[] {item.getCustomer().getId()},
                                    new AccountResultSetExtractor()));

            return item;
        }
}
```

Obviously, the loop you added to launch a number of threads and calculate all the prime numbers between zero and one million as shown in Listing 11-1 is unlikely to end up in your code. But it's exactly the type of accidental looping that could cause a catastrophic impact on the performance of a batch job over the course of processing millions of transactions. Figure 11-8 shows the impact this small loop makes on CPU utilization, according to VirtualVM.

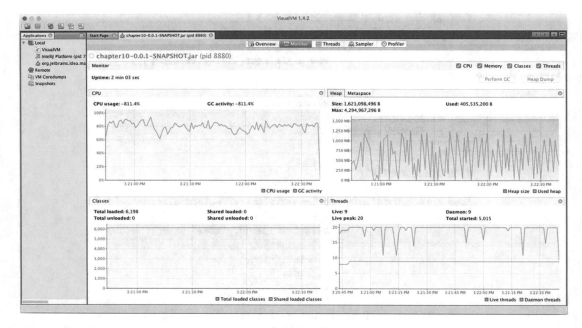

Figure 11-8. *Resource utilization for the updated statement job*

That code, as expected, sent our job into quite a frenzy. The first three steps of the job execute so fast, they don't even show up in the graph, but once the job gets to the last step, it consumes many more threads, memory, and CPU. But if you didn't know what caused this spike, where would you look next?

With a spike identified like this, the next place to look is in the Sampler tab. By rerunning the job under the same conditions, you can see what individual methods show up as hot spots in the job's execution.In this case, the lambda we created to run our calculations in is number three on the list by CPU time. By the end of the job, this method has taken up 24.2% of all the CPU time required to execute this job, as shown in Figure 11-9.

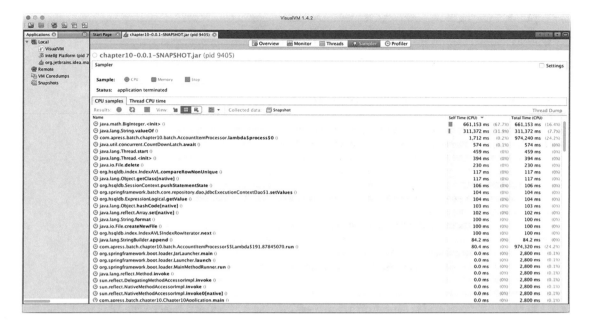

Figure 11-9. *The* `AccountItemProcessor` *has taken up quite a bit of CPU*

When you come across a scenario like this, a better way to view what is eating up CPU execution time is to filter the list by the package name you're using for your code. In this case, you can filter the list on `com.apress.batch.chapter10.*` to see what classes take up what percentage of the total CPU utilization. Under this filter, the culprit becomes crystal clear in this example: the `AccountItemProcessor.lambda$process$0` method and the 24.2% of the CPU time it takes up. The next highest on the list takes up 0%. At this point, you have all the information you can get from the tool, and it's time to begin digging through the code to determine what in `AccountItemProcessor` is using so much CPU.

Simple, isn't it? Not really. Although the process used here is what you would use to narrow down an issue in any system, the issue is rarely this easy to track down. However, using VisualVM you can progressively narrow down where the issue is in your job. CPU utilization isn't the only piece of performance. The next section looks at how to profile memory using VisualVM.

Memory Profiling

Although CPU utilization may seem like the place you're most likely to see issues, the truth is that it is my experience that memory issues are more likely to pop up in your software. The reason is that you use a number of frameworks that do things behind the scenes. When you use these frameworks incorrectly, large numbers of objects can be created without any indication that it has occurred until you run out of memory completely. This section looks at how to profile memory usage using VisualVM.

To look at how to profile memory, let's tweak what you did previously. However, this time instead of taking up processing time, you update it to simulate creating a `String` that is out of control. Although the code example may not be what you see in real-world systems, excess `String` manipulation is a common reason for memory issues. Listing 11-2 shows the code for the updated `AccountItemProcessor`.

Listing 11-2. PricingTierItemProcessor with a Memory Leak

```
@Component
public class AccountItemProcessor implements ItemProcessor<Statement, Statement> {

        @Autowired
        private final JdbcTemplate jdbcTemplate;

        public AccountItemProcessor(JdbcTemplate jdbcTemplate) {
                this.jdbcTemplate = jdbcTemplate;
        }

        @Override
        public Statement process(Statement item) throws Exception {

                String memoryBuster = "memoryBuster";

                for (int i = 0; i < 200; i++) {
                        memoryBuster += memoryBuster;
                }

                item.setAccounts(this.jdbcTemplate.query("select a.account_id," +
                                "       a.balance," +
                                "       a.last_statement_date," +
                                "       t.transaction_id," +
                                "       t.description," +
                                "       t.credit," +
                                "       t.debit," +
                                "       t.timestamp " +
                        "from account a left join " +   //HSQLDB
                                "   transaction t on a.account_id = t.account_account_id " +
//                        "from account a left join " +   //MYSQL
//                        "   transaction t on a.account_id = t.account_account_id " +
                        "where a.account_id in " +
                                "       (select account_account_id " +
                                "       from customer_account " +
                                "       where customer_customer_id = ?) " +
                        "order by t.timestamp",
                        new Object[] {item.getCustomer().getId()},
                                new AccountResultSetExtractor()));

                return item;
        }
}
```

In the version shown in Listing 11-2, you are creating a String that is out of control. By doing something like what you have in this example, you would expect the memory footprint to grow out of control as well.

When you run the statement job with this bug and profile it using VisualVM, you can see that things quickly get out of hand from a memory perspective; an OutOfMemoryException is thrown midway through the step. Figure 11-10 shows the VisualVM Monitor tab during a run of the statement job with the memory leak.

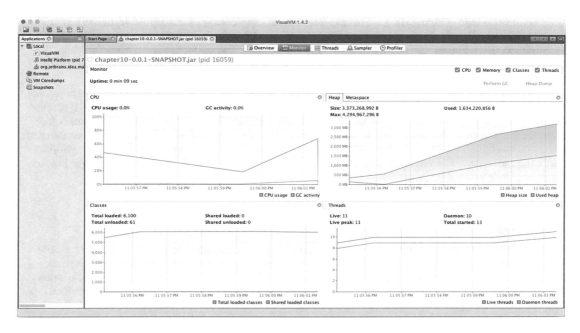

Figure 11-10. *Monitoring results of the statement job with a memory leak*

Notice at the very end of the memory graph in the upper-right corner of Figure 11-10 that memory usage spikes, causing the `OutOfMemoryException`. But how do you know what caused the spike? If you didn't know, the Sampler tab might be able to shed some light.

You've seen before that the Sampler tab can show what method calls are using up CPU, but it can also tell you what objects are taking up precious memory. To see that, begin by executing your job as you have previously. When it's running, connect to the process using VisualVM and go to the Sampler tab. To determine the cause of a memory leak, you need to determine what changes as the memory usage occurs. For example, in Figure 11-11, each block represents a class instance. The higher the blocks are stacked in each column; the more instances are in memory. Each column represents a snapshot in time within the JVM. When the program begins, the number of instances created is small (one in this case); this number slowly rises over time, occasionally declining when garbage collection occurs. Finally, it spikes at the end to nine instances. This is the type of increase in memory usage you look for with VisualVM.

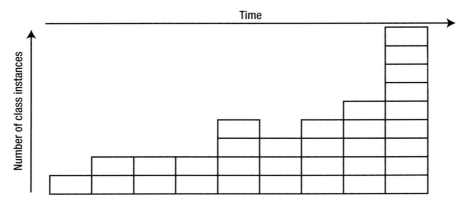

Figure 11-11. *Memory utilization over the life of a program*

To view this type of change in your batch jobs, you can use VisualVM's snapshot feature. As a job runs, click the Snapshot button in the middle of the screen. VisualVM records the exact state of the JVM when you take that snapshot. You can compare this with other snapshots to determine what changes. Typically, the change indicates the location of the issue. If it isn't the smoking gun, it's definitely where you should start looking.

The ability to scale batch jobs isn't a requirement to be able to address performance bugs as discussed in the previous sections of this chapter. On the contrary, jobs that have bugs like those discussed typically don't scale no matter what you do. Instead, you need to address the issues within your application before applying the scalability features that Spring Batch or any framework provides. When you have a system with none of these issues, the features that Spring Batch offers to scale it beyond a single-threaded, single-JVM approach are some of the strongest of any framework. You spend the rest of this chapter looking at how to use Spring Batch's scalability features.

Scaling a Job

In an enterprise, when things are going well, data gets big. More customers. More transactions. More site hits. More, more, more. Your batch jobs need to be able to keep up. Spring Batch was designed from the ground up to be highly scalable, to fit the needs of both small batch jobs and large enterprise-scale batch infrastructures. This section looks at the four different approaches Spring Batch takes for scaling batch jobs beyond the default flow: multithreaded steps, parallel steps, remote chunking, and partitioning.

Multithreaded Steps

When a step is processed, by default it's processed in a single thread. Although a multithreaded step is the easiest way to parallelize a job's execution, as with all multithreaded environments there are aspects you need to consider when using it. This section looks at Spring Batch's multithreaded step and how to use it safely in your batch jobs.

Spring Batch's multithreaded step concept allows a batch job to use Spring's `org.springframework.core.task.TaskExecutor` abstraction to execute each chunk in its own thread. Figure 11-12 shows an example of how processing works when using the multithreaded step.

As Figure 11-12 shows, any step in a job can be configured to perform within a threadpool, processing each chunk independently. As chunks are processed, Spring Batch keeps track of what is done accordingly. If an error occurs in any one of the threads, the job's processing is rolled back or terminated per the regular Spring Batch functionality.

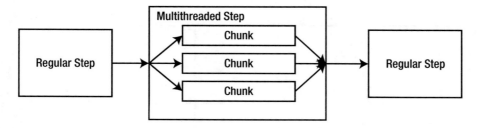

Figure 11-12. *Multithreaded step processing*

To configure a step to execute in a multithreaded manner, all you need to do is configure a reference to a `TaskExecutor` for the given step. If you use the statement job as an example, Listing 11-3 shows how to configure a single step job to use a multithreaded step.

Listing 11-3. `MultithreadedJobApplication` Using a Multithreaded Step

```
...
@EnableBatchProcessing
@SpringBootApplication
public class MultithreadedJobApplication {

    @Autowired
    private JobBuilderFactory jobBuilderFactory;

    @Autowired
    private StepBuilderFactory stepBuilderFactory;

    @Bean
    @StepScope
    public FlatFileItemReader<Transaction> fileTransactionReader(
                    @Value("#{jobParameters['inputFlatFile']}") Resource resource) {

            return new FlatFileItemReaderBuilder<Transaction>()
                            .name("transactionItemReader")
                            .resource(resource)
                            .saveState(false)
                            .delimited()
                            .names(new String[] {"account", "amount", "timestamp"})
                            .fieldSetMapper(fieldSet -> {
                                    Transaction transaction = new Transaction();

                                    transaction.setAccount(fieldSet.
                                    readString("account"));
                                    transaction.setAmount(fieldSet.
                                    readBigDecimal("amount"));
                                    transaction.setTimestamp(fieldSet.
                                    readDate("timestamp", "yyyy-MM-dd HH:mm:ss"));

                                    return transaction;
                            })
                            .build();
    }

    @Bean
    @StepScope
    public JdbcBatchItemWriter<Transaction> writer(DataSource dataSource) {
            return new JdbcBatchItemWriterBuilder<Transaction>()
                            .dataSource(dataSource)
                            .sql("INSERT INTO TRANSACTION (ACCOUNT, AMOUNT, TIMESTAMP)
                             VALUES (:account, :amount, :timestamp)")
                            .beanMapped()
                            .build();
    }
```

```
@Bean
public Job multithreadedJob() {
        return this.jobBuilderFactory.get("multithreadedJob")
                        .start(step1())
                        .build();
}

@Bean
public Step step1() {
        return this.stepBuilderFactory.get("step1")
                        .<Transaction, Transaction>chunk(100)
                        .reader(fileTransactionReader(null))
                        .writer(writer(null))
                        .taskExecutor(new SimpleAsyncTaskExecutor())
                        .build();
}

public static void main(String[] args) {
        String [] newArgs = new String[] {"inputFlatFile=/data/csv/bigtransactions.csv"};

        SpringApplication.run(MultithreadedJobApplication.class, newArgs);
}
}
```

As Listing 11-3 shows, all that is required to add the power of Spring's multithreading capabilities to a step in your job is to define a TaskExecutor implementation (you use org.springframework.core.task. SimpleAsyncTaskExecutor in this example) and reference it in your step. When you execute this job, Spring creates a new thread for each chunk executed within the step, executing each chunk in parallel. As you can imagine, this can be a powerful addition to most jobs.

But there is a catch when working with multithreaded steps. Most ItemReaders provided by Spring Batch are stateful. Spring Batch uses this state when it restarts a job, so it knows where processing left off. However, in a multithreaded environment, objects that maintain state in a way that is accessible to multiple threads (not synchronized, etc.) can run into issues of threads overwriting each other's state. Because of this, we turn the state saving feature of the reader off, preventing this job from being able to restart.

By adding a task executor can be a great first step in to improve performance. However, there are plenty of cases where this will not actually change the performance dynamic (if the input mechanism is already saturating resources like network, disk bus, etc). The next mechanism we will look at for scaling Spring Batch jobs is parallel steps.

Parallel Steps

Multithreaded steps provide the ability to process chunks of items within the same step of a job in parallel, but sometimes it's also helpful to be able to execute entire steps in parallel. Take for example importing multiple files that have no relationship to each other. There is no reason for one import to need to wait for the other import to complete before it begins. Spring Batch's ability to execute steps and even flows (reusable groups of steps) in parallel allows you to improve overall throughput on a job. This section looks at how to use Spring Batch's parallel steps and flows to improve the overall performance of your jobs.

If you consider the use case where you accept files from multiple sources, say each customer provides a file that you import into your system. Some clients prefer CSVs. Some clients prefer XML. The data is the same but the format is different. In this case, we can accomplish this a couple different ways; however, since each file is independent, one easy way to accomplish this is by executing parallel steps. Figure 11-13 illustrates the way parallel steps execute.

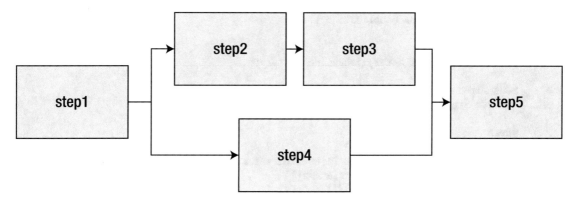

Figure 11-13. *Process flow for an order-processing job*

The job in Figure 11-14 begins with a single step. It then splits into two flows processing in parallel. Flow 1 (the top flow) executes step2 and on its completion step3. Flow 2 (the bottom flow) executes step4. Once flow 1 and 2 are both complete, then step 5 executes.

The job we will be looking at is a simple two-step job. Each step is responsible for importing data from a different input format. Step1 reads from XML files. Step2 reads from flat files. Both steps will be executed in parallel. The data from each format is the same. A transaction object consisting of an account, timestamp, and amount of the transaction is all that it contains. It's important to note that we're using a simplistic domain model here so that we can focus on the components of scaling and not the domain model itself.

Configuring the Parallel Steps

To execute steps in parallel, Spring Batch again uses Spring's TaskExecutor. In this case, each flow is executed in its own thread, allowing you to execute multiple flows in parallel. To configure this, you use the FlowBuilder's split() method. The split() method takes a TaskExecutor as its argument and returns a SplitBuilder. With the SplitBuilder you can add as many flow objects as you want. Each of those flows will be executed in its own thread (based on the rules of the underlying TaskExecutor) providing you the ability to execute steps or flows of steps in parallel.

It's important to note that the execution order of a job using split is similar to that of a regular job. In a regular job, a step doesn't complete until all the items are processed for the step and the next step doesn't begin until the previous one is completed. Using split, the step after the split isn't executed until all the flows configured within the split have been completed.

■ **Note**   The step after a split isn't executed until all the flows within a split are completed.

For each of these input sources we will configure a step to execute in parallel. Listing 11-4 shows the configuration of the entire job.

Listing 11-4. Configuration of Parallel Steps

```
@EnableBatchProcessing
@SpringBootApplication
public class ParallelStepsJobApplication {

        @Autowired
        private JobBuilderFactory jobBuilderFactory;

        @Autowired
        private StepBuilderFactory stepBuilderFactory;

        @Bean
        public Job parallelStepsJob() {
                Flow secondFlow = new FlowBuilder<Flow>("secondFlow")
                                .start(step2())
                                .build();

                Flow parallelFlow = new FlowBuilder<Flow>("parallelFlow")
                                .start(step1())
                                .split(new SimpleAsyncTaskExecutor())
                                .add(secondFlow)
                                .build();

                return this.jobBuilderFactory.get("parallelStepsJob")
                                .start(parallelFlow)
                                .end()
                                .build();
        }

        @Bean
        @StepScope
        public FlatFileItemReader<Transaction> fileTransactionReader(
                        @Value("#{jobParameters['inputFlatFile']}") Resource resource) {

                return new FlatFileItemReaderBuilder<Transaction>()
                                .name("flatFileTransactionReader")
                                .resource(resource)
                                .delimited()
                                .names(new String[] {"account", "amount", "timestamp"})
                                .fieldSetMapper(fieldSet -> {
                                        Transaction transaction = new Transaction();

                                        transaction.setAccount(fieldSet.
                                        readString("account"));
                                        transaction.setAmount(fieldSet.
                                        readBigDecimal("amount"));
                                        transaction.setTimestamp(fieldSet.
                                        readDate("timestamp", "yyyy-MM-dd HH:mm:ss"));
```

```java
                                return transaction;
                        })
                        .build();
}

@Bean
@StepScope
public StaxEventItemReader<Transaction> xmlTransactionReader(
                @Value("#{jobParameters['inputXmlFile']}") Resource resource) {
        Jaxb2Marshaller unmarshaller = new Jaxb2Marshaller();
        unmarshaller.setClassesToBeBound(Transaction.class);

        return new StaxEventItemReaderBuilder<Transaction>()
                        .name("xmlFileTransactionReader")
                        .resource(resource)
                        .addFragmentRootElements("transaction")
                        .unmarshaller(unmarshaller)
                        .build();
}

@Bean
@StepScope
public JdbcBatchItemWriter<Transaction> writer(DataSource dataSource) {
        return new JdbcBatchItemWriterBuilder<Transaction>()
                        .dataSource(dataSource)
                        .beanMapped()
                        .sql("INSERT INTO TRANSACTION (ACCOUNT, AMOUNT, TIMESTAMP)
                         VALUES (:account, :amount, :timestamp)")
                        .build();
}

@Bean
public Step step1() {
        return this.stepBuilderFactory.get("step1")
                        .<Transaction, Transaction>chunk(100)
                        .reader(xmlTransactionReader(null))
                        .writer(writer(null))
                        .build();
}

@Bean
public Step step2() {
        return this.stepBuilderFactory.get("step2")
                        .<Transaction, Transaction>chunk(100)
                        .reader(fileTransactionReader(null))
                        .writer(writer(null))
                        .build();
}

public static void main(String[] args) {
        String [] newArgs = new String[] {"inputFlatFile=/data/csv/
        bigtransactions.csv",
```

379

```
                                                "inputXmlFile=/data/xml/
                                                bigtransactions.xml"};

            SpringApplication.run(ParallelStepsJobApplication.class, newArgs);
        }
}
```

Listing 11-4 shows all that is needed to configure our job. The key items for this example are bolded in the listing where we use the `FlowBuilder` to create two flows. The first creates a flow to execute a step that ingests a CSV file containing the three fields previously talked about (`account`, `amount`, and `timestamp`). The second flow creates the actual split that will execute the step used to ingest XML as well as run the previously defined flow in parallel.

The rest of the listing defines the readers and writers for each of the steps as well as the steps themselves. Finally, the main method used by Spring Boot appears at the end of the listing. In order to run this job, you'll need input. Listing 11-5 shows the sample input for the CSV file.

Listing 11-5. bigtransactions.csv

```
5113971498870901,-546.68,2018-02-08 17:46:12
4041373995909987,-37.06,2018-02-02 21:10:33
3573694401052643,-784.93,2018-02-04 13:01:30
3543961469650122,925.44,2018-02-05 23:41:50
3536921428140325,507.57,2018-02-13 02:09:08
490516718399624409,-575.81,2018-02-15 20:43:12
201904179222112,-964.21,2018-02-08 15:50:21
5602221470889083,23.71,2018-02-14 10:23:41
5038678280559913,979.94,2018-02-05 04:28:31
```

And Listing 11-6 shows sample XML for the XML file input.

Listing 11-6. bigtransactions.csv

```xml
<transactions>
        <transaction>
                <account>633110684460535475</account>
                <amount>961.93</amount>
                <timestamp>2018-02-03 18:30:51</timestamp>
        </transaction>
        <transaction>
                <account>3555221131716404</account>
                <amount>759.62</amount>
                <timestamp>2018-02-12 20:02:01</timestamp>
        </transaction>
        <transaction>
                <account>30315923571992</account>
                <amount>648.92</amount>
                <timestamp>2018-02-12 23:16:45</timestamp>
        </transaction>
        <transaction>
                <account>5574851814767258</account>
                <amount>-90.11</amount>
```

```
                <timestamp>2018-02-04 10:01:04</timestamp>
        </transaction>
</transactions>
```

When you execute this job, you can follow along in the logs that both steps begin at the same time and the job completes once the two steps are complete. Listing 11-7 shows the snip of logs that illustrates this.

Listing 11-7. parallelStepsJob Logs

```
2018-12-03 15:46:09.575  INFO 44705 --- [            main] o.s.b.c.l.support.
SimpleJobLauncher      : Job: [FlowJob: [name=parallelStepsJob]] launched with the following
parameters:
[{inputXmlFile=/data/xml/bigtransactions.xml, inputFlatFile=/data/csv/bigtransactions.csv}]
2018-12-03 15:46:09.661  INFO 44705 --- [cTaskExecutor-2] o.s.batch.core.job.
SimpleStepHandler      : Executing step: [step1]
2018-12-03 15:46:09.670  INFO 44705 --- [cTaskExecutor-1] o.s.batch.core.job.
SimpleStepHandler      : Executing step: [step2]
2018-12-03 15:46:09.819  INFO 44705 --- [cTaskExecutor-2] o.s.oxm.jaxb.Jaxb2Marshaller
: Creating JAXBContext with classes to be bound [class io.spring.batch.scalingdemos.domain.
Transaction]
2018-12-03 15:46:29.960  INFO 44705 --- [            main] o.s.b.c.l.support.
SimpleJobLauncher      : Job: [FlowJob: [name=parallelStepsJob]] completed with the
following parameters: [{inputXmlFile=/data/xml/bigtransactions.xml, inputFlatFile=/data/csv/
bigtransactions.csv}] and the following status: [COMPLETED]
```

Executing steps in parallel can be a very useful tool when you have independent steps that you need to execute and want to improve performance. There is one other mechanism within Spring Batch that relies solely on threading within a single JVM for scaling and that is using the AsyncItemProcessor and AsyncItemWriter combination. This next section will take a look at how those can improve the performance of your ItemProcessor phase.

AsyncItemProcessor and AsyncItemWriter

There are certain processes where the ItemProcessor is the bottleneck of a step. Say, for example, the ItemProcessor has a complex calculation that needs to occur that slows down the overall execution of the step. A way to improve performance is to execute just the ItemProcessor phase of the step in a new thread. The AsyncItemProcessor and AsyncItemWriter allow you to do just that.

The AsycnItemProcessor is a decorator that will wrap whatever ItemProcessor implementation you have. When an item is passed to the decorator, the call to the underlying delegate is executed in a new thread. The Future that is returned, representing the results of the ItemProcessor's execution, is then passed to the AsyncItemWriter. Just like the AsyncItemProcessor, the AsyncItemWriter is also a decorator for a provided ItemWriter. The AsyncItemWriter finally unwraps the Future and passes the result to the delegate ItemWriter. It is important to note that the AsycnItemProcessor and AsyncItemWriter should be used together. Otherwise, you will be responsible for unwrapping the Futures returned by the AsyncItemProcessor yourself.

■ **Note**   AsyncItemProcessor and AsyncItemWriter should be used together.

Before we can use the `AsyncItemProcessor` and `AsyncItemWriter`, we need to import a new module into our project, the spring-batch-integration module. Listing 11-8 has the Maven configuration to add to your pom.xml.

Listing 11-8. spring-batch-integration

```
...
<dependency>
        <groupId>org.springframework.batch</groupId>
        <artifactId>spring-batch-integration</artifactId>
</dependency>
...
```

For this example, we will use the same use case we did in the parallel steps. However, this time we'll only import the CSV file. However, the difference is that we will add an `ItemProcessor` that does a `Thread.sleep(5)` for each item. While 5 milliseconds may not seem like a lot, if you process one million records in sequence, that can add over an hour of processing time to your job.[1] Not a trivial amount of time. However, if that is parallelized, you can see immediate benefits.

Once we've defined our `ItemProcessor`, we will define the `AsyncItemProcessor` to decorate ours. This will take a `TaskExecutor` to launch the underlying `ItemProcessor#process` call in another thread. Listing 11-9 shows the configuration of the `ItemProcessor` used for this example.

Listing 11-9. Async ItemProcessor

```
...
        @Bean
        public AsyncItemProcessor<Transaction, Transaction> asyncItemProcessor() {
                AsyncItemProcessor<Transaction, Transaction> processor = new
                AsyncItemProcessor<>();

                processor.setDelegate(processor());
                processor.setTaskExecutor(new SimpleAsyncTaskExecutor());

                return processor;
        }

        @Bean
        public ItemProcessor<Transaction, Transaction> processor() {
                return (transaction) -> {
                        Thread.sleep(5);
                        return transaction;
                };
        }
...
```

With the `ItemProcessors` defined, we can add the `AsyncItemWriter` and configure our `Step` to use them. Listing 11-10 begins with the writer that does the actual work. In this case, it's the same `ItemWriter` configuration that we used in the parallel step example. The second bean in the listing is the `AsyncItemWriter` that decorates the `JdbcBatchItemWriter` we are delegating to. After that, we configure the

---

[1]One million records x 5 milliseconds = 1 hour 23.34 minutes

step to use the AsyncItemProcessor and AsyncItemWriter instead of the delegates we also configured. One other detail about the step configuration to note is the types used on the chunk method call. Instead of the generics being <Transaction, Transaction> as you'd normally expect, since the second generic indicates the input of the ItemWriter, we need to update that to be <Transaction, Future<Transaction>> since a Future<Transaction> is what the AsycItemProcessor will actually return. The final piece of the listing is the configuration of our job to use the step we just configured.

Listing 11-10. Async ItemWriter

```
...
@Bean
public JdbcBatchItemWriter<Transaction> writer(DataSource dataSource) {
        return new JdbcBatchItemWriterBuilder<Transaction>()
                        .dataSource(dataSource)
                        .beanMapped()
                        .sql("INSERT INTO TRANSACTION (ACCOUNT, AMOUNT, TIMESTAMP) " +
                                "VALUES (:account, :amount, :timestamp)")
                        .build();
}

@Bean
public AsyncItemWriter<Transaction> asyncItemWriter() {
        AsyncItemWriter<Transaction> writer = new AsyncItemWriter<>();

        writer.setDelegate(writer(null));

        return writer;
}

@Bean
public Step step1async() {
        return this.stepBuilderFactory.get("step1async")
                        .<Transaction, Future<Transaction>>chunk(100)
                        .reader(fileTransactionReader(null))
                        .processor(asyncItemProcessor())
                        .writer(asyncItemWriter())
                        .build();
}

@Bean
public Job asyncJob() {
        return this.jobBuilderFactory.get("asyncJob")
                        .start(step1async())
                        .build();
}
...
```

If you run the job now with the AsyncItemProcessor and AsyncItemWriter in place, you can see that even with a million records, there is a significant performance improvement. One thing to note, the example here uses the SimpleAsyncTaskExecutor which uses a new thread per request. In a production environment, you'll want to use something safer like the ThreadPoolTaskExecutor.

Up to this point, all of the scaling options have been tied to using threads within a single JVM. However, not all workloads can fit into a single JVM. The next option for scaling Spring Batch workloads allows you to choose whether to use it via threads in a single JVM or via remote worker JVMs. Let's take a look at partitioning.

Partitioning

The majority of batch based workloads are I/O bound. Interacting with a database or reading files typically is where performance and scalability concerns come into play. To help with that, Spring Batch provides the ability for multiple workers to execute complete steps. The entire ItemReader, ItemProcessor, and ItemWriter interaction can be offloaded to workers. This section looks at what partitioning is and how to configure jobs to take advantage of this powerful Spring Batch feature.

Partitioning is a concept where a master step farms out work to any number of worker steps for processing. In a partitioned step, a large data set (say a database table with a million rows in it) is divided into smaller partitions. Each of those partitions is processed in parallel by the workers. Each worker is a complete Spring Batch step that is responsible for its own reading, processing, writing, and so on. There are great advantages to this model. For example, you get all of the features like restartability out of the box with this model. The implementation of the workers feels natural because it's just another step.

Using a partitioned step within Spring Batch requires the understanding of two main abstractions. The first is the Partitioner interface. This interface is responsible for understanding the data to be partitioned and how to divide it up into partitions. Going back to the example of a database table with a million rows in it, a Partitioner implementation may execute queries to determine what ids are part of each partition. Spring Batch provides one Partitioner implementation out of the box. The MultiResourcePartitioner looks at an array of Resources and creates a partition per Resource.

We should pause here and answer the question "What is a partition" within Spring Batch? How is it represented? It's actually very simple. A partition is represented by an ExecutionContext that contains the pertitinent data to identify what the partition consists of. When using the MultiResourcePartitioner, Spring Batch sets the name of the Resource in the ExecutionContext for each partition. That is stored in the job repository for later reference by the workers.

The Partitioner interface consists of a single method, partition(int gridSize). It returns a Map<String, ExecutionContext>. The gridSize is nothing more than a hint to the implementation as to how many workers are there to be able to divide the data in a way that is efficient for the overall cluster. That being said, there is nothing within Spring Batch that dynamically determines that value. It's up to you to calculate or set. The Map that the method returns should consist of key value pairs where the key is the name of the partition and should be unique. The ExecutionContext, as mentioned previously, is the representation of the partition metadata identifying what to process.

The other key abstraction in a partitioned step in Spring Batch is the PartitionHandler. This interface is one that understands how to communicate with the workers. How to tell each worker what to work on and how to identify when all the work is complete. While you will probably write your own Partitioner implementations when using Spring Batch, you probably will not write your own PartitionHandler.

Within the Spring portfolio, there are three implementations of this interface. The two in Spring Batch are the TaskExecutorPartitionHandler and the MessageChannelPartitionHandler. The TaskExecutorPartitionHandler launches the workers as threads within the same JVM allowing you to use the partitioning concept within a single JVM. The MessageChannelPartitionHandler uses Spring Integration to send the metadata to remote JVMs for processing. The last implementation of the PartitionHandler within Spring is provided by the Spring Cloud Task project. That project provides the DeployerPartitionHandler implementation. This implementation delegates to a Spring Cloud Deployer implementation to launch the workers on a supported platform[2] on demand. These workers start, execute their partition, then shut down providing dynamic scaling at runtime. We'll take a look at all three as we explore partitioning within Spring Batch. Figure 11-14 illustrates the relationships between the various components of a partitioned Spring Batch step.

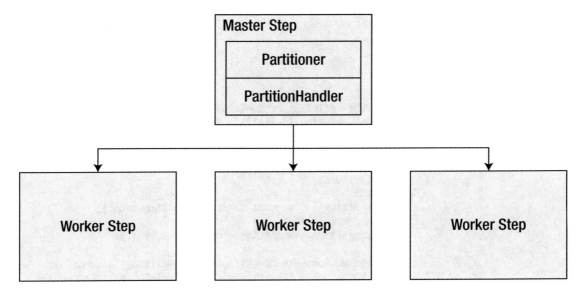

Figure 11-14. *A partitioned Spring Batch step*

There are some considerations that need to be taken when using a partitioned step. For example, the state of the step is maintained within the job repository. That is across both the master and all workers, so all components of the cluster must be configured to talk to the same job repository database instance. Another consideration is that if you are to use the `MessageChannelPartitionHandler`, you'll need to be able to communicate with the remote JVMs. This means setting up messaging middleware or some other mechanism supported by Spring Integration for that communication. Let's begin to dig into partitioning from a code perspective by using a single JVM.

TaskExecutorPartitionHandler

The `TaskExecutorPartitionHandler` is a component that allows a partitioned step to execute the workers using threads within a single JVM. For many use cases, this is a good way to start with partitioning before taking on the added complexities of orchestrating remote JVMs. The limitations of this approach center around the limits of a single JVM. There are limits on what you can accomplish on a single box (how fast you can get data off the disk, how fast and how many network connections you can have at once, etc.) which are what you are trying to exercise when using threads in this manner. You are really trying to push a single JVM to its max performance.

The use case we are going to look at for using a partitioned step is the ingestion of multiple files into a database. Each file can be processed independently. This style of use case (I/O bound with no dependencies between the input) is a classic example of when to use partitioning. We'll use the same input CSV files

---

[2]As of this writing, CloudFoundry, Kubernetes, and local implementations are maintained by the Spring team.

we used in our split configuration. The domain object will be the same as well. In fact, the code for the worker step we'll run in our job is also the same as step2 in the split job in Listing 11-4 with one very minor difference. The method signature for our fileTransactionReader obtains the name of the file to read in via the job parameters. When using partitioning, we obtain any partition specific information (like the file we are going to process) from the step execution context. Listing 11-11 shows the updated listing for the reader in our worker step.

Listing 11-11. fileTransactionReader

```
...
@Bean
@StepScope
public FlatFileItemReader<Transaction> fileTransactionReader(
                @Value("#{stepExecutionContext['file']}") Resource resource) {

        return new FlatFileItemReaderBuilder<Transaction>()
                        .name("flatFileTransactionReader")
                        .resource(resource)
                        .delimited()
                        .names(new String[] {"account", "amount", "timestamp"})
                        .fieldSetMapper(fieldSet -> {
                                Transaction transaction = new Transaction();

                                transaction.setAccount(fieldSet.readString("account"));
                                transaction.setAmount(fieldSet.readBigDecimal("amount"));
                                transaction.setTimestamp(fieldSet.readDate("timestamp",
                                                        "yyyy-MM-dd HH:mm:ss"));

                                return transaction;
                        })
                        .build();
}
...
```

One of the great advantages to how Spring Batch's scalability features work is that you can add them iteratively without need for heavy rewrites each time. Partitioning is just another example of that. The only difference between the single threaded version of this job and the partitioned version is the addition of the partitioned step configuration and the change to the reader to obtain the file location from the step execution context instead of the job parameters.

Now it is important to note that the step we defined previously and are reusing is now the worker step and not the master step in our job. Because of that, our job will not directly reference that step. Instead, we need to define our partitioned step with two components: A Partitioner implementation and a PartitionHandler implementation. For this example, both are available out of the box from Spring Batch. All we need to do is to configure them. Let's start with the Partitioner.

The Partitioner interface, as previously discussed, is responsible for understanding the data set and how to divide it up into partitions. Spring Batch only provides one implementation out of the box, the MultiResourcePartitioner. This implementation takes an array of resources and creates a new partition for each resource. Listing 11-12 illustrates our configuration of our MultiResourcePartitioner.

Listing 11-12. MultiResourcePartitioner

```
...
@Bean
@StepScope
public MultiResourcePartitioner partitioner(
        @Value("#{jobParameters['inputFiles']}") Resource[] resources) {

    MultiResourcePartitioner partitioner = new MultiResourcePartitioner();

    partitioner.setKeyName("file");
    partitioner.setResources(resources);

    return partitioner;
}
```

The bean definition takes in a single parameter, the array of `Resource` objects provided by Spring when we pass in the path to the input files. The `MultiResourcePartitioner` takes two values to be set. The first is the key name. This is the value that the workers will use to look in the provided `ExecutionContext` for the name of the resource to read. In this example, that value "file" matches the key we used in the defintion of the `FlatFileItemReader` in Listing 11-11. The other value we need to set on this `Partitioner` is that array of `Resources`. Once that is done, we can return the instance.

The other component we'll use in our partitioned step is the `PartitionHandler` implementation. For this first look at partitioning, we're going to use the `TaskExecutorPartitionHandler`. Here, we set two values on that implementation. The first is the step we want to execute; in this case, it is called step1. The second value is a `TaskExecutor`. By default, if you either do not provide a `PartitionHandler` to a `PartitionStepBuilder` or do not provide a `TaskExecutor` to the `TaskExecutorPartitionHandler`, a `SyncTaskExecutor` is used as the default. While that may seem like it defeats the purpose if you're looking for parallelism, there really isn't another reasonable, production grade option. So you'll always want to be sure to set the `TaskExecutor` on this implementation to something that uses multiple threads. For testing, the `SimpleAsyncTaskExecutor` will do.[3] Listing 11-13 shows the configuration of the `TaskExecutorPartitionHandler` for our job.

Listing 11-13. TaskExecutorPartitionHandler

```
...
@Bean
public TaskExecutorPartitionHandler partitionHandler() {
    TaskExecutorPartitionHandler partitionHandler =
            new TaskExecutorPartitionHandler();

    partitionHandler.setStep(step1());
    partitionHandler.setTaskExecutor(new SimpleAsyncTaskExecutor());

    return partitionHandler;
}
...
```

---

[3]Do not use the `SimpleAsyncTaskExecutor` in production as it does not recycle threads and does not have any kind of limit on how many it creates.

With both of our new beans defined, we can create our new partitioned step and update our job to use it. The partitionedMaster step uses the normal StepBuilderFactory to obtain a builder. It then sets the partitioner providing both the step name to execute and the Partitioner instance. It needs the step name to be able to create the step execution contexts for each partition. The second thing we set is the partition handler. Listing 11-14 illustrates both the partitionedMaster step as well as the job updated to reference it.

Listing 11-14. partitionedMaster Step and Job

```
...
@Bean
public Step partitionedMaster() {
        return this.stepBuilderFactory.get("step1")
                        .partitioner(step1().getName(), partitioner(null))
                        .partitionHandler(partitionHandler())
                        .build();
}

@Bean
public Job partitionedJob() {
        return this.jobBuilderFactory.get("partitionedJob")
                        .start(partitionedMaster())
                        .build();
}
...
```

With everything defined, we can build and execute the job. Once the jar is built, we can execute the job using the command java -jar partition-demo-0.0.1-SNAPSHOT.jar inputFiles=/data/csv/ transactions*.csv. When we look at the output from the logs, not much will have changed. However, there are two key differences. The first is that all the files located in /data/csv with the prefix transactions and a .csv extension will have been imported into the TRANSACTION table. Second, if we take a look at the BATCH_STEP_ EXECUTION table, there is one record for the partitioned step, then one additional record for each partition that was executed as well. Listing 11-15 shows the results of the BATCH_STEP_EXECUTION table after running three files.

Listing 11-15. BATCH_STEP_EXECUTION After a Partitioned Step

```
mysql> select step_name, status, commit_count, read_count, write_count from SCALING.BATCH_
STEP_EXECUTION;
+------------------+-----------+--------------+------------+-------------+
| step_name        | status    | commit_count | read_count | write_count |
+------------------+-----------+--------------+------------+-------------+
| step1            | COMPLETED |          303 |      30000 |       30000 |
| step1:partition1 | COMPLETED |          101 |      10000 |       10000 |
| step1:partition2 | COMPLETED |          101 |      10000 |       10000 |
| step1:partition0 | COMPLETED |          101 |      10000 |       10000 |
+------------------+-----------+--------------+------------+-------------+
4 rows in set (0.01 sec)
```

The TaskExecutorPartitionHandler is the simplest way to add partitioning to a Spring Batch step. However, it's also the most restrictive when you consider that it is tied to a single JVM. In the next section, we will take a look at distributed batch processing by using the MessageChannelPartitionHandler to distribute the partition workload over multiple JVMs.

MessageChannelPartitionHandler

Spring Integration is a project in the Spring portfolio that implements the enterprise integration patterns presented in the book "Enterprise Integration Patterns" by Gregor Hohpe and Bobby Woolf. When we are looking for ways to communicate between JVMs on a Spring project, the components in this framework normally come to mind, which brings us to the `MessageChannelPartitionHandler`.

The `MesssageChannelPartitionHandler` is a `PartitionHandler` implementation that uses Spring Integration's `MessageChannel` abstraction to communicate with external JVMs via some means. For this example, we will be using RabbitMQ, an open source message broker that is easy to use both in production environments as well as locally.

When we're looking at a partitioned step in a distributed fashion, the topolgoy obviously changes a bit. Instead of our worker step implementations running within the same JVM via threads, we'll have a listener in each of the worker JVMs listening for a request to execute the worker step. Figure 11-15 illustrates the topology of a remote partitioned step.

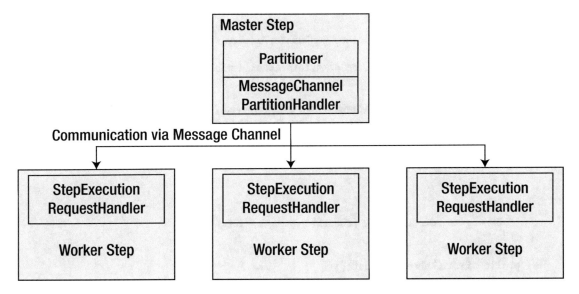

Figure 11-15. *A remote partitioned Spring Batch step*

As you can see, the master communicates with the workers via messages to execute the worker step. Each of the worker JVMs has a listener listening on the queue for those messages. When the request comes over the wire, the listener executes the step and returns the results. It's important to note a few details of this architecture.

The first is that all JVMs involved in this process need to be configured to use the same job repository. Since that's where the state is maintained, each step is responsible for maintaining the results as it processes. Without this shared state, restartability of a partitioned job would not be possible. The second is that each of the worker steps is executed outside of the context of a job. So anything that you'd need from the `JobExecution` or job's `ExecutionContext` will not be available in the workers.

Now the configuration of this type of partitioned step is the same on the master side except for three things. First, we'll need to configure the communication mechanisms for our app to talk to RabbitMQ. That is easy thanks to Spring Boot and Spring Integration. The second piece we need to do is to configure our new `PartitionHandler`, the `MessageChannelPartitionHandler`. Fortunately, Spring Batch provides a builder that helps with all of that as well. Finally, to simplify the deployment, we use Spring profiles to group the components for the worker applications and for the master application. This allows us to use a single JAR for both and just specific which we are running via a simple parameter.

Let's start the configuration with the master. We'll create a new configuration class called MasterConfiguration. We will add a new annotation to it, @EnableBatchIntegration. This annotation provides the required builders for building remote partitioned steps in a simple way. On the master side of the configuration, there are really two flows that need to be configured, the outbound flow (sending the messages to the workers) and the inbound flow (receiving messages from the workers). Listing 11-16 begins showing the configuration by configuring the outbound flow.

Listing 11-16. MasterConfiguration and Outbound Flow

```
...
@Configuration
@Profile("master")
@EnableBatchIntegration
public class MasterConfiguration {

        private final JobBuilderFactory jobBuilderFactory;

        private final RemotePartitioningMasterStepBuilderFactory
                masterStepBuilderFactory;

        public MasterConfiguration(JobBuilderFactory jobBuilderFactory,
                        RemotePartitioningMasterStepBuilderFactory masterStepBuilderFactory) {

                this.jobBuilderFactory = jobBuilderFactory;
                this.masterStepBuilderFactory = masterStepBuilderFactory;
        }

        /*
         * Configure outbound flow (requests going to workers)
         */
        @Bean
        public DirectChannel requests() {
                return new DirectChannel();
        }

        @Bean
        public IntegrationFlow outboundFlow(AmqpTemplate amqpTemplate) {
                return IntegrationFlows.from(requests())
                                .handle(Amqp.outboundAdapter(amqpTemplate)
                                                .routingKey("requests"))
                                .get();
        }
...
```

The block of code in Listing 11-16 begins with the annotations to identify the class as a configuration class, that it is to be used with the master profile and that we will be using the Spring Batch Integration components in it. The first builder factory within the class is the same as we've used to define every job so far. However, the second builder factory (RemotePartitioningMasterStepBuilderFactory) is new. This is a special builder factory for getting step builders that know how to build a master remote partitioned step.

The code in this listing continues with a constructor allowing for both builder factories to be injected. The next pieces are the configuration of the outbound flow, connecting one end of a direct channel to an AMQP Template (provided by Spring Boot's autoconfiguration). The `IntegrationFlow` uses Spring Integration's Java DSL to configure that flow. In English, it says "when a message comes in on the requests channel, pass it to the handler configured. The handler configured is an AMQP outbound adapter that will send the messages to a RabbitMQ queue named requests."

Once we have the outbound plumbing configured, we should configure the inbound plumbing. Now we should note that there are two options for obtaining the results from the workers with remote partitioning. The first (and the option we will use here) is to receive messages sent back to the master from each worker, aggregate them, and evaluate the results to determine if the step was successful or not. The alternative is to poll the job repository to check the status of each `StepExecution` we sent out. Once they are all updated in the database as being complete, we can then evaluate the status from there. As I mentioned previously, we'll use the mechanism of receiving result messages from each worker in this example. Listing 11-17 illustrates the configuration of the master flow to receive those resulting requests.

Listing 11-17. Inbound Flow

```
...
        /*
         * Configure inbound flow (replies coming from workers)
         */
        @Bean
        public DirectChannel replies() {
                return new DirectChannel();
        }

        @Bean
        public IntegrationFlow inboundFlow(ConnectionFactory connectionFactory) {
                return IntegrationFlows
                                .from(Amqp.inboundAdapter(connectionFactory,"replies"))
                                .channel(replies())
                                .get();
        }
...
```

The configuration of the inbound flow is essentially the opposite of the outbound one. We again, begin with the definition of a direct channel. From there, the `IntegrationFlow` we configure using Spring Integration's Java DSL reads in English like this "When we receive a message on the replies queue in RabbitMQ, take that message and put it on the replies channel."

The last pieces on the master side to configure are the `Step` and `Job`. I should note that you want the job to be configured in the master profile when using remote partitioning because if you use the same Spring Boot uber jar for both the master and the worker (as we recommend), Spring Boot will automatically execute the job in the worker application as well as the master if you don't put the job exclusively in the master profile. Listing 11-18 illustrates the creation of the `Partitioner` (also required in the master step), the `Step`, and `Job` for the remote partitioning job.

Listing 11-18. Inbound Flow

```
...
        @Bean
        @StepScope
        public MultiResourcePartitioner partitioner(
                @Value("#{jobParameters['inputFiles']}") Resource[] resources) {
```

```
                MultiResourcePartitioner partitioner = new MultiResourcePartitioner();

                partitioner.setKeyName("file");
                partitioner.setResources(resources);

                return partitioner;
        }

        @Bean
        public Step masterStep() {
                return this.masterStepBuilderFactory.get("masterStep")
                                .partitioner("workerStep", partitioner(null))
                                .outputChannel(requests())
                                .inputChannel(replies())
                                .build();
        }

        @Bean
        public Job remotePartitioningJob() {
                return this.jobBuilderFactory.get("remotePartitioningJob")
                                .start(masterStep())
                                .build();
        }

}
```

Listing 11-18 begins with the definition of the MultiResourcePartitioner. This is the same as it was in the TaskExecutorPartitionHandler section. Our data hasn't changed, so how we divide it up shouldn't. Next we define our remote partitioned step. We begin by obtaining the builder from the factory. From there, we provide our Partitioner, an output channel (requests), and an input channel (replies). With those components configured, Spring Batch will wire up all the components needed to send the outbound requests, receive the replies, and aggregate them into a single result for the master step.

With the master side configured, we can take a look at the worker configuration. Again, we'll use another configuration class for this called WorkerConfiguration. It will actually look very similar to the master in that we will again configure two flows, an inbound flow and an outbound flow, and a Step. However, instead of configuring a Job, we will just need to configure the reader and writer that our worker step use (hint: it will be the same as we used in the TaskExecutorPartitionHandler section). Listing 11-19 begins the listing for the WorkerConfiguration class.

Listing 11-19. WorkerConfiguration and inboundFlow

```
...
@Configuration
@Profile("!master")
@EnableBatchIntegration
public class WorkerConfiguration {

        private final RemotePartitioningWorkerStepBuilderFactory
                workerStepBuilderFactory;
```

```
        public WorkerConfiguration(
                RemotePartitioningWorkerStepBuilderFactory workerStepBuilderFactory) {

                this.workerStepBuilderFactory = workerStepBuilderFactory;
        }

        /*
         * Configure inbound flow (requests coming from the master)
         */
        @Bean
        public DirectChannel requests() {
                return new DirectChannel();
        }

        @Bean
        public IntegrationFlow inboundFlow(ConnectionFactory connectionFactory) {
                return IntegrationFlows
                                .from(Amqp.inboundAdapter(connectionFactory, "requests"))
                                .channel(requests())
                                .get();
        }

        @Bean
        public DirectChannel replies() {
                return new DirectChannel();
        }
}
...
```

Similar to the `MasterConfiguration` in Listing 11-16, we begin with the `@Configuration` annotation, a `@Profile` annotation (in this case indicating that we want to use this configuration any time we are not using the master profile), and the `@EnableBatchIntegration` annotation. To prevent duplication, you can also move this annotation to the main class if you wanted.

The class then begins with another new factory. This time we have the `RemotePartitioningWorker StepBuilderFactory`. This is the counterpart to the `RemotePartitioningMasterStepBuilderFactory`. This builder factory provides a step builder that can build the components needed for the worker side of a remote partitioned step. Those components consist of a `StepExecutionRequestHandler` and the other components of our remote step. The `StepExecutionRequestHander` is responsible for receiving the message from the master step and executing it in the remote JVM. The code continues with the definition of our inbound channel, `requests`, and the integration flow used to define how our request is routed. In plain words, the `inboundFlow` is configured to take each request it receives from the AMQP queue named `requests` and pass it to the `requests` channel. The last piece of the worker side plumbing is the definition of the `replies` channel. This channel will serve as the return route for the results of each partition.

With the inbound plumbing constructed, the worker step is constructed with the new builders as shown in Listing 11-20.

Listing 11-20. workerStep Configuration

```
...
public Step workerStep() {
        return this.workerStepBuilderFactory.get("workerStep")
                        .inputChannel(requests())
```

```
                            .outputChannel(replies())
                            .<Transaction, Transaction>chunk(100)
                            .reader(fileTransactionReader(null))
                            .writer(writer(null))
                            .build();
}
...
```

Just like the master step begins with specifying the name, input channel, and output channel, so does the worker. However, from there, the rest of the configuration for this step is exactly as you'd expect for any other Spring Batch step. In this case, we specify the chunk size, the reader and writer to use, and call `build()`. The reader and writer for this job are the same we used in the thread based partitioned example.

Launching this example is a bit more complex. Now, we are dealing with multiple JVMs. Before we execute our job, we need our messaging middleware up and running. In our case, we'll be using RabbitMQ, so we need to be sure that is running locally. For OS X users who installed RabbitMQ via brew, the command is typically `rabbitmq-server`. Once Rabbit is up and running, we can start our worker JVMs. For this example, we'll execute three worker JVMs. To do this, we'll need to execute the same command in three different shell windows. Each one will be executed with the command `java -jar target/partitioned-demo-0.0.1-SNAPSHOT.jar --spring.profiles.active=worker`. Once each of those applications has started, you'll notice that they just sit and wait. Since there is no `Job` bean defined in the `worker` profile, Spring Boot doesn't automatically start our processing like it has in our other examples. To launch our job, we'll need to launch the `master` profile with the command `java -jar target/partitioned-demo-0.0.1-SNAPSHOT.jar --spring.profiles.active=master`. This will launch our master application, and Spring Boot will launch the job. If we monitor the logs in the workers, we'll see as each worker picks up the request from the queue and executes it as well as when they all finish in the master. On a clean database, the output in the database for this execution should match that of what we saw from the `TaskExecutorPartitionHandler` example.

We've looked at two different mechanisms for executing partitioned workloads in Spring Batch so far. The first was within a single JVM with threads and the second was with static worker JVMs waiting for work in a classic messaging workload style. However, neither of these options lend themselves well to the benefits that can be had in the cloud. The third option is designed to take advantage of the dynamic resources available in the cloud when using partitioning in Spring Batch.

DeployerPartitionHandler

The biggest driver to the cloud is the abilty to get compute on demand. If I need two servers 80% of the time, but I need 100 servers that other 20%, I can easily accomplish both with out paying for 100 servers 100% of the time. That ability to scale on demand is a huge benefit for many workloads. Batch processing is probably one of the best examples of where that kind of elasticity is ideal. Think about it. Most batch processes are run on a schedule of some kind, execute within a finite timeframe, then are no longer needed until the next window. In traditional Java deployments, these applications would sit idle in application containers 24/7, wasting resources. In the cloud, it doesn't have to be that way.

The `DeployerPartitionHandler` is the last `PartitionHandler` implementation provided by the Spring portfolio out of the box. This `PartitionHandler` utilizes another abstraction, a `TaskLauncher`, to execute the workers on demand. The flow goes like this. The master application is executed with no workers running. When the partitioned step begins, the master determines how many partitions there are. Once that is known, the `PartitionHandler` launches a new instance of the application on platform to execute a partition. It will execute as many as required (up to a configured max) to get the work done. Say the master determines there are ten partitions and the `DeployerPartitionHandler` is configured to use at most four workers. The `DeployerPartitionHandler` will keep four workers busy until all ten partitions have been executed.

With this approach, you get the same benefits of partitioning in Spring Batch (restartability, higher throughput, etc.) with the dynamic scaling of the cloud. Now in order for this to work, you need a

TaskLauncher that is supported for your platform as well as Spring Cloud Task. As of the writing of this book, there are TaskLaunchers for CloudFoundry, Kubernetes, and a local version. Listing 11-21 has the Maven imports for running this example locally.

Listing 11-21. Maven Dependencies for Spring Cloud Task and the Local Deployer

```
...
<dependency>
        <groupId>org.springframework.cloud</groupId>
        <artifactId>spring-cloud-starter-task</artifactId>
</dependency>
<dependency>
        <groupId>org.springframework.cloud</groupId>
        <artifactId>spring-cloud-deployer-local</artifactId>
</dependency>
...
```

For our look into this PartitionHandler implementation, we will use the same job we have used up to now. The general configuration looks like a combination of the two previous options. It will have two profiles since we'll have independent JVMs, each with their own responsibility (one for running the master and multiple for running the worker); however, it won't have the Spring Integration configuration because we won't need any form of messaging middleware in between our master and worker applications. Listing 11-22 begins the configuration for this job with the real differences.

Listing 11-22. BatchConfiguration

```
...
@Configuration
public class BatchConfiguration {

        @Autowired
        private JobBuilderFactory jobBuilderFactory;

        @Autowired
        private StepBuilderFactory stepBuilderFactory;

        @Autowired
        private JobRepository jobRepository;

        @Autowired
        private ConfigurableApplicationContext context;

        @Bean
        @Profile("master")
        public DeployerPartitionHandler partitionHandler(TaskLauncher taskLauncher,
                        JobExplorer jobExplorer,
                        ApplicationContext context,
                        Environment environment) {
                Resource resource =
                        context.getResource("file:///path-to-jar/partitioned-demo-0.0.1-
                        SNAPSHOT.jar");
```

```
                DeployerPartitionHandler partitionHandler =
                        new DeployerPartitionHandler(taskLauncher, jobExplorer, resource,
                        "step1");

                List<String> commandLineArgs = new ArrayList<>(3);
                commandLineArgs.add("--spring.profiles.active=worker");
                commandLineArgs.add("--spring.cloud.task.initialize.enable=false");
                commandLineArgs.add("--spring.batch.initializer.enabled=false");
                commandLineArgs.add("--spring.datasource.initialize=false");
                partitionHandler.setCommandLineArgsProvider(
                        new PassThroughCommandLineArgsProvider(commandLineArgs));
                partitionHandler.setEnvironmentVariablesProvider(
                        new SimpleEnvironmentVariablesProvider(environment));
                partitionHandler.setMaxWorkers(3);
                partitionHandler.setApplicationName("PartitionedBatchJobTask");

                return partitionHandler;
        }

        @Bean
        @Profile("worker")
        public DeployerStepExecutionHandler stepExecutionHandler(JobExplorer jobExplorer) {
                return new DeployerStepExecutionHandler(this.context, jobExplorer, this.
                jobRepository);
        }
...
```

Starting at the top, we create the class definition with the @Configuration annotation. After that, we autowire in the normal Spring Batch step and job builder factories. Since there is nothing "special" about these steps, we don't need special builders to create our Step. We also autowire in the JobRepository and the current app's context which are used in the worker profile. We'll get to their use in a bit.

The first bean configured here is the DeployerPartitionHandler. There is a lot going on in there, so let's break down what it's going to do before we walk through configuring it. The DeployerPartitionHandler is going to launch a new instance of an application on the given platform. Think about deploying a new application on Kubernetes. You create a docker image, publish it to a docker registry, then use Kubernetes's tools to download and push that to a Kubernetes cluster. The DeployerPartitionHandler does the same thing. So you create a docker image, publish it to a docker registry, then the DeployerPartitionHandler downloads it and pushes it automatically to the Kubernetes cluster (assuming you are using Kubernetes for this). Now let's get back to the configuration and see how what it has to do maps to how we configure it.

The configuration for the deployerPartitionHandler begins by obtaining a Resource that is the artifact that will be executed as the worker. So if you are using the Kubernetes deployer under the hood, this would point to a docker resource. In our case, since we are going to use the local deployer (the default), we are pointing that resource to our Spring Boot über jar. When the workers are launched, the DeployerPartitionHandler will delegate to the TaskLauncher to create the command to execute (in the local's case, it will generate the java -jar command executed from a shell).

Once we've obtained a reference to the Resource we will run, we create the DeployerPartitionHandler instance. It's constructor takes three arguments, a taskLauncher, jobExplorer, the Resource we just obtained, and the step name to be executed. The taskLauncher is the platform specific piece that knows how to launch an application on the platform. Spring provides three options as of the writing of this book: Local, CloudFoundry, and Kubernetes. The JobExplorer reference is used to poll the job repository to see if the workers are complete. The Resource is the resource to launch as previously mentioned. Finally, the name of the step to execute is what the worker will launch once the app is running.

Once we have the `DeployerPartitionHandler` created, there are two main abstractions related to it that we'll need to work with. The first is the `CommandLineArgsProvider`. This is a strategy interface that allows users to customize what command line arguments are passed to the uber jar when it is executed. In our case, we're going to use the one provided out of the box from Spring, the `PassThroughCommandLineArgsProvider` to pass in a list of args. The args we are passing are as follows:

- `--spring.profiles.active=worker`: This arg is used to set the profile on each worker launched to be a `worker`.

- `--spring.cloud.task.intialize.enable=false`: This tells Spring Cloud Task to not run the database initialization code for each worker. Since it will be initialized on startup of the master (or by other means) we don't want each worker to reinitialize it.

- `--spring-batch.initializer.enabled=false`: Same concept as the previous argument, only for the Spring Batch tables instead of the Spring Cloud Task ones.

- `--spring.datasource.initialize=false`: Again, same concept as the previous two arguments, only for any other database scrips declared in our über jar.

Once we have defined the `CommandLineArgsProvider`, the other abstraction related to this `PartitionHandler` is the `EnvironmentVariablesProvider`. This is also a strategy interface that is used to set any environment variables within the shell the worker apps are running in. In our case, we are going to copy the current environment over to the workers by using the `SimpleEnvironmentVariablesProvider`.

The last two pieces we need to configure are the max number of workers we will allow to be running at once and the application name. We configure the max workers so that Spring doesn't launch an unknown number of workers. If we do not set that value, Spring will launch one worker per partition. If your `Partitioner` determines you have 1000 partitions, it will launch 1000 workers ... probably not what you want. By setting it to three, if we have less than three partitions, the `PartitionHandler` will launch one per partition. If we have more than three partitions, the `PartitionHandler` will launch three, then as each one finishes, it will launch more up to the max three until all partitions are complete. The application name we need to configure is used by Spring Cloud Task in the task repository for tracking the execution of the workers.

Now that the `DeployerPartitionHandler` is configured, we need a mechanism within the workers to launch the step that is requested. When we are using the `MessageChannelPartitionHandler`, there is a listener that monitors a channel for the requests and launches them as they come in (`StepExecutionRequestHandler`). When using the `DeployerPartitionHandler`, we need another mechanism to launch the worker step, the `DeployerStepExecutionHandler`. This handler, instead of obtaining its information from a message via a channel, pulls the values of the job execution id, step execution id, and step name out of the environment via well known properties. It then executes the step just as the `StepExecutionRequestHandler` in the previous example did. To configure the `DeployerStepExecutionHandler`, we simply need to provide a context (for the handler to obtain a handle on the step to execute), a `jobExplorer` (to obtain the `StepExecution` to execute from the job repository), and a `JobRepository` to update the step's execution upon a step failure that is not handled by the step implementation itself.

While a lot is going on there, that is all that is different from the previous example ... except for how to execute it. In the previous example, we had to manually launch the workers and make sure RabbitMQ was running before they were launched. With this approach, we only need to worry about launching the master. It will handle launching the workers for us. So once we have built our project, all we need to do is `java -jar target/partitioned-demo-0.0.1-SNAPSHOT.jar --spring.profiles.active=master`. We can monitor the progress of the app in the job repository but that doesn't illustrate to us the fact that the other workers are being launched. When they are, you'll see in the logs of the master the locations of the log files as shown in Listing 11-23.

Listing 11-23. Launching Worker JVMs

```
2019-01-05 10:34:16.533  INFO 67745 --- [           main] o.s.c.t.b.l.TaskBatchExecutionList
ener    : The job execution id 1 was run within the task execution 1
2019-01-05 10:34:16.562  INFO 67745 --- [           main] o.s.batch.core.job.
SimpleStepHandler      : Executing step: [step1]
2019-01-05 10:34:16.640 DEBUG 67745 --- [           main] o.s.c.t.b.p.DeployerPartitionHandl
er      : 3 partitions were returned
2019-01-05 10:34:16.684  INFO 67745 --- [           main] o.s.c.d.spi.local.
LocalTaskLauncher      : launching task PartitionedBatchJobTask-a5e75fd0-0c90-49b3-9e2b-
428d5182765c
    Logs will be in /var/folders/6s/2mwfrcbx5tg1mxr251bbl44m0000gn/T/spring-cloud-
dataflow-9037584903989022167/PartitionedBatchJobTask-1546706056645/PartitionedBatchJobTask-
a5e75fd0-0c90-49b3-9e2b-428d5182765c
2019-01-05 10:34:16.697  INFO 67745 --- [           main] o.s.c.d.spi.local.
LocalTaskLauncher      : launching task PartitionedBatchJobTask-4a4c7152-8f3c-48ce-84d3-
cac0919d4385
    Logs will be in /var/folders/6s/2mwfrcbx5tg1mxr251bbl44m0000gn/T/spring-cloud-
dataflow-9037584903989022167/PartitionedBatchJobTask-1546706056689/PartitionedBatchJobTask-
4a4c7152-8f3c-48ce-84d3-cac0919d4385
2019-01-05 10:34:16.709  INFO 67745 --- [           main] o.s.c.d.spi.local.
LocalTaskLauncher      : launching task PartitionedBatchJobTask-705cf4dd-b708-491f-8191-
f6f520cc018c
    Logs will be in /var/folders/6s/2mwfrcbx5tg1mxr251bbl44m0000gn/T/spring-cloud-
dataflow-9037584903989022167/PartitionedBatchJobTask-1546706056699/PartitionedBatchJobTask-
705cf4dd-b708-491f-8191-f6f520cc018c
```

As you can see in Listing 11-23, when the worker JVM is launched using the LocalTaskLauncher (the default), the location of the log files is presented in this log. If we open one of those directories up, we'll see a stdout.log and a stderr.log for standard out and standard error, respectively. The stdout.log file will look just like any other Spring Boot log file assuming all goes well. There is another way to tell that the extra JVMs are launching (besides the tell tail sign of your laptop fan spooling up). We can also monitor the execution of the extra JVMs via the jps command. This command is like the Unix ps command, only it lists all the Java virtual machines running. Listing 11-24 shows the list I have when I execute this job.

Listing 11-24. jps Output

```
→  ~ jps
68944 RemoteMavenServer
42899
88027 partitioned-demo-0.0.1-SNAPSHOT.jar
88045 partitioned-demo-0.0.1-SNAPSHOT.jar
88044 partitioned-demo-0.0.1-SNAPSHOT.jar
88047 Jps
88046 partitioned-demo-0.0.1-SNAPSHOT.jar
```

You can see that we end up with four Java processes, one for the master and three workers. With just like all the other examples, the output is the same as seen in Listing 11-25.

Listing 11-25. Partitioned Job Output

```
mysql> select step_name, status, commit_count, read_count, write_count from SCALING.BATCH_
STEP_EXECUTION;
+------------------+-----------+--------------+------------+-------------+
| step_name        | status    | commit_count | read_count | write_count |
+------------------+-----------+--------------+------------+-------------+
| step1            | COMPLETED |          303 |      30000 |       30000 |
| step1:partition1 | COMPLETED |          101 |      10000 |       10000 |
| step1:partition0 | COMPLETED |          101 |      10000 |       10000 |
| step1:partition2 | COMPLETED |          101 |      10000 |       10000 |
+------------------+-----------+--------------+------------+-------------+
4 rows in set (0.01 sec)
```

Partitioning a workload, whether it be within a single JVM or spread across a cluster, is a powerful tool. When working with a use case that is bound by IO, it can have profound impacts on the performance of a job.

However, not all processes are bound by IO. If you have a use case where the processor piece is the bottle neck and you need more than what a single JVM can handle, remote chunking may be the right scaling option for you. The next section will take a look at this last option for scaling Spring Batch jobs.

Remote Chunking

Distributed computing where the processing of data is offloaded to the cluster is a common pattern seen in many verticles. One of the more extreme examples is the BOINC system developed by Berkley. This framework was originally developed for use with the SETI@Home project which searches for evidence of extraterrestrial intelligence by having unused personal computers process radio signals recorded from radio telescopes. BOINC was later extracted from the SETI@Home project and is now a more general use framework used in the scientific community for a number of @Home projects (Folding@Home which looks at protein folding, Einstein@Home which looks for pulsars, Rosetta@Home which performs protein structure prediction for disease research, etc.). The idea around BOINC, however, is actually quite simple. It consists of a master command server that receives requests for data to be processed. It replies with input to be processed by the requesting worker. The worker downloads the input, performs the required processing, then uploads the results.

Remote chunking is similar to how BOINC works only in a push instead of pull model. Unlike remote partitioning where metadata is sent over the wire to the workers, remote chunking (like BOINC) sends the actual data to be processed over the wire. The master reads the data, sends it to the workers for processing, then the workers write the output. This type of scaling outside of a single JVM is useful only when item processing is the bottleneck in your process. If input or output is the bottleneck, this type of scaling only makes things worse. There are a couple things to consider before using remote chunking as your method for scaling batch processing:

- *Processing needs to be the bottleneck:* Because reading and writing are completed in the master JVM, in order for remote chunking to be of any benefit, the cost of sending data to the slaves for processing must be less than the benefit received from parallelizing the processing.

- *Guaranteed delivery is required:* Because Spring Batch doesn't maintain any type of information about who is processing what, if one of the slaves goes down during processing, Spring Batch has no way to know what data is in play. Thus a persisted form of communication (typically a persistent messaging based solution) is required.

To configure a job using remote chunking, you begin with a normally configured job that contains a step that you want to execute remotely. Spring Batch allows you to add this functionality with no changes to the configuration of the job itself. Instead, you hijack the ItemProcessor of the step to be remotely processed and insert an instance of a ChunkHandler implementation (provided by Spring Batch Integration). The org.springframework.batch.integration.chunk.ChunkHandler interface has a single method, handleChunk, that works just like the ItemProcessor interface. However, instead of actually doing the work for a given item, the ChunkHandler implementation sends the item to be processed remotely and listens for the response. When the item returns, it's written normally by the local ItemWriter. Figure 11-16 shows the structure of a step that is using remote chunking.

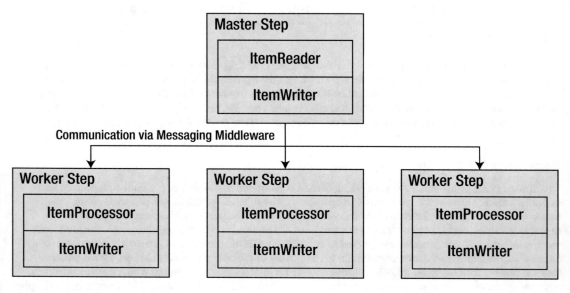

Figure 11-16. *The structure of a step using remote chunking*

As Figure 11-16 shows, any one of the steps in a job can be configured to do its processing via remote chunking. When you configure a given step, that step's ItemProcessor is replaced with a ChunkHandler, as mentioned previously. That ChunkHandler's implementation uses a special writer (org.springframework. batch.integration.chunk.ChunkMessageChannelItemWriter) to write the items to the queue. The workers are nothing more than message-driven POJOs that execute your business logic. When the processing is completed, the output of the ItemProcessor is persisted via the ItemWriter.

For this example, we will just demonstrate the flow of data through the process doing the same import job we did in the partitioned examples before. Just like the partitioned job that uses the MessageChannelPartitionHandler, we'll be working with two profiles and Spring Integration for configuring the communication between the master and the worker nodes. Let's begin with the master configuration. Listing 11-26 has the configuration for the master step.

Listing 11-26. Master Remote Chunked Step

```
...
@EnableBatchIntegration
@Configuration
public class BatchConfiguration {
```

```java
@Configuration
@Profile("!worker")
public static class MasterConfiguration {

        @Autowired
        private JobBuilderFactory jobBuilderFactory;

        @Autowired
        private RemoteChunkingMasterStepBuilderFactory
                remoteChunkingMasterStepBuilderFactory;

        @Bean
        public DirectChannel requests() {
                return new DirectChannel();
        }

        @Bean
        public IntegrationFlow outboundFlow(AmqpTemplate amqpTemplate) {
                return IntegrationFlows.from(requests())
                                .handle(Amqp.outboundAdapter(amqpTemplate)
                                                .routingKey("requests"))
                                .get();
        }

        @Bean
        public QueueChannel replies() {
                return new QueueChannel();
        }
        @Bean
        public IntegrationFlow inboundFlow(
                ConnectionFactory connectionFactory) {

                return IntegrationFlows
                                .from(Amqp.inboundAdapter(connectionFactory,
                                "replies"))
                                .channel(replies())
                                .get();
        }

        @Bean
        @StepScope
        public FlatFileItemReader<Transaction> fileTransactionReader(
                @Value("#{jobParameters['inputFlatFile']}") Resource resource) {

                return new FlatFileItemReaderBuilder<Transaction>()
                                .saveState(false)
                                .resource(resource)
                                .delimited()
                                .names(new String[] {"account",
                                                "amount",
                                                "timestamp"})
```

```
                                    .fieldSetMapper(fieldSet -> {
                                        Transaction transaction = new Transaction();

                                        transaction.setAccount(
                                                fieldSet.readString("account"));
                                        transaction.setAmount(
                                                fieldSet.readBigDecimal("amount"));
                                        transaction.setTimestamp(
                                                fieldSet.readDate("timestamp",
                                                        "yyyy-MM-dd HH:mm:ss"));

                                        return transaction;
                                    })
                                    .build();
        }
        @Bean
        public TaskletStep masterStep() {
                return this.remoteChunkingMasterStepBuilderFactory.get("masterStep")
                                .<Transaction, Transaction>chunk(100)
                                .reader(fileTransactionReader(null))
                                .outputChannel(requests())
                                .inputChannel(replies())
                                .build();
        }

        @Bean
        public Job remoteChunkingJob() {
                return this.jobBuilderFactory.get("remoteChunkingJob")
                                .start(masterStep())
                                .build();
        }
    }
...
```

Just like the remote partitioning example earlier in this chapter, the remote chunking configuration begins with the @EnableBatchIntegration annotation. Spring Batch provides special builder factories to build a step that uses remote chunking and this annotation enables that functionality. Once in the class we break the master/worker configuration up using inner classes. The master configuration is first.

The master's configuration begins with the autowiring of both a jobBuilderFactory (like all Spring Batch configuration classes) and a new builder factory called the RemoteChunkingMasterStepBuilderFactory. Just as the RemotePartitioningMasterStepBuilderFactory provides a special builder for creating the master step for remote partitioning, the RemoteChunkingMasterStepBuilderFactory will provide a builder for creating the master step for a remote chunking step.

After the two autowired factories, we create a channel for the outbound flow. This channel will be used to send data to RabbitMQ for processing by the workers. The flow using that channel is the next bean definition. You'll notice that this outbound flow is the same as what is used in the outbound flow in our remote partitioning example. Read in English, the Spring Integration DSL states to take each message from the requests channel and pass it to the AMQP outbound adapter which will send it to the RabbitMQ queue named requests.

Once we have the outbound flow configured, we configure the inbound flow. Now you may be wondering, since the writing is going to happen on the worker side of this step, what replies are we expecting. The replies that come from the worker side are actually the StepContribution for the master to apply. This allows the stats in the job repository to be accurate even if this is a distributed step. To configure the inbound flow, we configure a direct channel and then the inbound Spring Integration flow. This flow reads from the replies queue in RabbitMQ, take the messages and put them in the replies channel.

The next component configured in the master is the FlatFileItemReader we are using to read the transaction input file. This is the same as the reader we have configured previously. The reason I call it out here is that it is the only part of a chunked step that is configured in the master side. The ItemProcessor and ItemWriter are both configured in the worker side of the application.

Once we have the components of the step configured, we can configure the Step itself. The configuration of the masterStep reads just like any other chunk based step to begin with. We specify the name, chunk size, and the reader. However, instead of either an ItemProcessor or an ItemWriter, we specify an input channel and an output channel. These are used to configure the step to send and receive the messages to the workers. The last bean in the master configuration is the configuration for the Job. We put this bean in the master profile so Boot only executes it on the master and not once per JVM.

Seven beans for the master and three of them would be required even if we were not using remote chunking to scale the step. Not bad. Let's take a look at the worker's configuration. It is found in Listing 11-27.

Listing 11-27. Worker Remote Chunked Step

```
...
        @Configuration
        @Profile("worker")
        public static class WorkerConfiguration {

                @Autowired
                private RemoteChunkingWorkerBuilder<Transaction, Transaction> workerBuilder;

                @Bean
                public DirectChannel requests() {
                        return new DirectChannel();
                }

                @Bean
                public DirectChannel replies() {
                        return new DirectChannel();
                }

                @Bean
                public IntegrationFlow inboundFlow(ConnectionFactory connectionFactory) {
                        return IntegrationFlows
                                        .from(Amqp.inboundAdapter(connectionFactory,
                                        "requests"))
                                        .channel(requests())
                                        .get();
                }
                @Bean
                public IntegrationFlow outboundFlow(AmqpTemplate template) {
                        return IntegrationFlows.from(replies())
```

```
                                        .handle(Amqp.outboundAdapter(template)
                                                    .routingKey("replies"))
                        .get();
        }

        @Bean
        public IntegrationFlow integrationFlow() {
                return this.workerBuilder
                                .itemProcessor(processor())
                                .itemWriter(writer(null))
                                .inputChannel(requests())
                                .outputChannel(replies())
                                .build();
        }

        @Bean
        public ItemProcessor<Transaction, Transaction> processor() {
                return transaction -> {
                        System.out.println("processing transaction = " + transaction);
                        return transaction;
                };
        }
        @Bean
        public JdbcBatchItemWriter<Transaction> writer(DataSource dataSource) {
                return new JdbcBatchItemWriterBuilder<Transaction>()
                                .dataSource(dataSource)
                                .beanMapped()
                                .sql("INSERT INTO TRANSACTION (ACCOUNT, AMOUNT,
                                TIMESTAMP) " +
                                        "VALUES (:account, :amount, :timestamp)")
                                .build();
        }
    }
}
```

Again using an inner class to group the beans makes it easier to rationalize about them. The class begins with the autowiring of the RemoteChunkingWorkerBuilder. You'll note that this is not a FactoryBean bean like the others. The reason for this is that no factory is needed. When working with the other builders, the factory you interact with does some Spring "magic" behind the scenes. In the case of this builder, there is no magic needed so you just get the builder itself.

Moving down the configuration, we again have an inbound flow and an outbound flow that do the same things that they did on the master side of the configuration. Things get different when we realize that we have three integration flows instead of the two we had in the master. On the worker side, instead of configuring a step, we actually configure an integration flow when using remote chunking. Behind the scenes, this builder will create a chain that accepts the incoming requests, passes them to a service activator for handling the batch related processing, and the result is returned to the outbound flow. To configure that flow, we use the new builder, the RemoteChunkingWorkerBuilder to configure the ItemProcessor, the ItemWriter, the input channel, and the output channel.

The configuration completes when we configure our ItemProcessor (a simple lambda that passes the transaction to System.out) and the ItemWriter which stores the items in a database table.

Once we have our configuration, we can build our project and execute it. The mechanisms for launching this are actually the same as they were for the remote partitioning example where we used RabbitMQ. We begin by starting RabbitMQ (if it's not already running) via `rabbitmq-server`. From there, we can launch our workers each with the command `java -jar target/chunking-demo-0.0.1-SNAPSHOT.jar`. Finally we can launch the master with the command `java -jar target/chunking-demo-0.0.1-SNAPSHOT.jar --spring.profiles.active=master`.

With all of the components running, we can see the output of our job in the database just as we have up to now as shown in Listing 11-28.

Listing 11-28. Remote Chunking Job Output

```
mysql> select step_name, status, commit_count, read_count, write_count from SCALING.BATCH_
STEP_EXECUTION;
+------------------+-----------+--------------+------------+-------------+
| step_name        | status    | commit_count | read_count | write_count |
+------------------+-----------+--------------+------------+-------------+
| step1            | COMPLETED |          303 |      30000 |       30000 |
+------------------+-----------+--------------+------------+-------------+
1 rows in set (0.01 sec)
```

Summary

One of the primary reasons for using Spring Batch is its ability to scale without having a large impact on your existing code base. Although you could write any of these features yourself, none of them are easy to implement well, and it doesn't make sense to reinvent the wheel. Spring Batch provides an excellent set of ways to scale jobs with a number of options.

This chapter looked at how to profile jobs to obtain information about where bottlenecks exist. You then worked through examples of each of the five options Spring Batch provides for scalability: parallel steps, multithreaded steps, asynchronous item processing, remote chunking, and partitioning.

CHAPTER 12

■ ■ ■

Cloud Native Batch

Batch processing has been around for a long time. Since the dawn of automated computing, gathering data, running a process over it, and generating output from it have been a fundamental piece of it. As enterprises transition to the cloud, it is just natural that batch processing also migrates there.

However, what makes an application ready for the cloud or "cloud native"? Can we just pick up our Spring Batch application and drop it in the cloud? In short, probably not. Even if it does "work" there are additional concerns about running an application in the cloud. A concept called the "twelve factor application" was devised explicitly for this purpose. A twelve factor application addresses the extra concerns related to running an application within the cloud.

The Spring Cloud project is a portfolio of projects that build on Spring Boot to enable cloud native development. Things like circuit breakers, service discovery, configuration management, and job orchestration (all related to twelve factor applications) all fall into this bucket of functionality. Don't know what any of those are? Don't worry, we'll walk through them in this chapter.

In this chapter we will iterate over a very simple Spring Batch application, taking it from a traditional Spring Boot based Spring Batch application, to a cloud native one. Each iteration we'll add new features to it to take advantage of cloud native features until we have an application that utilizes features of Spring Cloud. Specifically

- We'll walk through what are the tenants of a twelve factor application and how do they apply to batch processing.

- We'll look at a very simple Spring Batch application that we'll migrate to a cloud native application.

- We will add additional resiliency to our batch application by using the circuit breaker pattern to handle interacting with a faulty REST API.

- We'll then externalize our configuration with both Spring Cloud Config Server and Spring Cloud Eureka.

- Finally, we'll orchestrate our batch job with Spring Cloud Data Flow.

Before we can get started on looking at the details, let's take a look at what a twelve factor application is.

© Michael T. Minella 2019

M. T. Minella, *The Definitive Guide to Spring Batch*, https://doi.org/10.1007/978-1-4842-3724-3_12

Twelve Factor Applications

The concept of a twelve factor application came out of Heroku and their work with cloud computing. The goals of a twelve factor application are to develop patterns that can be used to develop applications as services. The twelve factors are[1]

1. Codebase: One codebased tracked in revision control with many deploys

2. Dependencies: Explicitly declare and isolate dependencies

3. Config: Store configuration in the environment

4. Backing services: Treat backing services as attached resources

5. Build, release, run: Strictly separate build and run stages

6. Processes: Execute the app as one or more stateless processes

7. Port binding: Export services via port binding

8. Concurrency: Scale out via the process model

9. Disposability: Maximize robustness with fast startup and graceful shutdown

10. Dev/prod parity: Keep development, staging, and production as similar as possible

11. Logs: Treat logs as event streams

12. Admin processes: Run admin/management tasks as one-off processes

Let's walk through each of those, define what they really mean as well as talk about how they each apply to batch processing.

Codebase

The codebase of your application should be consolidated in a single version control repository. The idea here is that if you need to break your codebase up into multiple repositories, you probably don't have an application … you have a distributed system. Twelve factor applications make up a distributed system, but each of these apps are self-contained and independent. From the perspective of batch processing, in most of our book we've been following this model of a single batch job per application. In legacy environments, however, it is not uncommon to see codebases where there is a monolithic WAR or EAR file that contains multiple batch jobs. In a cloud native environment, you'd want to break that up into multiple applications.

Dependencies

Dependencies are a part of developing in java. You use your build system (Maven or Gradle) to download and include your dependencies in your Spring Boot über jar as we have in all the examples to date. This model is a specific requirement for twelve factor applications. The idea here is that you don't want your application depending on something external to it. All dependences should be encapsulated within the application via some mechanism. Spring Boot handles this for us.

---

[1]https://12factor.net/

Config

Configuration must be separated from the code in a twelve factor application. Why is that? Because the same artifact will work through multiple environments and it must be independent of those environments. While even Spring provides mechanisms for configuring environments independently via features like profiles, that model doesn't scale as cleanly as using something like environment variables or a centralized configuration server. In the batch world, we have a number of things we need to have configured. Whether it be the database for the job repository or access to other systems for our inputs and outputs, Spring Batch applications will need to be adapted to meet this principle.

Backing Services

A backing service is any service that the application depends on. This may be a RDBMS, SMTP server, S3, third party APIs, and so on. The key to this tenant though is that all of these should be referenceable via a URL or other locator that is made available via the configuration. There should be no direct dependence on a particular service instance within the code. For example, whether you use a local MySQL or Amazon's RDS in the cloud, your code should not change. Only the configuration should need to change (changing the URL, username, password, etc.). In a batch application, this means structuring our code and configuration separately in ways where our application can be configured this way. Fortunately, using good development practices with Spring Boot already enforces these practices.

Build, Release, Run

The twelve factor tenants strictly divide the process of building an application, releasing it, and running it. Building an application is about compiling and testing the application's code. Releasing the artifact is not only about creating the artifact but providing a unique identifier for that version and stored in a place that it cannot be modified (a Maven repository, for example). Running the application is the taking of that released artifact and executing it in an environment. With Spring Batch, you'll want to setup a continuous integration system like Jenkins or Concourse to run your build pipeline. From there, tools like Spinnaker can handle the release and run pieces in very robust ways, providing for things like rollbacks if a deployment fails, and so on.

Processes

Twelve factor applications are stateless and share nothing. This means that if I execute an application once, it will not expect any preexisting data either locally on the file system, or in memory. The classic example of how to break this tenant in the web world is building your application to require sticky sessions. This design implies that there is data that persists in memory from request to request, is therefore stateful, and will scale poorly. In the batch world, Spring Batch has been designed from the beginning to be stateless. I bet you're thinking, but wait, batch jobs are stateful. They can be restarted and have state in the ExecutionContext. While that is all true, the state is maintained within the job repository, a relational database that is cloud friendly. I can execute a properly designed batch job on one node, have it fail, then restart it on a completely different node with no other preexisting state defined besides the ability to connect to the same database for the job repository.

Port Binding

A twelve factor application does not require the runtime addition of a server of some kind to expose it as a service. It is self-contained and will bind to a port itself if that is required to execute the functionality it is designed for. An example of this would be in a non-twelve factor application, you would take a WAR file and deploy it on Tomcat. In a twelve factor application, it is completely self-contained and so Tomcat is embedded in the application (as is done in Spring Boot). In Spring Batch or batch jobs in general, this isn't as big of a concern given that batch jobs are by definition, self-contained. That being said, Spring does provide the ability to handle scenarios where you can have a Spring Batch job open a port to expose itself for various other reasons.

Concurrency

Twelve factor applications must be able to scale via processes. This does not mean that the use of threads is forbidden or even discouraged. Threads as found in the JVM are great tools for scaling an application and should be used in the right scenarios. However, a JVM can only grow to a certain point and so using multiple instances must be an option. As we discussed in Chapter 11, we can scale Spring Batch applications either via threads within a JVM or externally with remote chunking and remote partitioning. Both are examples of how Spring Batch fits well into this model.

Disposability

In a legacy world, the idea that a process is going to go down is a scary one. The idea of pager duty in a world where applications must stay up or customers are impacted send shivers down the spine of any developer. However, in a cloud native world, processes are disposable. They can be stopped and restarted without notice and need to be architected as such. Processes should start as quickly s possible and shut down gracefully when a request to stop is received. Spring Batch naturally falls into this. Due to how it works with the JobRepository, it has the ability to be shut down and restarted on demand. That being said, the process itself is still stateful and on a restart, state does need to be restored. Spring Batch optimizes this by doing things like skipping records that have already been processed on a restart.

Dev/Prod Parity

One of the goals of the cloud is business agility. Remove the delays of things like getting servers racked and databases provisioned. Speed from a business perspective is a key driver to the cloud. The ultimate goal of that speed is continuous deployment. Going from deploying to production once a quarter or once a month, to multiple times a day. This can only occur when there is parity between environments. If you are developing against MySQL but deploying in production against Oracle, there may be issues that are not found. Parity between environments is key to minimizing issues that can be exposed by these differences. Spring Batch doesn't address these kinds of nonfunctional demands but doing so will make your release and deployment process much smoother.

Logs

We said earlier that processes must be disposable and stateless. This includes their log files as well. Applications in a cloud native environment will ideally write all log output to stdout to be consumed either by a developer as they work at their terminal window or via a log ingestion system like Splunk or related system. This allows for the aggregation across instances to better understand what is happening within your system. Spring integrates well with all major Java logging frameworks and as so handles this requirement well.

Admin Processes

Administration processes like database migrations should be run as one-off processes. The funny thing about this tenant of the twelve factor application is that it's not uncommon for batch processes *to be the* administration processes themselves.

That is it for the twelve factors in cloud native application development. However, how do they apply from a tactical perspective to our Spring Batch applications. In the next section, we'll begin exploring that by looking at the Spring Batch application we are going to evolve from a basic application to a cloud native one.

A Simple Batch Job

For the rest of this chapter, we will take a simple Spring Batch application and evolve it to use the capabilities in Spring to make it cloud native. The use case for this job is intended to be unusually simplistic. We want to focus not on the domain model, but on the additional capabilities we are adding with each iteration. So for our use, a single step batch job that downloads files from Amazon's S3 to a local directory, it will read in each file line by line, calling a REST API to enrich the data, then store the results a database.

Before we create our batch job we will create our project from Spring Initializr. We will bring in the usual suspects for dependencies plus one we have not looked at. We can start with batch, jdbc, and MySQL. However we will add the AWS dependency as well to be able to interact with Amazon's S3 store. We will be adding dependencies as we add new features; however, for our base application, that should suffice.

To start looking at our job, we'll begin with the main class. This is nothing more than the normal Spring Boot main class we have used in every other example in this book. Listing 12-1 shows the main class.

Listing 12-1. CloudNativeBatchApplication.java

```
...
@EnableBatchProcessing
@SpringBootApplication
public class CloudNativeBatchApplication {

        public static void main(String[] args) {
                SpringApplication.run(CloudNativeBatchApplication.class, args);
        }
}
```

The class begins with two normal annotations, the `@EnableBatchProcessing` annotation used to bootstrap all of the Spring Batch infrastructure, and `@SpringBootApplication` to bootstrap classpath scanning and autoconfiguration. The main method on the class simply points Spring Boot as to where to begin looking for configurations via the classpath scanning. The rest resides in other configuration classes.

This leads us to the `JobConfiguration` class. This class contains all of the configuration for all of the components required for our Spring Batch job to execute. Listing 12-2 provides the code for the `JobConfiguration` class.

Listing 12-2. JobConfiguration.java

```
...
@Configuration
public class JobConfiguration {

        @Autowired
        private StepBuilderFactory stepBuilderFactory;
```

```
@Autowired
private JobBuilderFactory jobBuilderFactory;

@Bean
public DownloadingJobExecutionListener downloadingStepExecutionListener() {
        return new DownloadingJobExecutionListener();
}

@Bean
@StepScope
public MultiResourceItemReader reader(
        @Value("#{jobExecutionContext['localFiles']}")String paths) throws Exception {

        System.out.println(">> paths = " + paths);
        MultiResourceItemReader<Foo> reader = new MultiResourceItemReader<>();

        reader.setName("multiReader");
        reader.setDelegate(delegate());

        String [] parsedPaths = paths.split(",");
        System.out.println(">> parsedPaths = " + parsedPaths.length);
        List<Resource> resources = new ArrayList<>(parsedPaths.length);

        for (String parsedPath : parsedPaths) {
                Resource resource = new FileSystemResource(parsedPath);
                System.out.println(">> resource = " + resource.getURI());
                resources.add(resource);
        }
        reader.setResources(resources.toArray(new Resource[resources.size()]));

        return reader;
}

@Bean
@StepScope
public FlatFileItemReader<Foo> delegate() throws Exception {
        FlatFileItemReader<Foo> reader = new FlatFileItemReaderBuilder<Foo>()
                        .name("fooReader")
                        .delimited()
                        .names(new String[] {"first", "second", "third"})
                        .targetType(Foo.class)
                        .build();

        return reader;
}
@Bean
@StepScope
public EnrichmentProcessor processor() {
        return new EnrichmentProcessor();
}
```

```
@Bean
public JdbcBatchItemWriter<Foo> writer(DataSource dataSource) {

        return new JdbcBatchItemWriterBuilder<Foo>()
                        .dataSource(dataSource)
                        .beanMapped()
                        .sql("INSERT INTO FOO VALUES (:first, :second, :third,
                        :message)")
                        .build();
}

@Bean
public Step load() throws Exception {
        return this.stepBuilderFactory.get("load")
                        .<Foo, Foo>chunk(20)
                        .reader(reader(null))
                        .processor(processor())
                        .writer(writer(null))
                        .build();
}

@Bean
public Job job(JobExecutionListener jobExecutionListener) throws Exception {
        return this.jobBuilderFactory.get("s3jdbc")
                        .listener(jobExecutionListener)
                        .start(load())
                        .build();
}

@Bean
public RestTemplate restTemplate() {
        return new RestTemplate();
}
}
```

There is quite a bit of code in this but the vast majority of it should be familiar. Let's start at the top. After the @Configuration and class definition, we're autowiring in two fields. Both are the builders used to create our steps and jobs.

The first bean definition in our JobConfiguration class is one for a DownloadingJobExecutionListener. This is a JobExecutionListener that we will implement in just a bit, but it is responsible for downloading the files from S3 to be imported into our database. In the beforeJob method is where we will do the download. This design allows us to be compliant with the Processes tenant of the twelve factor application. If our job fails and is restarted on a new node or container in the cloud, the beforeJob(JobExecution jobExecution) method will be re-executed and the files will be redownloaded.

Since we are processing multiple files, the next bean is the MultiResourceItemReader. This factory method looks at the directory containing the files downloaded by the previous listener and configures the MultiResourceItemReader to read all the files in it. You'll notice this method is marked as step scoped via the @StepScope annotation. This is due to the fact that the listener places a list of the files it downloaded into job's ExecutionContext to be pulled out by this method.

413

The MultiResourceItemReader needs another ItemReader to delegate to during the reading. The next bean in our configuration, the delegate, is the one we'll use. This factory method provides a FlatFileItemReader configured to read a CSV with three values in each record. The values are originally named "first," "second," and "third" (real inventive). The values for each of these are mapped to a domain object called Foo that contains a setFirst(int value), setSecond(int value), and a setThird(String value) methods on it for Spring to automatically call.

Once the readers have been defined, the ItemProcessor that will be calling the REST API will be defined. The factory method for this component simply returning a new instance of the EnrichmentItemProcessor. This custom ItemProcessor we'll look at later, but all it does is call a REST API that returns the number of times it's been called. The EnrichmentItemProcessor then set the field message with the result.

The final component for the step is the ItemWriter. For this, we're using the JdbcBatchItemWriter to write our enriched items. Using the builder for that ItemWriter, we configure the DataSource, identify that we'll be using the BeanPropertyItemSqlParameterSourceProvider via the beanMapped() method call, provide our SQL insert statement, and call build.

That is all of our components for the load step. Now to build the step, we use the autowired StepBuilderFactory to get StepBuilder. We identify that we're going to use a chunk based step passing it in the MultiResourceItemReader we configured previously, the EnrichmentItemProcessor, and the JdbcBatchItemWriter. This constructs the step we'll use to import the data.

In order to user our step, we need to configure our job. Using the autowired JobBuilderFactory, we configure the DownloadingJobExecutionListener as a JobExecutionListener on the job, identify that the job will start with our load step, and call build on the builder.

The last bean in this class may not seem like it's needed. We explicitly configure a RestTemplate bean. Spring Boot will automatically add one to your context with the right starter. However, we are going to customize this component in a future iteration of the batch job, so we'll explicitly configure one for now.

That's all for the JobConfiguration class. We identified two custom classes we created for this job, the DownloadingJobExecutionListner and the EnrichmentProcessor. We'll take a look at the DownloadingJobExecutionListner first. In the DownloadingJobExecutionListener, we use Spring's S3 resource capabilities from the Spring AWS project to navigate the resources stored in the configured S3 bucket and download it in the beforeJob(JobExecution jobExecution) method. Listing 12-3 displays the code for this listener.

Listing 12-3. DownloadingJobExecutionListner.java

```
...
public class DownloadingJobExecutionListener extends JobExecutionListenerSupport {

        @Autowired
        private ResourcePatternResolver resourcePatternResolver;

        @Value("${job.resource-path}")
        private String path;

        @Override
        public void beforeJob(JobExecution jobExecution) {

                try {
                        Resource[] resources =
                                this.resourcePatternResolver.getResources(this.path);

                        StringBuilder paths = new StringBuilder();
```

```
                for (Resource resource : resources) {

                        File file = File.createTempFile("input", ".csv");

                        StreamUtils.copy(resource.getInputStream(),
                                            new FileOutputStream(file));

                        paths.append(file.getAbsolutePath() + ",");
                        System.out.println(">> downloaded file : " +
                                    file.getAbsolutePath());
                }

                jobExecution.getExecutionContext()
                                    .put("localFiles",
                                    paths.substring(0, paths.length() - 1));
            }
            catch (IOException e) {
                e.printStackTrace();
            }
        }
    }
}
```

The DownloadingJobExecutionListener extends JobExecutionListenerSupport which provides a no-op implementation of the JobExecutionListener for us to override just the methods we care about. In this case, our implementation only requires us to implement the beforeJob method. In there, we begin by getting a list of all the resources found in the S3 bucket we configured via the job.resource-path application parameter. Once we've obtained the array of Resources (files), we create a temporary file for each one, then use Spring's StreamUtils to download the file saving off the absolute path for each file downloaded. Once all the files have been downloaded, the list of paths to them is saved in the job's ExecutionContext with the key localFiles.

One key point about this that makes this work in a cloud native environment is that each time the job runs, this listener will re-execute, redownloading each of the files needed. It will also overwrite the values in the job's ExecutionContext from the last time, preventing the job from looking for files that most likely won't be there if the job is being run in a new container.

The last piece of code for this example job is the EnrichmentProcessor. This processor makes a simple GET call to a REST API using the RestTemplate we defined in our JobConfiguration. Listing 12-4 shows the code for the EnrichmentProcessor.

Listing 12-4. EnrichmentProcessor.java

```
...
public class EnrichmentProcessor implements ItemProcessor<Foo, Foo> {

    @Autowired
    private RestTemplate restTemplate;

    @Override
    public Foo process(Foo foo) throws Exception {
            ResponseEntity<String> responseEntity =
                this.restTemplate.exchange(
                        "http://localhost:8080/enrich",
                        HttpMethod.GET,
```

```
                                 null,
                                 String.class);
                    foo.setMessage(responseEntity.getBody());

                    return foo;
          }
}
```

Once the response is received from the `RestTemplate` call, it is stored in the item. In our case, it will just be a message stating "Enriched X" where X is the number of times the controller in the REST API has been called.

With all of the code for our job defined, our job is in need of a bit of configuration. To begin, we'll use Spring Boot's application.yml to configure our application. Listing 12-5 provides an example of the application.yml required for this application.

Listing 12-5. application.yml

```
spring:
  datasource:
    driverClassName: org.mariadb.jdbc.Driver
    url: jdbc:mysql://localhost:3306/cloud_native_batch
    username: 'root'
    password: 'password'
    schema: schema-mysql.sql
job:
  resource-path: s3://def-guide-spring-batch/inputs/*.csv
cloud:
  aws:
    credentials:
      accessKey: 'OPAR802SSRDI9NIGDBWA'
      secretKey: 'SDKEjF9IqNOIjTKIJVaEOG9UwI+=DOEFTjOkS2B4'
    region:
      static: us-east-1
      auto: false
```

■ **Note** You will need to configure the database url, username, password, `job.resource-path`, AWS credentials, and region to match your own values. The preceding ones are provided as samples and will not work for you.

The application.yml begins by configuring our `DataSource` with the normal values required by Spring Boot. The only additional piece is the schema value which points to the database table for our `Foo` item. Once the database is configured, the `job.resource-path` points to the S3 bucket where our input files are located on AWS. The last configuration pieces are our AWS credentials and region configuration. The `cloud.aws.region.auto` is an important value. When you run your code on AWS, Spring will automatically configure the region to be the same as the region the code is running in. For example, if you are in the US East 1 region, Spring will automatically configure your S3 region to be the same. The code, as configured, is not configured to be run on AWS, so we turn off the autoconfiguration of our region and configure it explicitly.

That's it for the batch job. If you tried to compile and run it now, however, you'd get an exception. We haven't gone over the REST API the `EnrichmentProcessor` will be calling.

The REST API we'll build will be a web module created via Spring Initializr. Once we've downloaded and imported our project, we will only need to add a single class, our controller. Listing 12-6 shows the code for that controller.

Listing 12-6. EnrichmentController.java

```
...
@RestController
public class EnrichmentController {

        private int count = 0;

        @GetMapping("/enrich")
        public String enrich() {
                this.count++;

                return String.format("Enriched %s", this.count);
        }
}
```

This controller is annotated with the `@RestController` which indicates to Spring that the values returned from the methods will be returned to the client in its raw form. Its sole field is the counter used to keep track of how many times the controller is called.[2] The `enrich` method is annotated with @GetMapping mapping that method to the /enrich URL. All our method does is increment our counter and format the message, returning the formatted message to the caller.

If we build and run the REST API and the job, you should see that all the data stored in S3 will be imported successfully into your database. This job and REST API is the baseline for what we will iterate on for the rest of the chapter.

It is important to note that while this chapter discusses the building of a cloud native application, we will not be running it on any given cloud. Given the number of different options on the market at the time of this writing and no clear winner at this point, we will leave out the issues of deploying these on a specific cloud provider. That being said, everything in this chapter is intended to be cloud agnostic and can be run on any of the major cloud providers (CloudFoundry, Kubernetes, Google Cloud Platform, Amazon Web Services, etc.).

The first cloud native feature we'll take a look at with Spring Batch is one that is common in cloud native web applications, the circuit breaker.

Circuit Breaker

One interesting aspect of batch processes is that they are very efficient. If you need to make a large number of API calls, a batch job will do them efficiently. If you need to read messages from a queue, batching them up is a known way to improve performance. The same goes for writing to databases. The list goes on. However, with that efficiency can come trouble.

Say that REST API becomes overloaded. Does calling it over and over mercilessly in a batch process make sense? Einstein defined insanity as doing the same thing over and over again and expecting different results. Yet this is exactly what can happen.

---

[2]We are ignoring the thread safety issues of using an `int` for this use case for the sake of simplicity.

Instead, what if we gave the API a chance to catch up before hitting it with another request? Netflix popularized this technique in their microservices architecture via a framework called Hystrix. The idea is simple. Identify a method that is surrounded by a circuit breaker. When a threshold of exceptions is surpassed, the circuit breaker trips stopping calls to that method and routing traffic to an alternative method. This alternative method usually handles things in a different way. For example, returning a default value instead of one the REST API returns. The circuit breaker will, based on some algorithm, trickle traffic back through to the original method to test to see if it's back online. Once it is, the circuit breaker resets and traffic is restored as normal to the original method.

Netflix has since deprecated Hystrix and resilience4j has taken its place, but Spring Batch actually needs neither of them to implement the circuit breaker pattern. Spring Batch depends on a library called Spring Retry. This little known library is actually heavily utilized throughout the Spring portfolio. It provides the fault tolerant capabilities in the fault tolerant step in Spring Batch. However, for our purposes, we're going to use a feature added in recent versions … a basic circuit breaker.

Our use case will work like this. Our REST API will be configured to return random exceptions. We'll wrap our `EnrichmentProcessor#process` method in the circuit breaker and have the message field of our item set to "error" by the alternative method.

Given Spring Batch's fault tolerant capabilities, you may wonder why we would even do this instead of using those facilities. There are really two reasons to use a circuit breaker in this instance instead of a fault tolerant step. The first is performance. When an operation is retried in Spring Batch, as we've identified in previous chapters, the framework will roll back the transaction, set the commit count to one, and then retry each item in its own transaction. This can be a very performance killing operation. If you can just flag items as errors and can rerun them later, that's a much more efficient mechanism for dealing with errors. The second reason is due to the use case. Spring Batch will let you retry items but it doesn't offer the ability to back off the pressure on the offending code. If your service needs time to recover, Spring Batch doesn't have a good facility built in to handle that. Enter Spring Retry's circuit breakers.

The key components to Spring Retry's circuit breakers are two annotations. The first, not surprisingly, is `@CircuitBreaker`. This method level annotation indicates that something should be wrapped in a circuit breaker. By default, the circuit breaker will be closed until three exceptions of the configured type (all exceptions by default) are thrown from the method in 5 seconds. The circuit breaker, once tripped will remain open for 20 seconds by default before attempting the main path again. All of this is configurable via the annotation parameters as shown in Table 12-1.

Table 12-1. *CircuitBreaker Attributes*

Name	Description	Default
exclude	Array of exceptions to exclude. Useful for excluding specific subclasses of an exception for example.	Empty (all exceptions if the include attribute is also empty)
include	Array of exceptions to retry with	Empty (all exceptions if the exclude attribute is also empty)
label	A unique tag for circuit breaker reporting	The method signature where the annotation is declared
maxAttempts	Max number of attempts before opening the circuit breaker (including the first failure)	3
openTimeout	The interval over which the maxAttempts must occur before tripping the circuit breaker	5000
resetTimeout	Number of milliseconds before trying the main path again	20000 (20 seconds)
value	Array of exceptions to retry with	Empty (all exceptions)

The other annotation used by the Spring Retry is the @Recover annotation. This method level annotation indicates the method to call when the retryable method fails or when the circuit breaker is flipped. The method annotated with @Recover is required to have the same method signature as that of the method annotated with the @CircuitBreaker annotation associated with it.

The final piece of the puzzle for adding the circuit breaker functionality to our ItemProcessor is the @EnableRetry annotation. This annotation bootstraps the Spring mechanisms for proxying the retryable method calls. By adding this to our main class along with the @EnableBatchProcessing, we'll have the infrastructure we need. To see the code update we need to make, take a look at Listing 12-7.

Listing 12-7. EnrichmentProcessor with Circuit Breaker

```
...
public class EnrichmentProcessor implements ItemProcessor<Foo, Foo> {

        @Autowired
        private RestTemplate restTemplate;

        @Recover
        public Foo fallback(Foo foo) {
                foo.setMessage("error");
                return foo;
        }

        @CircuitBreaker(maxAttempts = 1)
        @Override
        public Foo process(Foo foo) {
                ResponseEntity<String> responseEntity =
                        this.restTemplate.exchange(
                                "http://localhost:8080/enrich",
                                HttpMethod.GET,
                                null,
                                String.class);
                foo.setMessage(responseEntity.getBody());

                return foo;
        }
}
```

The two differences between the original version in Listing 12-6 and here are the addition of the @CircuitBreaker annotation to the process method and the addition of the fallback method annotated with the @Recover annotation. Our fallback method has the same method signature as the process method. However, instead of making the remote call, we provide a default message. The @CircuitBreaker annotation on the process method is configured to trip after a single attempt within the default 5 seconds and reset after 20 seconds.

To test this configuration, we're going to update the EnrichmentController to throw a random exception about 50% of the time. This will simulate flakiness in our system and will cause our circuit breaker to trip. Listing 12-8 shows the code for the updated controller.

Listing 12-8. EnrichmentController Updated to Throw Random Exceptions

```
...
@RestController
public class EnrichmentController {

        private int count = 0;

        @GetMapping("/enrich")
        public String enrich() {
                if(Math.random() > .5) {
                        throw new RuntimeException("I screwed up");
                }
                else {
                        this.count++;

                        return String.format("Enriched %s", this.count);
                }
        }
}
```

With the updates to our controller as shown in Listing 12-8, if we run our job again, this time we'll see some records imported with the message "Enriched X" with X being the number of times the controller has been called without throwing an exception. However, we will also see "error" in some of the messages. If we compare the number of error messages in the database with the number of stack traces in the logs from the REST application, we should see that there are more error messages than stack traces. That confirms that our circuit breaker was tripped and our fallback method was called instead of trying the REST API.

Now that we've added some additional resiliency to our application, we're going to work on externalizing the configuration for our application. We'll do that in two ways. First using Spring Cloud's Config Server and second by using service discovery provided by Eureka. The next section will look at the mechanisms for how to make this work.

Externalizing Configuration

Every Spring Boot application up to this point in the book has used the application.properties or application.yml to configure our applications. However, that poses a problem in a cloud native environment in that since our configuration is bundled with our application in the jar file, we cannot change it easily as we move from environment to environment. There are also security concerns with the current approach in that we have sensitive secrets (username and password for our database as well as Amazon credentials) stored in plain text in our artifact that is typically published to a public repository of some kind.

There has to be a better way. And thanks to the Spring Cloud projects, there is. In this section, we will look at two mechanisms for externalizing our configuration. The first will be using Spring Cloud Config Server to provide and secure the values we are currently storing in our application.yml file. We will also secure them using encryption tools provided by the Spring Cloud CLI. Finally, we will use service binding to allow our batch job to locate and contact our REST API instead of having a URL, port, and so on hard coded or configured in our application directly.

Spring Cloud Config

Spring Cloud Config is a configuration server that provides a server for serving configuration stored in either a git repository or a database back end. In order to consume this configuration, an application uses the Spring Cloud Config client within their application. This client will call the server, obtain the configuration properties from the server, and populate a Spring Environment with those values so all of the normal property injection, and so on, that Spring Boot does for you when using your application.yml still works.

To enable this functionality in our application, we'll need to do two things. The first is to include the client in our application. This is done by adding the dependency for the spring-cloud-starter-config to our pom.xml. Listing 12-9 shows the dependency we'll need.

Listing 12-9. Spring Cloud Config Client Dependency

```
...
<dependency>
        <groupId>org.springframework.cloud</groupId>
        <artifactId>spring-cloud-starter-config</artifactId>
</dependency>
...
```

With that dependency added, the other change we'll need to make to our application is to replace the application specific configuration in our application.yml file with the configuration the config server client requires. For our purposes (running everything on our local machine), we'll only need to configure two properties: `spring.application.name` and `spring.cloud.config.failFast`. The first property `spring.application.name` is used by the client to ask for the correct configuration from the server. We'll specify the value for this property to be `cloud-native-batch`. Spring Cloud Config supports all of the normal features of Spring based properties including things like profiles; however, that is out of the scope of this book. The other property we are configuring, `spring.cloud.config.failFast` tells the client to throw an exception, preventing the application from starting, if it cannot retrieve the configuration from the config server. By default, the client will ignore the inability to retrieve the configuration and use whatever is locally configured; however, in our case we want to be sure we are reading our configuration from the config server. With those two properties configured in our batch application, all other configuration can be deleted.

With the client side configured we need to provide our configuration to the config server. The easiest way to get up and running with the config server is with the Spring Cloud CLI.[4] This CLI provides the ability to launch the various Spring Cloud server components with a single command as well as some useful utilities for things like encrypting values. The project page for it has instructions for its installation.

Once the Spring Cloud CLI is installed on your machine, we need to do some configuration for the config server. We will use a git repository for our mechanism for storing our configuration. In order for the config server to use our repository, we need to tell it we are going to use git and the location of the repository. We configure this in a configserver.yml file located in ~/.spring-cloud. Listing 12-10 shows what our configserver.yml looks like.

---

[3]`https://cloud.spring.io/spring-cloud-cli/`

Listing 12-10. configserver.yml

```
spring:
  profiles:
    active: git
  cloud:
    config:
      server:
        git:
          uri: file:///Users/mminella/.spring-cloud/config/
```

With the configuration in Listing 12-10, we can add a git repository to the `~/.spring-cloud/config` directory. This is where we will store our configuration. We can begin by just copying over our old application.yml file to this directory, and committing it into a new git repository (via `git init`, `git add`, and `git commit`). However, we would also like to secure our secrets in this file so that we don't need to worry about attackers getting hold of this file. To do that, we'll use Spring Cloud CLI's encrypt functionality.

Spring Cloud CLI's encrypt functionality supports encrypting values using either a String as a key or as a file (an RSA public key, for example). We'll keep things simple here and use the `String` key. Listing 12-11 shows the results of using Spring's encrypt utilities.

Listing 12-11. Encryption Round Trip

```
➜  config git:(master) ✗ spring encrypt mysecret --key foo
ea48c11ca890b7cb7ffb37de912c4603d97be9d9b1ec05c7dbd3d2183a1da8ee
➜  config git:(master) ✗ spring decrypt --key foo
ea48c11ca890b7cb7ffb37de912c4603d97be9d9b1ec05c7dbd3d2183a1da8ee
mysecret
```

Using this technique, we can encrypt all of our secrets and past the encrypted values in the configuration file being served by our config server. The format for pasting the secret is to prepend the value with {cipher} and wrap the entire value in single quotes. Listing 12-12 shows the full cloud-native-batch.yml file with sample values for the encrypted values.

Listing 12-12. cloud-native-batch.yml

```
spring:
  datasource:
    driverClassName: org.mariadb.jdbc.Driver
    url: jdbc:mysql://localhost:3306/cloud_native_batch
    username: '{cipher}19775a12b552cd22e1530f745a7b842c90d903e60f8a934b072c21454321de17'
    password: '{cipher}abcdefa44d2db148cd788507068e770fa7b64c4d1980ef6ab86cdefabc118def'
    schema: schema-mysql.sql
  batch:
    initalizr:
      enabled: false
job:
  resource-path: s3://def-guide-spring-batch/inputs/*.csv
cloud:
  aws:
    credentials:
      accessKey: '{cipher}a7201398734bcd468f5efab785c2b6714042d62844e93f4a436bc4fd2e95fa4bcd
      26e8fab459c99807d2ef08a212018b'
```

```
secretKey: '{cipher}40a1bc039598defa78b3129c878afa0d36e1ea55f4849c1c7b92e809416737
de05dc45b7eafce3c2bc184811f514e2a9ad5f0a8bb3e503282158b577d27937'
region:
  static: us-east-1
  auto: false
```

With our configuration file created and the config server pointed to the right place, we can start our config server and test our batch job using it. Start by launching the REST API via the normal Spring boot `java -jar rest-service/target/rest-service-0.0.1-SNAPSHOT.jar` command we've used previously. With the REST API running, we can launch the config server with the command `spring cloud configserver`. This will launch the config server and point it to the configuration file we just committed in our git repository. Once the config server is running we can launch our batch job. With these updates, you should see no difference in the output of the job since the behavior is the same. The only difference is the mechanism for obtaining the configuration for our job.

The second part of externalizing our configuration is using service binding via Eureka to connect our batch job to the REST API. We'll take a look at how that works in the next section.

Service Binding via Eureka

The other piece of externalizing configuration is the use of service binding. Spring Cloud Netflix contains an implementation of Eureka, a service discovery tool provided by Netflix's open source initiatives. Eureka provides the ability for services to register themselves so that they can be discovered dynamically by other services. There is also a client that allows services to be discovered by other services. In this section, we'll look at how to use service discovery via Eureka to allow our batch job to connect to the REST API without explicitly configuring our job to do so.

Enabling service discovery works in a similar manner as cloud configuration did in the previous section. Eureka has a client and a server. An application registers with the server identifying that it is available to be discovered. Our REST API will register itself to be discoverable. Our batch job will then obtain the information on how to communicate with the REST API from Eureka when it launches. This means that all we will need to provide our job is where Eureka is and what services we need to talk to and Eureka will handle the rest of the configuration for us. To make this happen, there are only a couple lines of code that need to change.

We'll start with the new dependency that is required. Eureka, like Cloud Config has a client component and a server component. We will need to include the client dependency into our REST API and our batch job. The dependency we will be adding is found in Listing 12-13.

Listing 12-13. Eureka Dependency

```
...
<dependency>
        <groupId>org.springframework.cloud</groupId>
        <artifactId>spring-cloud-starter-eureka</artifactId>
</dependency>
...
```

With that dependency added to both our batch job and our REST API, we'll be able to configure each appropriately. To start, let's configure the REST API to register itself with Eureka on startup. In order for us to do that, we will make two small changes. The first is adding the `@EnableDiscoveryClient` annotation to our main class as shown in Listing 12-14.

Listing 12-14. Adding @EnableDiscoveryClient

```
...
@EnableDiscoveryClient
@SpringBootApplication
public class RestServiceApplication {

        public static void main(String[] args) {
                SpringApplication.run(RestServiceApplication.class, args);
        }
}
```

With the new annotation added, our REST API will automatically register with Eureka on the localhost. You can configure a remote instance via standard properties in a production use case or even use Spring Cloud Config to specify its location.

The final change needed for the REST API is we need to add a boostrap.yml file to our project. This file has the same format as an application.yml file; however, there is a minor difference. application.yml is loaded as your ApplicationContext is loaded. However, for some Spring Cloud features, that is too late in the process, so Spring Cloud works by creating a bootstrap ApplicationContext, which serves as a parent context to your ApplicationContext. It is this bootstrap context that will load the bootstrap.yml. For our use case, we need to configure the name of the application in the bootstrap.yml. This name is the name that our application will register with Eureka. Listing 12-15 shows the bootstrap.yml for the REST API.

Listing 12-15. bootstrap.yml

```
spring:
  application:
    name: rest-service
```

With those changes, the REST API is ready to be used via Eureka. Now it's onto our batch job. There are four updates we'll need to make to our batch job in order to consume the REST API via Eureka's provided configuration. The first is adding the client dependency. This is the same dependency that was provided in Listing 12-13.

Once we've added the dependency, we can update our main class. Again, we'll add the @EnableDiscoveryClient annotation, however with a small modification. We wanted our REST API to register with Eureka as a service. We don't want to register our job as one, we just want to obtain the configuration details for other services. Because of this, we'll set the autoRegister value on the annotation to false as shown in Listing 12-16.

Listing 12-16. CloudNativeBatchApplication

```
@EnableRetry
@EnableBatchProcessing
@SpringBootApplication
@EnableDiscoveryClient(autoRegister = false)
public class CloudNativeBatchApplication {

        public static void main(String[] args) {
                SpringApplication.run(CloudNativeBatchApplication.class, args);
        }
}
```

Once we've enabled the Eureka client within our batch application, we need to configure our RestTemplate to use it. This occurs via another annotation called @LoadBalanced. This annotation, added to the bean definition for our RestTemplate, autoconfigures our RestTemplate to be able to use the configurations provided via Eureka including things like client side load balancing. Listing 12-17 shows what the updated bean definition for our RestTemplate looks like with this annotation applied.

Listing 12-17. LoadBalanced RestTemplate

```
...
@Bean
@LoadBalanced
public RestTemplate restTemplate() {
        return new RestTemplate();
}
...
```

The final update we need to apply to our batch job is to actually reference the service by name. In the previous iteration of our EnrichmentProcessor, we specified the host and port of the REST API directly in our code. This obviously is not an ideal in a cloud environment since you may not know the host name. The benefit of using Eureka is that we just specify the name of the service and Spring Cloud handles the rest. Our updated EnrichmentProcessor calls a service name (rest-service as specified in our bootstrap. yml) instead of a host and port. Listing 12-18 shows the updated EnrichmentProcessor.

Listing 12-18. EnrichmentProcessor

```
...
public class EnrichmentProcessor implements ItemProcessor<Foo, Foo> {

        @Autowired
        private RestTemplate restTemplate;

        @Recover
        public Foo fallback(Foo foo) {
                foo.setMessage("error");
                return foo;
        }

        @CircuitBreaker
        @Override
        public Foo process(Foo foo) {
                ResponseEntity<String> responseEntity = this.restTemplate.exchange(
                                "http://rest-service/enrich",
                                HttpMethod.GET,
                                null,
                                String.class);
                foo.setMessage(responseEntity.getBody());

                return foo;
        }
}
```

That's all there is. Now to run these components, we'll need to start Eureka locally like we did Spring Cloud Config. To do that, we can use the Spring Cloud CLI and the command `spring cloud eureka`. If you are starting both the config server and Eureka at the same time, you can simplify their startup via the command `spring cloud configserver eureka` and both servers will be launched locally.

Once Eureka is up and running, you can navigate to its web dashboard via the URL provided in the logs (defaults to `http://localhost:8761`) to see what services have been registered. On startup, none will be registered. With Eureka started up, you can launch the REST API and monitor the dashboard for its registration. Figure 12-1 illustrates what you should see once the REST API has registered with Eureka.

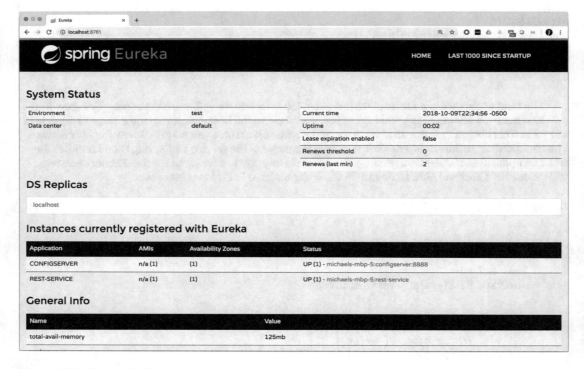

Figure 12-1. *Eureka dashboard*

You'll notice that the `ConfigServer` also registers with Eureka by default. If you click on the status URL for the REST API, you'll get a whitelabeled error page because we haven't configured anything for the Spring Boot `/info` endpoint. However, if you modify the URL to be `/enrich` instead of `/info` at the end, you'll see the result of our API being called.

The last step is to run our job. You can do that via the same java commands we've done up to now. Verifying the results should show the same behaviors in our previous run.

Externalizing configuration is one of the most important pieces of making cloud native processes resilient to their new, dynamic environment. However, given the dynamic nature of the cloud, orchestration mechanisms need to be compatible with them. In the next section, we'll take a look at another Spring Cloud project that handles the orchestration pieces of cloud native batch processing.

Orchestrating Batch Processes

Spring Batch, by design, does not handle orchestration. There is no scheduler or other mechanism for launching batch jobs at a given time within the framework. The framework delegates this responsibility to other mechanisms to allow it to integrate with whatever orchestration tool makes the most sense for your enterprise. Whether it's a large scale enterprise scheduler like Control-M or something as simple as cron, Spring Batch is able to run with it.

That being said, the Spring portfolio does have a tool for orchestrating data processing applications. That tool is called Spring Cloud Data Flow. Unlike most of the Spring portfolio which consists of frameworks and libraries you use to build custom applications with, Spring Cloud Data Flow is a fully built tool that you use to orchestrate applications for either streaming or task based workloads.

An entire book could be written on the Spring Cloud Data Flow ecosystem.[4] We are only going to cover a very small subsection related to batch processing. I encourage you to explore books, online documentation, and talks that have been given on the subject to learn more. For our needs, we'll begin by answering the question, what is Spring Cloud Data Flow?

Spring Cloud Data Flow

At its core, Spring Cloud Data Flow is an orchestration tool. Up to this point, if we've wanted to launch a Spring Batch job using our Spring Boot über jar, we would type at the command line java -jar, the name of the jar file, and any parameters required to run our job. And that works fine in development. You could even argue that it may be okay to run an ad hoc job once in a while. But what about in a cloud environment? How does your application get deployed to the cloud? How does it get launched with the correct parameters? How do you monitor your batch job? How do you manage dependencies between batch jobs?

These are all use cases that Spring Cloud Data Flow solves. Spring Cloud Data Flow is a server application that launches your batch jobs for you on the platform you are using. Spring Cloud Data Flow's server supports three different platforms: CloudFoundry, Kubernetes, and local. Each is capable of deploying and launching your batch jobs.

Spring Cloud Data Flow consists of a Spring Boot application that serves as a server responsible for deploying and launching your batch jobs on the given platform. You interact with that server via either an interactive shell or a web based user interface. Both of those communicate with the server via a set of REST APIs that you can also consume directly. Figure 12-2 shows a diagram of the architecture involved when using Spring Cloud Data Flow.

---

[4]In fact, one has. Spring Cloud Data Flow by Felipe Gutierrez.

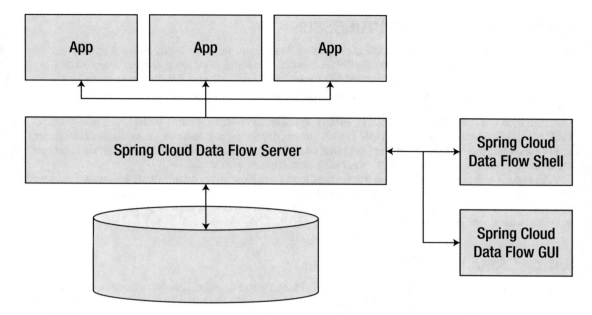

Figure 12-2. *Spring Cloud Data Flow architecture*

It's important to note that with Spring Cloud Data Flow, everything is Spring Boot based. The server is a Spring Boot application, the shell is a Spring Boot application, and the apps Spring Cloud Data Flow deploys are all typically Spring Boot applications.

To get started using Data Flow, we'll need to download it. We'll begin by downloading the server and the shell via the wget commands:

```
wget https://repo.spring.io/milestone/org/springframework/cloud/spring-cloud-dataflow-
server-local/1.7.0.M1/spring-cloud-dataflow-server-local-1.7.0.M1.jar
```

```
wget https://repo.spring.io/milestone/org/springframework/cloud/spring-cloud-dataflow-
shell/1.7.0.M1/spring-cloud-dataflow-shell-1.7.0.M1.jar
```

With those jar files downloaded, you'll need to configure the Spring Batch job repository. Fortunately, Data Flow's server comes preconfigured with the Spring Cloud Config client, so it will pick up the same values for our job repository that we've already configured for our batch job. And that is important. In order for Spring Cloud Data Flow to be useful from a monitoring perspective, it needs to be looking at the same Job Repository tables that our batch job is writing to.

So with our config server already running from the previous section, all we need to do is start the Data Flow server via the command java -jar spring-cloud-dataflow-server-local-1.7.0.M1.jar. This will launch the Data Flow server and point it to our previously configured database. Once the server is up and running, we can connect to it either via the web based user interface or the interactive shell.

We'll begin by using the shell. To launch the shell, you can execute the command java -jar spring-cloud-dataflow-shell-1.7.0.M1.jar. By default, it will automatically connect to our locally running Spring Cloud Data Flow server. Once it is running, you should be presented with a prompt. This is an interactive shell so if at any time you don't know what to do, press tab and the shell will give you guidance.

As we said earlier, Spring Cloud Data Flow is responsible for the orchestration of applications. When dealing with Spring based applications, there are two main types of workloads Spring Cloud Data Flow orchestrates: streams and tasks. Let's take a look at tasks and how they work in this context.

Spring Cloud Task

When we think about most cloud native applications, our minds naturally go to things like REST APIs or integration applications. All of these workloads have one key thing in common…they all are not meant to end. If the REST API that powers your website goes down in the middle of the night, you are probably going to get a page or call to fix it. However, as we well know, not all workloads fit nicely into that never-ending model. That's where tasks come in.

Spring Cloud Task is a framework for building microservices that have an expected end. Database migrations, batch jobs, data science batch training of models. These all are workloads that we would like to run in the cloud with a bit more robustness than just throwing a script up in a cloud and hoping it was run successfully. Spring Cloud Task provides a series of functional and non functional features that allow you to run finite workloads in the cloud in a production hardened manor. Spring Cloud Task includes the following features:

- *A Task Repository* – Modeled after Spring Batch's Job Repository, Spring Cloud Task provides a repository backed by a database that stores a task's start time, end time, results, parameters that were passed to it, and any error that was thrown in the course of the task's execution.

- *Listeners* – Spring Cloud Task provides the ability to hook into the various execution points of a task just like Spring Batch provides the ability to hook into the various stages of a job's lifecycle.

- *Integration with Spring Cloud Stream* – Spring Cloud Stream is a framework for building message based microservices. Spring Cloud Task provides integration with Spring Cloud Stream by implementing listeners for Spring Cloud Task and Spring Batch that emit informational messages via messaging middleware. Things like a task started or finished, a job started or finished, a step started or finished, etc.

- *Integration with Spring Batch* – Spring Cloud Task provides two main integrations with Spring Batch. The first being the informational messages Spring Cloud Task can enable as mentioned in the previous bullet. The other is the `DeployerPartitionHandler` as discussed in Chapter 11.

For Spring Cloud Data Flow to have the visibility into a task the way it needs, the task must be a Spring Cloud Task[5]. To make this happen, we will need to tweak the application we have been working on throughout this chapter in two minor ways. First, we will need to add the dependency for the Spring Cloud Task starter. This Maven entry can be found in Listing 12-19.

Listing 12-19. Spring Cloud Task starter dependency

```
<dependency>
    <groupId>org.springframework.cloud</groupId>
    <artifactId>spring-cloud-starter-task</artifactId>
</dependency>
```

---

[5]Or emulate the same functionality in non-Spring or polyglot situations.

With Spring Cloud Task added to our project, we need to enable its functionality. To do this, the Spring developers have made it really hard to do…you just add a single annotation. The @EnableTask annotation bootstraps the Spring Cloud Task functionality we will need to interact with Spring Cloud Data Flow. We can add that to our main class as shown in Listing 12-20.

Listing 12-20. Spring Cloud Task's @EnableTask annotation

```
@EnableTask
@EnableRetry
@EnableBatchProcessing
@EnableDiscoveryClient(autoRegister = false)
public class CloudNativeBatchApplication {

    public static void main(String[] args) {
        SpringApplication.run(CloudNativeBatchApplication.class, args);
    }
}
```

Now that our application is a task, we can register it with Spring Cloud Data Flow which is what we will look at next.

Registering and Running a Task

In order for Data Flow to be able to orchestrate the applications, it needs to know where the bits are. To tell it, we register our applications, providing Data Flow with a name and coordinates of where the executable is. This executable can be a jar file somewhere (Maven repository, hosted via http, etc.) or a Docker image in a registry (Dockerhub, etc.). In our case, we'll specify the Maven coordinates for the jar file we've been creating up to now to register our application with Data Flow. To do this, we use the app register command. It takes three arguments: name which is the name we want to give our application, type which is the type of application (source, processor, sink, or task), and the URI specifying the location of the bits. In our case, we'll register the application using Maven coordinates since we can install it conveniently into our local Maven repository. So the command we'll use is app register --name fileImport --type task --uri "maven://io.spring.cloud-native-batch:batch-job:0.0.1-SNAPSHOT". If you follow that command with app list, you will see the fileImport task has been added.

Once we have our application registered, we need to create a task definition. A definition is like a template for launching tasks. It's a combination of the task name along with any properties that need to be set in order to run it. A task definition uses a pipes and filters syntax similar to a normal UNIX shell so it should feel very familiar. We'll use the shell to also create our task definition. To do this, we'll use the command task create myFileImport --definition "fileImport". If we had any parameters that needed to be set for our fileImport task, we could configure them in the definition in a format similar to "fileImport --foo=bar". However, since we are going to use the config server to obtain our configuration, it simplifies our definition greatly.

The last step for launching a job with Spring Cloud Data Flow is to determine how you want to launch it. There are a few different mechanisms possible. You can launch a task on demand via the shell, GUI, or REST API. You can schedule the task to be run if you are running Spring Cloud Data Flow on a platform that supports it.[6] Or you can take an event based approach by defining a stream that will launch a task when something particular happens (the download of a file for example). In our case, we are just going to launch the task ad hoc. To do this, we'll use the shell to launch our task. The command for launching a task is task

---

[6]CloudFoundry and Kubernetes are the two platforms that support task scheduling as of this writing.

launch `myFileImport`. If we had any additional command line arguments we wanted to add at runtime, we could append them via the `--arguments` parameter (task launch `myFileImport –arguements foo=bar`) and properties we can add via the `--propeties` parameter.

With the job running, we can monitor it via the data in the job repository. Again, this can be done via the shell, REST API, or the GUI; however, it's better represented via the GUI, so we'll take a look there. To go to the GUI, open a browser and navigate to `http://localhost:9393/dashboard`. Along the left is a set of tabs for the various features of Spring Cloud Data Flow. Apps has a list of all the registered applications in the system as well as the ability to register more of your own. Runtime illustrates the status of all the running applications deployed via Spring Cloud Data Flow. Streams provides the ability to define and launch message based microservices based on Spring Cloud Stream as a stream. Tasks provides the ability to define and execute tasks as well as provides a view into the task repository used by Spring Cloud Task. Jobs is an extension of the Task tab that provides the ability to browse the Spring Batch job repository. Analytics is a tab for doing basic visualizations using Spring Cloud Data Flow's built in analytics capabilities. Finally, Audit Records is a way to view the audit flow that Spring Cloud Data Flow provides for security and compliance use cases.

Let's go to the Task tab. Within the task tab, you will see two additional tabs: Task and Executions. The Task tab will list all the task definitions as well as provide the ability to create, launch, schedule, and delete task definitions. However, we've already run our task from the shell, so we want to go over to the Executions tab. Here, you'll find a list of all the task executions that are in the task repository. In our case, we'll have an entry for the `myFileImport` task that we just ran. Figure 12-3 shows the dashboard with the task execution data.

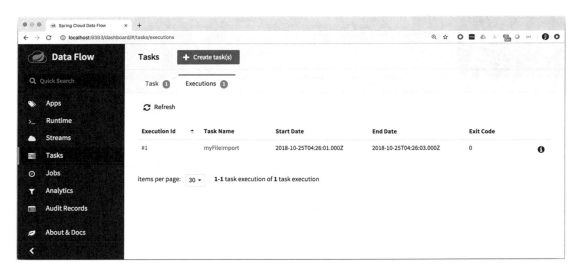

Figure 12-3. *Task Executions in Spring Cloud Data Flow*

Within the dashboard, we can see the name, start and end times, and the exit code from the task once it's been run. By clicking on the #1 (the task's execution id), we can see the details for this run including any arguments that were passed, the external execution id (the id for the underlying system), if the task contained a batch job and a link to its execution id, as well as the data that was on the original list. If an exception had occurred while our job was running, its stack trace would appear in the exit message field on this page. Figure 12-4 illustrates the details of a task execution.

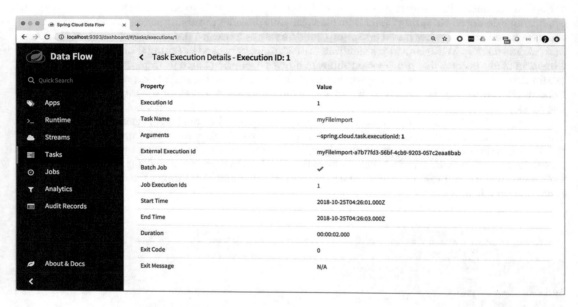

Figure 12-4. *Task execution detail in Spring Cloud Data Flow*

In our case, we not only care about the task execution but what happened with our batch job. We can find out in one of two ways. Either clicking the Jobs tab on the left, or, since we're already on the task execution detail page, we can just click the link to the job execution id and it will take us directly to the detail page for the job execution. On this page you'll see all the same fields you'd see in the BATCH_JOB_EXECUTION table as well as any job parameters. It also includes a summary view of each step executed within this job execution. Figure 12-5 shows what to expect.

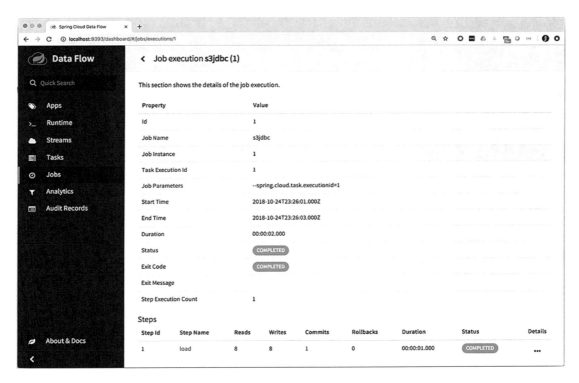

Figure 12-5. *Job execution detail in Spring Cloud Data Flow*

The last page of interest to us when monitoring our batch jobs is the step execution detail page. If we click on the load link under Step Name at the bottom, we will see all the details of the step execution and its related step execution context. Figure 12-6 illustrates the step execution detail page.

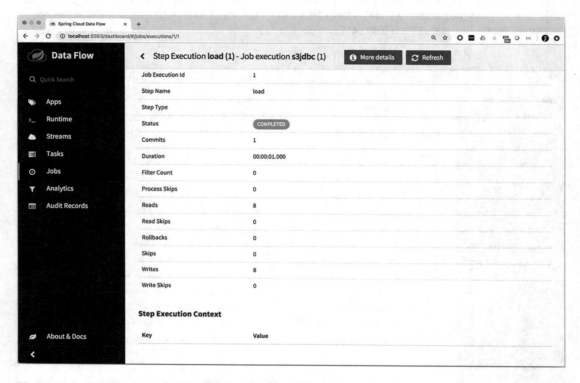

Figure 12-6. *Step execution detail in Spring Cloud Data Flow*

Spring Cloud Data Flow provides a robust solution for orchestrating and monitoring batch jobs on a cloud platform. It's an important piece of the cloud native story for batch processing.

Summary

While the cloud is all the rage, batch processing may not have the sex appeal many other areas of modern development have. However, as we've walked through in this chapter, running batch applications on a modern cloud platform is not only possible, but provides real benefits. We'll continue to expand on those benefits when we look at scaling batch applications in the next chapter.

CHAPTER 13

■ ■ ■

Testing Batch Processes

Testing is everyone's lease favorite part of programming. The funny thing is, like most things in life, once you get good at it, testing actually is fun. It allows you to be more productive. It provides a safety net for you to try new things. Programmatic tests also give you a test bed to try new technologies (most companies don't mind if you want to try something new in the tests but mind greatly if you try it in code that's going to production). You've spent the previous ten chapters writing code without the ability to prove that any of it works. This chapter looks at how to exercise your code in a variety of ways so you can not only prove that it works as designed, but also provide a safety net for when you change it.

This chapter covers the following topics:

- *Unit tests with JUnit and Mockito*: You begin with a high-level overview of the JUnit and Mockito frameworks. Although you move past JUnit's base functionality in the later parts of the chapter, the concepts that Spring has incorporated into its testing apparatus are based in the JUnit conventions, so knowing them helps you understand what is going on in the more advanced tests. The chapter also covers how the mock object framework Mockito can help you unit-test the components you develop for your batch processes.

- *Integration testing using Spring Batch's utilities*: Batch processing is a specific domain and has its own execution requirements. This section will cover how to test batch processes using some of the tools that Spring Batch provides.

The most fundamental aspect of testing begins with unit testing, so the discussion begins there.

Unit Tests with JUnit and Mockito

Probably the easiest to write and perhaps the most valuable, unit tests are the most overlooked type of testing. Although the development done in this book hasn't taken a test-driven approach for a number of reasons, you're encouraged to do so in your own development. As a proven way to improve not only the quality of the software you produce but also the overall productivity of any individual developer and a team as a whole, the code encased in these tests is some of the most valuable you can produce. This section looks at how to use JUnit and Mockito to unit-test the components you develop for your batch processes.

What is a unit test? It's a test of a single, isolated component in a repeatable way. Let's break down that definition to understand how it applies to what you're trying to do:

- *A test of a single:* One. Unit tests are intended to test the smallest building blocks of your application. A single method is typically the scope of a unit test.

- *Isolated:* Dependencies can wreak havoc on the testing of a system. Yet all systems have dependencies. The goal of a unit test isn't to test your integration with each of these dependencies, but to instead test how your component works by itself.

- *In a repeatable way:* When you fire up a browser and click through your application, it isn't a repeatable exercise. You may enter different data each time. You may click the buttons in a slightly different order. Unit tests should be able to repeat the exact same scenario time and time again. This allows you to use them to regression-test as you make changes in your system.

The frameworks you use to execute the isolated testing of your components in a repeatable way are JUnit, Mockito, and the Spring framework. The first two are common, multipurpose frameworks that are useful for creating unit tests for your code. The Spring test utilities are helpful for testing more broad concerns including the integration of the different layers and even testing job execution from end to end (from a service or Spring Batch component to the database and back).

JUnit

Considered the gold standard for testing frameworks in Java,[1] JUnit is a simple framework that provides the ability to unit-test Java classes in a standard way. Whereas most frameworks you work with require add-ons to things like your IDE and build process, Maven and most Java IDEs have JUnit support built in with no additional configuration required. Entire books have been written on the topic of testing and using frameworks like JUnit, but it's important to quickly review these concepts. This section looks at JUnit and its most commonly used features.

The current version of JUnit as of the writing of this book is JUnit 5.2.0. Although each revision contains marginal improvements and bug fixes, the last major revision of the framework was the move from JUnit 4 to JUnit 5, was a major overhaul to the API used to create test cases. Test cases? Let's step back a minute and go over how JUnit tests are structured.

JUnit Lifecycle

JUnit tests are broken down into what are called *test cases*. Each test case is intended to test a particular piece of functionality, with the common divisor being at the class level. The common practice is to have at least one test case for each class. A test case is nothing more than a Java class configured with JUnit annotations to be executed by JUnit. In a test case exist both test methods and methods that are executed to set preconditions and clean up post conditions after each test or group of tests. Listing 13-1 shows a very basic JUnit test case.

Listing 13-1. A Basic JUnit Test Case

```
package com.apress.springbatch.chapter13;

import org.junit.jupiter.api.Test;
import static org.junit.jupiter.api.Assertions.*;

public class StringTest {

    @Test
    public void testStringEquals() {
        String michael = "Michael";
        String michael2 = michael;
        String michael3 = new String("Michael");
        String michael4 = "Michael";
```

---

[1] Or at least it won the Betamax vs. VHS wars against frameworks like TestNG and others.

```
        assertTrue(michael == michael2);
        assertFalse(michael == michael3);
        assertTrue(michael.equals(michael2));
        assertTrue(michael.equals(michael3));
        assertTrue(michael == michael4);
        assertTrue(michael.equals(michael4));
    }
}
```

There is nothing fancy about the unit test in Listing 13-1. All it does is prove that using == when comparing Strings isn't the same as using the .equals method. However, let's walk through the different pieces of the test. First, a JUnit test case is a regular POJO. You aren't required to extend any particular class, and the only requirement that JUnit has for your class is that it has a no argument constructor.

In each test, you have one or more test methods (one in this case). Each test method is required to be public, to be void, and to take zero arguments. To indicate that a method is a test method to be executed by JUnit, you use the @Test annotation. JUnit executes each method annotated with the @Test annotation once during the execution of a given test.

The last piece of StringTest is the assert methods used in the test method. The test method has a simple flow. It begins by setting up the conditions required for this test, and then it executes the tests and validates the results at the same time using JUnit's assert methods. The methods of the org.junit.Assert class are used to validate the results of a given test scenario. In the case of StringTest in Listing 13-1, you're validating that calling the .equals method on a String object compares the contents of the String, whereas using == to compare two Strings verifies that they're the same instance only.

Although this test is helpful, there are a couple other useful annotations that you should know about when using JUnit. The first two are related to the JUnit test lifecycle. JUnit allows you to configure methods to run before each test method and after each test method so that you can set up generic preconditions and do basic cleanup after each execution. To execute a method before each test method, you use the @BeforeEach annotation; @AfterEach indicates that the method should be executed after each test method. Just like any test method, the @BeforeEach and @AfterEach marked methods are required to be public, be void, and take no arguments. Typically, you create a new instance of an object to be tested in the method marked with @BeforeEach to prevent any residual effects from one test having an effect on another test. Figure 13-1 shows the lifecycle of a JUnit test case using the @BeforeEach, @Test, and @AfterEach annotations.

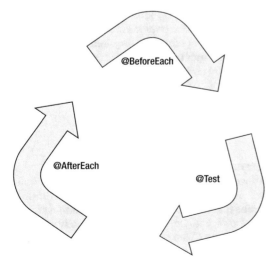

Figure 13-1. *JUnit lifecycle*

437

As Figure 13-1 shows, JUnit executes those three methods in sequence for each method identified with the @Test annotation until all the test methods in the test case have been executed. Listing 13-2 shows an example test case using all three of the discussed annotations.

Listing 13-2. Test of Foo

```
...
public class FooTest {

    private Foo fooInstance;

    @BeforeEach
    public void setUp() {
        fooInstance = new Foo();
    }

    @Test
    public void testBar() {
        String results = fooInstance.bar();

        assertNotNull("Results were null", results);
        assertEquals("The test was not a success", "success", results);
    }

    @AfterEach
    public void tearDown() {
        fooInstance.close();
    }
}
```

JUnit provides a number of other variants on these features, including @BeforeAll to execute one-time setup for all the test methods in a given test class, @Ignore to indicate test methods and classes to skip, and @RunWith to indicate a class to run your test case other than the default used by JUnit. However, those are outside of the scope of this book. The goal of this section is to give you the tools required to be able to test your batch processes. Using just the @BeforeEach, @Test, and @AfterEach annotations along with the assert methods available on JUnit's Assert class, you can test the vast majority of scenarios required.

But there is a small catch. The earlier unit test definition said that unit testing is the testing of components in isolation. How can you test a data access object (DAO) using JUnit when it depends on JDBC and a database? How about testing an ItemStream, which requires you to use Spring Batch components as some of its parameters? Mock objects fill this void, and you look at those next.

Mock Objects

It would be very easy to write software like the String object tested earlier, which has no dependencies. However, most systems are complex. Batch jobs can require dozens or more classes and depend on external systems including application servers, messaging middleware, and databases, just to name a few. All these moving parts can be difficult to manage and provide interactions that are outside the scope of a unit test. For example, when you want to test the business logic for one of your ItemProcessors, do you really need to test that Spring Batch is saving the context correctly to your database? That is outside the scope of a unit test. Don't get this wrong—that does need to be tested, and you look at it later in this chapter. However, to test your business logic, you don't need to exercise the various dependencies that your production system interacts with. You use mock objects to replace these dependencies in your testing environment and exercise your business logic without being effected by outside dependencies.

▪ **Note** Stubs are not mock objects. Stubs are hard coded implementations that are used in testing where mock objects are reusable constructs that allow the definition of the required behavior at run time.

Let's take a minute to call out that mock objects aren't stubs. *Stubs* are implementations that you write to replace various pieces of an application. Stubs contain hard coded logic intended to mimic a particular behavior during execution. They aren't mock objects (no matter what they're named in your project)!

How do mock objects work? There are essentially two different approaches most mock object frameworks take: proxy based and class remapping. Because proxy-based mock objects are not only the most popular but the easiest to use, let's look at them first.

A *proxy object* is an object that is used to take the place of a real object. In the case of mock objects, a proxy object is used to imitate the real object your code is dependent on. You create a proxy object with the mocking framework and then set it on the object using either a setter or constructor. This points out an inherent issue with mocking using proxy objects: you have to be able to set up the dependency through an external means. In other words, you can't create the dependency by calling new MyObject() in your method, because there is no way to mock the object created by calling new MyObject().[2] This is one of the reasons Dependency Injection frameworks like Spring have taken off—they allow you to inject your proxy objects without modifying any code.

The second form of mocking is to remap the class file in the class loader. The mocking framework JMockit is the only framework I'm aware of that currently exploits this ability for mock objects. The concept is provided by the java.lang.instrument.Insturmentation interface. You tell the classloader to remap the reference to the class file it loads. So, let's say you have a class MyDependency with the corresponding .class file MyDependency.class, and you want to mock it to use MyMock instead. By using this type of mock objects, you actually remap in the classloader the reference from MyDependency to MyMock.class. This allows you to mock objects that are created by using the new operator. Although this approach provides more power than the proxy-object approach because of its ability to inject literally any implementation into the classloader, it's also harder and more confusing to get going given the knowledge of classloaders you need in order to be able to use all its features.

Mockito is a popular proxy-based mock object framework that provides a large amount of flexibility coupled with an expressive syntax. It allows you to create easy-to-understand unit tests with relative ease. Let's take a look.

Mockito

Mockito allows you to mock the behaviors you care about and verify only the behaviors that matter. In this section, you look at some of the functionality available with Mockito and use it to test Spring Batch components.

Both Junit and Mockito are included in the spring-boot-starter-test dependency, so you don't need to do anything once you've created your project from Spring Initializr in order to get started writing your tests.

To see how Mockito works, let's look at one of the classes you developed for the statement job in Chapter 10. The CustomerItemValidator you created to validate that the customer existed is a prime candidate to use mock objects, with its dependencies on an external JdbcTemplate. To refresh your memory, Listing 13-3 shows the code from that ItemProcessor.

---

[2]This isn't 100% true. PowerMock lets you mock the new operator. You can find more information on PowerMock at http://code.google.com/p/powermock/.

Listing 13-3. `CustomerItemValidator`

```
...
@Component
public class CustomerItemValidator
implements Validator<CustomerUpdate> {

private final NamedParameterJdbcTemplate jdbcTemplate;

public static final String FIND_CUSTOMER =
        "SELECT COUNT(*) FROM CUSTOMER WHERE customer_id = :id";

public CustomerItemValidator(NamedParameterTemplate template) {
        this.jdbcTemplate = template;
}

@Override
public void validate(CustomerUpdate customer)
        throws ValidationException {

        Map<String, Long> parameterMap =
                Collections.singletonMap("id", customer.getCustomerId());

        Long count =
                jdbcTemplate.queryForObject(FIND_CUSTOMER,parameterMap,Long.class);

        if(count == 0) {
                throw new ValidationException(
                        String.format("Customer id %s was not able to be found",
                                customer.getCustomerId()));
        }
}
}
```

The method you're testing for this class is obviously `validate()`. This method requires one external dependency: an instance of a `NamedParameterJdbcTemplate`. To test this method, you have two test methods, one for each of the method's two execution branches (one for when a customer is found and one for when it is not).

To start this test, let's create the test-case class and the `@BeforeEach` method so your objects are built for later use. Listing 13-4 shows the test case with the `setUp()` method identified with the `@BeforeEach` annotation and three class attributes.

Listing 13-4. `CustomerStatementReaderTest`

```
...
public class CustomerItemValidatorTests {

        @Mock
        private NamedParameterJdbcTemplate template;

        private CustomerItemValidator validator;
```

```
        @BeforeEach
        public void setUp() {
                MockitoAnnotations.initMocks(this);
                this.validator = new CustomerItemValidator(this.template);
        }
...
}
```

The two attributes of the test class are the class under test (CustomerItemValidator) and the one dependency (NamedParameterJdbcTemplate). By using the @Mock annotation, you tell Mockito to create a mock for the NamedParameterJdbcTemplate. When the test is executed, Mockito creates a proxy for each of these for your test to use.

In the setup method, you do two things. First, you initialize the mocks with Mockito's MockitoAnnotations.initMocks method. This method initializes all the objects you previously indicated with a mock object for you to use. This is a quick and easy way to create the mock objects you need in the future.

The next thing you do in the setUp() method is create a new instance of the class to test. By creating this class here, you can be sure that each test method contains a clean instance of the class under test. This prevents any residual state in the test object from one method from having an impact on your other test methods. After you create CustomerItemValidator, you inject the mock objects the same way Spring would do it for you on bootstrapping the application.

Because you now have a new instance of the object under test and a fresh set of mock objects to satisfy your dependencies on the Spring Batch framework as well as the database, you can write your test methods. The first one, which tests when the customer is found by our query, is very easy; see Listing 13-5.

Listing 13-5. testValidCustomer()

```
...
@Test
public void testValidCustomer() {

        // given

        CustomerUpdate customer = new CustomerUpdate(5L);

        // when
        ArgumentCaptor<Map<String, Long>> parameterMap =
                        ArgumentCaptor.forClass(Map.class);
        when(this.template.queryForObject(eq(CustomerItemValidator.FIND_CUSTOMER),
                                        parameterMap.capture(),
                                        eq(Long.class)))
                .thenReturn(2L);

        this.validator.validate(customer);

        // then

        assertEquals(5L, (long) parameterMap.getValue().get("id"));
}
...
```

I've laid this test out in a behavior driven design style[3] with the comments //given, //when, //then. Given these inputs, when these actions occur, then these should be the results. In this example, given the CustomerUpdate object with an id of 5, when we execute the validator, no exception should be thrown.

However, our code is slightly more complex than that. In the when section we are using a feature of Mockito to capture the value passed to a mock. In this case, we're capturing the Map being passed to our NamedParameterJdbcTemplate, so we can assert that the parameter we're passing is as expected. We also tell our mock that when it receives a call with the parameters FIND_CUSTOMER, any Map (since we will be capturing it for further analysis), and the class Long, then to return the value 2. When our method under test is executed, when the correct parameters are passed to the mock, Mockito will return 2. Mockito returns type appropriate defaults for all values when a return is not specified. In our case, if the method is called with any other arrangement of parameters, it will return null (null is the default for an object) causing our test to fail. Once Mockito returns a 2, our logic validates that it is not equal to 0 and does not throw an exception.

The second test method we will write is actually almost the same as the first. The main difference is we will be expecting our code under test to throw an exception when the customer provided is not found so we can assert that condition. Listing 13-6 shows the code for the testInvalidCustomer method.

Listing 13-6. testInvalidCustomer

```
...
@Test
public void testInvalidCustomer() {

        // given

        CustomerUpdate customerUpdate = new CustomerUpdate(5L);

        // when
        ArgumentCaptor<Map<String, Long>> parameterMap =
                        ArgumentCaptor.forClass(Map.class);
        when(this.template.queryForObject(eq(CustomerItemValidator.FIND_CUSTOMER),
                                parameterMap.capture(),
                                eq(Long.class)))
                .thenReturn(0L);

        Throwable exception = assertThrows(ValidationException.class,
                                () -> this.validator.validate(customerUpdate));

        // then

        assertEquals("Customer id 5 was not able to be found",
                        exception.getMessage());
}
...
```

The testInvalidCustomer() method illustrates another new feature in JUnit 5, the assertThrows method. In previous versions of JUnit you needed to handle assertions on exceptions in various ways (using JUnit rules, catching the exception yourself, etc.). Now, we can validate that the type of exception that is thrown is correct via the org.junit.jupiter.api.Assertions.assertThrows method. This method takes

---

[3]https://dannorth.net/introducing-bdd/

the class of the exception to expect and a closure executing the code under test. The assertion will fail if no exception or the wrong type of exception is thrown. It will pass and return the exception thrown if the type is correct. From there, we can assert that the error message is what we expect in the // then section of our test.

These two tests provide us with the ability to reliably validate the behavior of our class as well as refactor without fear of impacting other parts of the codebase. Unit tests are the foundation of a solid system. They not only provide the ability to make the changes you need without fear, but also force you to keep your code concise and serve as executable documentation for your system. However, you don't build a foundation just to look at it. You build on top of it. In the next section, you expand your testing capabilities.

Integration Tests with Spring Classes

The previous section discussed unit tests and their benefits. Unit tests, however useful they are, do have their limits. Integration testing takes your automated testing to the next level by bootstrapping your application and running it with the same dependencies you tried so hard to extract previously. This section looks at how to use Spring's integration test facilities to test interactions with various Spring beans, databases, and finally batch resources.

General Integration Testing with Spring

Integration testing is about testing different pieces talking to each other. Does the DAO get wired correctly, and is the Hibernate mapping correct so you can save the data you need? Does your service retrieve the correct beans from a given factory? These and other cases are tested when you write integration tests. But how do you do that without having to set up all the infrastructure and make sure that infrastructure is available everywhere you want to run these tests? Luckily, you don't have to.

The two primary use cases for integration testing with the core Spring integration-testing facilities are to test database interaction and to test Spring bean interaction (was a service wired up correctly, etc.). To test this, let's look at the `NamedParameterJdbcTemplate` you mocked in the unit tests previously (`CustomerItemValidator`). However, this time, you let Spring wire up `CustomerItemValidator` itself, and you use an in-memory instance of HSQLDB as your database so that you can execute the tests anywhere, anytime. HSQLDB is a 100% Java implemented database that is great for integration testing because it's lightweight to spool up an instance. To get started, let's look at how to configure your test environment.

Configuring the Testing Environment

To isolate your test execution from external resource requirements (specific database servers, and so on), you should configure a couple of things. Specifically, you should use a test configuration for your database that creates an instance of HSQLDB for you in memory. To do that, you need to update your POM file to include the HSQLDB database drivers. The specific dependency you need to add is shown in Listing 13-7.

Listing 13-7. HSQLDB's Database Driver Dependency

```
...
<dependency>
    <groupId>org.hsqldb</groupId>
    <artifactId>hsqldb</artifactId>
    <scope>test</scope>
</dependency>
...
```

Once we have the additional dependency in place, we can write our integration test. Spring Boot provides a few facilities to make writing integration tests very easy. In fact, since we don't need to define any mock objects, our integration test will be shorter than our unit test.

To begin, we use the @ExtendWith(SpringExtension.class) annotation on our test class. This is the JUnit 5 equivalent to @RunWith(SpringRunner.class). By using this annotation, we get all the Spring goodness for our test. From there we'll want to use Spring Boot's autoconfiguration capabilities to create our database and populate it. Like most things when using Spring Boot, this is very easy and requires only a single annotation: @JdbcTest. This annotation will create an in-memory database for us and populate it with the data Spring Boot normally uses (via our initialization scripts).

With our database automatically created and populated for us by Spring Boot, all we'll need to do is autowire in the DataSource related to it. Listing 13-8 shows the initial class definition with the class level annotations on it and the autowired DataSource.

Listing 13-8. CustomerItemValidatorIntegrationTests

```
@ExtendWith(SpringExtension.class)
@JdbcTest
public class CustomerItemValidatorIntegrationTests {

        @Autowired
        private DataSource dataSource;

        private CustomerItemValidator customerItemValidator;
...
}
```

Once we have our DataSource, we can configure our setUp() method to create our validator. Listing 13-9 has the setUp() method.

Listing 13-9. CustomerItemValidatorIntegrationTests Setup

```
...
@BeforeEach
public void setUp() {
        NamedParameterJdbcTemplate template =
                new NamedParameterJdbcTemplate(this.dataSource);
        this.customerItemValidator = new CustomerItemValidator(template);
}
...
```

With our validator created, we can create our tests. However, unlike our unit tests that require us to mock the values, we can just make the calls and assert the results. Listing 13-10 lists functionaly the same tests as we did in our unit tests, just as integration tests.

Listing 13-10. CustomerItemValidatorIntegrationTests Tests

```
...
@Test
public void testNoCustomers() {
        CustomerUpdate customerUpdate = new CustomerUpdate(-5L);
```

```
        ValidationException exception =
                        assertThrows(ValidationException.class,
                                () -> this.customerItemValidator.validate(customerUpdate));

        assertEquals("Customer id -5 was not able to be found",
                        exception.getMessage());

}

@Test
public void testCustomers() {
        CustomerUpdate customerUpdate = new CustomerUpdate(5L);
        this.customerItemValidator.validate(customerUpdate);
}
...
```

This time, for our test with no customers, we create a `CustomerUpdate` object with an id that we know isn't in our test data. We can then capture the exception that is thrown and assert the results just as we did in our unit test. In our test that validates when a customer is found, we create a `CustomerUpdate` object with an id that is in our test data and call the validator. If no exception is thrown, the test passes.

Integration tests like those found in `CustomerItemValidatorIntegrationTests` can be hugely valuable when you're developing a system. The ability to determine if things are being wired correctly, if SQL is correct, and even if the order of operations between components of a system is correct can provide considerable security when you're dealing with complex systems.

The final piece of testing with Spring Batch is testing the Spring Batch components themselves. `ItemReaders`, steps, and even entire jobs can be tested with the tools provided by Spring. The final section of this chapter looks at how to use those components and test pieces of your batch process.

Testing Spring Batch

Although the ability to test components like a DAO or a service is definitely needed when you're working with robust batch jobs, working with the Spring Batch framework introduces a collection of additional complexities into your code that need to be addressed in order to build a robust test suite. This section looks at how to handle testing Spring Batch–specific components, including elements that depend on custom scopes, Spring Batch steps, and even complete jobs.

Testing Step and Job Scoped Beans

As you've seen in many examples throughout this book, the step and job scopes defined by Spring Batch for your Spring Beans is a very helpful tool. However, when you're writing integration tests for components that use the step scope, you run into an issue: If you're executing those components outside the scope of a step, how do those dependencies get resolved? In this section, you look at the two ways Spring Batch offers to simulate that a bean is being executed in the scope of a step.

You've seen in the past how using the step scope allows Spring Batch to inject runtime values from the job and/or step context into your beans. Previous examples include the injection of an input or output file name, or criteria for a particular database query. In each of those examples, Spring Batch obtains the values from the `JobExecution` or the `StepExecution`. If you aren't running the step in a job, you don't have either of those executions. Spring Batch provides two different ways to emulate the execution in a step so that those values can be injected. The first approach uses a `TestExecutionListener`.

TestExecutionListener is a Spring API that allows you to define things to occur before or after a test method. Unlike using JUnit's @BeforeEach and @AfterEach annotations, using Spring's TestExecutionListener allows you to inject behavior across all the methods in a test case in a more reusable way. Although Spring provides three useful implementations of the TestExecutionListener interface (DependencyInjectionTestExecutionListener, DirtiesContextTestExecutionListener, and TransactionalTestExecutionListener), Spring Batch provides two that handle what you're looking for: StepScopeTestExecutionListener and JobScopeTestExecutionListener. Since they both function the same, we'll use the StepScopeTestExecutionListener for the remainder of this section since it's more commonly used.

StepScopeTestExecutionListener provides two features you need. First, it uses a factory method from your test case to obtain a StepExecution and uses the one returned as the context for the current test method. Second, it provides a StepContext for the life of each test method. Figure 13-2 shows the flow of a test being executed using StepScopeTestExecutionListener.

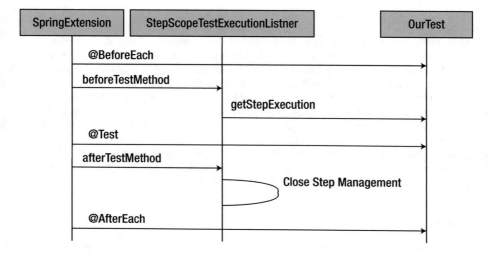

Figure 13-2. *Test execution using StepScopeTestExecutionListener*

As you can see, the factory method you create in the test case (getStepExecution) is called before each test method to obtain a new StepExecution. If there is no factory method, Spring Batch uses a default StepExecution.

To test this, you configure a FlatFileItemReader to obtain the location of the file to read from the jobParameters. The reader we'll use is the one from our sample application that reads in the customer update file. If you'll remember, it was step scoped and a job parameter was used to provide the location of the file to read. Listing 13-11 shows the configuration of that reader.

Listing 13-11. ImportJobConfiguration#customerUpdateItemReader

```
...
@Bean
@StepScope
public FlatFileItemReader<CustomerUpdate> customerUpdateItemReader(
        @Value("#{jobParameters['customerUpdateFile']}") Resource inputFile)
        throws Exception {
```

```
        return new FlatFileItemReaderBuilder<CustomerUpdate>()
                        .name("customerUpdateItemReader")
                        .resource(inputFile)
                        .lineTokenizer(customerUpdatesLineTokenizer())
                        .fieldSetMapper(customerUpdateFieldSetMapper())
                        .build();
}
...
```

We'll use our reader to read in a test file consisting of three records, one for each of the three record formats possible in the file. Our test will then assert that the reader is returning the correct type for each record. Listing 13-12 shows an example of what the test file would look like.

Listing 13-12. Test customerUpdateFile.csv

```
2,5,,,Montgomery,Alabama,36134
3,5,,,,316-510-9138,2
1,5,Rozelle,Heda,Farnill
```

The customerUpdateItemReader defined in Listing 13-11 requires additional dependencies but we won't need to worry about their configuration. We're actually just going to use the original configuration from our application in our test. This provides the most accurate way of testing that all components are working together as expected. This will require a bit of infrastructure on our test, so let's begin looking at our test with the infrastructure we need to put in place. Listing 13-13 lays out what we need.

Listing 13-13. FlatFileItemReaderTests Infrastructure

```
...
@ExtendWith(SpringExtension.class)
@ContextConfiguration(classes = {ImportJobConfiguration.class,
                                 CustomerItemValidator.class,
                                 AccountItemProcessor.class})
@JdbcTest
@EnableBatchProcessing
@SpringBatchTest
public class FlatFileItemReaderTests {

        @Autowired
        private FlatFileItemReader<CustomerUpdate> customerUpdateItemReader;
...
}
```

The first annotation in Listing 13-13 should be familiar since we used it in our integration test. The @ExtendWith(SpringExtension.class) is what triggers Spring's testing goodness. You'll also notice that this test is using the @JdbcTest annotation again. While we won't use the database it creates in our unit test, since we're recycling the configuration of our application and it requires a database, we'll need to provide one. The next annotation of interest is the @ContextConfiguration annotation. This annotation is where we specify the classes (or resources if you're using XML configuration) to build your ApplicationContext. In this case, we have three classes to provide. The ImportJobConfiguration class is annotated with @Configuration and is where we do all of our @Bean style configuration. However, we do have two other components defined that are used by beans in the ImportJobConfiguration. Those are the CustomerItemValidator and the AccountItemProcessor. Both of these classes are annotated with the @Component annotation and so also need to be included in the array of classes we provide.

All of the annotations up to this point on our test have been either Spring Framework or Spring Boot annotations. The last two are Spring Batch specific. The first one we've used in every example in this book, @EnableBatchProcessing. Since we're recycling the configuration from our application, our ApplicationContext will contain our job and steps (don't worry, they won't be executed). Spring Batch will require that those get wired up with a JobRepository which is why we're using that annotation.

The last annotation is a newer one from Spring Batch 4.1: @SpringBatchTest. This annotation provides a number of utilities for testing automatically to your ApplicationContext. Specifically it adds four beans:

- A JobLauncherTestUtils instance for launching jobs or steps

- A JobRepositoryTestUtils which can be used to create or JobExecutions from a JobRepository

- A StepScopeTestExecutionListner and a JobScopeTextExecutionListener to allow for testing of step and job scoped beans

The last beans are the ones of interest to us right now since our reader is step scoped. In order to use the StepScopeTestExecutionListener to handle step scoped dependencies we need to create a method that will provide a StepExecution populated with what we need. In our case, the reader we are testing requires that a job parameter named customerUpdateFile pointing to the file to be read is provided. This factory method is shown in Listing 13-14.

Listing 13-14. FlatFileItemReaderTests#getStepExecution

```
...
public StepExecution getStepExecution() {
        JobParameters jobParameters = new JobParametersBuilder()
                        .addString("customerUpdateFile", "classpath:customerUpdateFile.csv")
                        .toJobParameters();

        return MetaDataInstanceFactory.createStepExecution(jobParameters);
}
...
```

The getStepExecution() method in Listing 13-14 is pretty straightforward. We begin by creating a JobParameters object with the customerUpdateFile parameter pointing to a test version of the customerFile. csv. We then use the MetaDataInstanceFactory to create a StepExecution. The MetaDataInstanceFactory is a utility class for creating Step and Job Execution instances. It's different from the JobRepositoryTestUtils in that the resulting Step or Job Execution from the MetaDataInstanceFactory is not persisted in a JobRepository where as getting one from the JobRepositoryTestUtils is.

With the StepExecution factory method created, all we need to do now is write our test. Since we're really relying heavily on Spring Batch's normal behavior, our test isn't very complex. We will open the reader that is injected into our test and then read the three records in our test file and validate that they return the correct type. Listing 13-15 shows our test.

Listing 13-15. FlatFileItemReaderTests# testTypeConversion

```
...
@Test
public void testTypeConversion() throws Exception {
        this.customerUpdateItemReader.open(new ExecutionContext());

        assertTrue(this.customerUpdateItemReader.read() instanceof CustomerAddressUpdate);
        assertTrue(this.customerUpdateItemReader.read() instanceof CustomerContactUpdate);
```

```
        assertTrue(this.customerUpdateItemReader.read() instanceof CustomerNameUpdate);
}
...
```

Integration tests of this nature can be very useful to test custom developed components such as custom `ItemReaders` and `ItemWriters` or related components. However, as you can see, the value of testing Spring Batch's own components is minimal at best. Rest assured, it has test coverage for these very things. Instead, it may be more useful to test your batch jobs by executing an entire step. The next section looks at the tools Spring Batch provides to make that happen.

Testing a Step

Jobs are broken into steps. This book has established that. Each step is an independent piece of functionality that can be executed with minimal impact on other steps. Because of the inherent decoupling of steps with a batch job, steps become prime candidates for testing. In this section, you look at how to test a Spring Batch step in its entirety.

In the `step` scope–based examples in the previous section, you tested the `ItemReader` of a job that reads in a file. However let's take a look at testing it within the context of a step. For this test, we'll test the execution of the step that `ItemReader` is used in. So what will happen is that we'll execute the step (read our file and update the database as expected) then we'll be able to look at the results of the step to validate that it executed correctly.

To begin, we'll set up the infrastructure for our test. Listing 13-16 shows that the infrastructure for this test is very similar to that included in our step scoped test.

Listing 13-16. `ImportCustomerUpdatesTests` Infrastructure

```
...
@ExtendWith(SpringExtension.class)
@JdbcTest
@ContextConfiguration(classes = {ImportJobConfiguration.class,
                                 CustomerItemValidator.class,
                                 AccountItemProcessor.class,
                                 BatchAutoConfiguration.class})
@SpringBatchTest
@Transactional(propagation = Propagation.NOT_SUPPORTED)
public class ImportCustomerUpdatesTests {

        @Autowired
        private JobLauncherTestUtils jobLauncherTestUtils;

        @Autowired
        private DataSource dataSource;

        private JdbcOperations jdbcTemplate;

        @BeforeEach
        public void setUp() {
                this.jdbcTemplate = new JdbcTemplate(this.dataSource);
        }
...
}
```

The annotations at the start of this test match what we had in the step scoped test:

- `@ExtendWith(SpringExtension.class)`: This enables all the Spring goodness in JUnit 5.

- `@JdbcTest`: Provides facilities including an in-memory database for tests that do database testing.

- `@ContextConfiguration`: Provides the classes required to build our `ApplicationContext`.

- `@SpringBatchTest`: Provides utilities for testing Spring Batch jobs. The specific one we care about in this example is the `JobLauncherTestUtils`.

- `@Transactional(propagation = Propagation.NOT_SUPPORTED)`: By default, `@JdbcTest` will wrap each test method in a transaction and roll it back upon completion. In normal unit testing scenarios, this makes sense. However, in our case, Spring Batch manages the transactions for us and being wrapped in another transaction actually causes an error. This annotation turns off the transactional behavior of the `@JdbcTest` annotation.

With those annotations applied we can autowire in the facilities that are provided for us. In this case, it is the `JobLauncherTestUtils` and the `DataSource`. We can also create a `setUp()` method that will create a `JdbcTemplate` for us with the provided `DataSource`. We'll use that in our test to validate the results of our step's execution.

Next let's look at the starting data we'll be using for this test. Table 13-1 illustrates each database field and it's expected value before running the test and after.

Table 13-1. *Test Values*

Database Column Name	Initial Value	Final Value
customer_id	5	5
first_name	Danette	Rozelle
middle_name	null	Heda
last_name	Langelay	Farnill
address1	36 Ronald Regan Terrace	36 Ronald Regan Terrace
address2	P.O. Box 33	P.O. Box 33
city	Gaithersburg	Montgomery
state	Maryland	Alabama
postal_code	99790	36134
ssn	832-86-3661	832-86-3661
email_address	tlangelay4@mac.com	tlangelay4@mac.com
home_phone	240-906-7652	240-906-7652
cell_phone	907-709-2649	907-709-2649
work_phone	null	316-510-9138
notification_pref	3	2

The initial values are already in the database, loaded via a script. The final values are the results of executing our test with the file listed in Listing 13-17 as input.

Listing 13-17. Input File

```
2,5,,,Montgomery,Alabama,36134
3,5,,,,316-510-9138,2
1,5,Rozelle,Heda,Farnill
```

With the inputs and outputs established, we can write our test method. Surprisingly, the majority of the code in the method is actually the code required to query the database for the results. We begin our test method by defining the job parameters needed to run our step, which in this case are the same parameters we used in our last test. Once we have our parameters defined, we can execute our step via a call to this.jobLauncherTestUtils.launchStep("importCustomerUpdates", jobParameters);. This call will find a Step called importCustomerUpdates and run it providing the job parameters we pass in. The last part of our test is asserting that the data in the database is what we expect. We execute our query using a JdbcTemplate, map the results to a Map, and then assert that each one is the value we expect. Listing 13-18 shows the complete test method.

Listing 13-18. ImportCustomerUpdatesTests#test

```
...
        @Test
        public void test() {
                JobParameters jobParameters = new JobParametersBuilder()
                                .addString("customerUpdateFile", "classpath:customerFile.csv")
                                .toJobParameters();

                JobExecution jobExecution =
                        this.jobLauncherTestUtils.launchStep("importCustomerUpdates",
                                                                        jobParameters);

                assertEquals(BatchStatus.COMPLETED,
                                        jobExecution.getStatus());

                List<Map<String, String>> results =
                        this.jdbcTemplate.query("select * from customer where customer_id = 5",
                                                        (rs, rowNum) -> {

                        Map<String, String> item = new HashMap<>();
                        item.put("customer_id", rs.getString("customer_id"));
                        item.put("first_name", rs.getString("first_name"));
                        item.put("middle_name", rs.getString("middle_name"));
                        item.put("last_name", rs.getString("last_name"));
                        item.put("address1", rs.getString("address1"));
                        item.put("address2", rs.getString("address2"));
                        item.put("city", rs.getString("city"));
                        item.put("state", rs.getString("state"));
                        item.put("postal_code", rs.getString("postal_code"));
                        item.put("ssn", rs.getString("ssn"));
                        item.put("email_address", rs.getString("email_address"));
                        item.put("home_phone", rs.getString("home_phone"));
```

```
                        item.put("cell_phone", rs.getString("cell_phone"));
                        item.put("work_phone", rs.getString("work_phone"));
                        item.put("notification_pref", rs.getString("notification_pref"));

                        return item;
            });

            Map<String, String> result = results.get(0);

            assertEquals("5", result.get("customer_id"));
            assertEquals("Rozelle", result.get("first_name"));
            assertEquals("Heda", result.get("middle_name"));
            assertEquals("Farnill", result.get("last_name"));
            assertEquals("36 Ronald Regan Terrace", result.get("address1"));
            assertEquals("P.O. Box 33", result.get("address2"));
            assertEquals("Montgomery", result.get("city"));
            assertEquals("Alabama", result.get("state"));
            assertEquals("36134", result.get("postal_code"));
            assertEquals("832-86-3661", result.get("ssn"));
            assertEquals("tlangelay4@mac.com", result.get("email_address"));
            assertEquals("240-906-7652", result.get("home_phone"));
            assertEquals("907-709-2649", result.get("cell_phone"));
            assertEquals("316-510-9138", result.get("work_phone"));
            assertEquals("2", result.get("notification_pref"));
        }
}
```

The only thing left that you could possibly test is the entire job. In the next section, you move to true functional testing and test a batch job from end to end.

Testing a Job

Testing an entire job can be a daunting task. Some jobs, as you've seen, can be quite complex and require setup that isn't easy to do. However, the benefits of being able to automate the execution and result verification can't be ignored. Thus you're strongly encouraged to attempt to automate testing at this level whenever possible. This section looks at how to use `JobLauncherTestUtils` to execute an entire job for testing purposes. You'll soon find out, it's actually very similar to executing a step by itself.

For the sake of this test, we'll use a job that is a bit easier to test than our statement job. In this case, we'll have a single step job that reads items from a list and writes them to `System.out`. While it may not be fancy, it will allow us to focus on the elements of testing the job and not the job itself.

We'll begin creating our test the same way we've done all of the other integration tests, with the infrastructure. This test uses the same infrastructure we've used in the other integration tests. Listing 13-19 shows the annotations we use.

Listing 13-19. JobTests Infrastructure

```
...
@ExtendWith(SpringExtension.class)
@SpringBatchTest
@ContextConfiguration(classes = {JobTests.BatchConfiguration.class,
BatchAutoConfiguration.class})
```

```
public class JobTests {

        @Autowired
        private JobLauncherTestUtils jobLauncherTestUtils;
...
```

With our infrastructure in place including the annotations and the `JobLauncherTestUtils` injected into our test class, we need to define our job. For the sake of this example, we'll define it within a static class inside our test case. Listing 13-20 shows the code for our job.

Listing 13-20. Test Job

```
...
@Configuration
@EnableBatchProcessing
public static class BatchConfiguration {

        @Autowired
        private JobBuilderFactory jobBuilderFactory;

        @Autowired
        private StepBuilderFactory stepBuilderFactory;

        @Bean
        public ListItemReader<String> itemReader() {
                return new ListItemReader<>(Arrays.asList("foo", "bar", "baz"));
        }

        @Bean
        public ItemWriter<String> itemWriter() {
                return (list -> {
                        list.forEach(System.out::println);
                });
        }

        @Bean
        public Step step1() {
                return this.stepBuilderFactory.get("step1")
                                .<String, String>chunk(10)
                                .reader(itemReader())
                                .writer(itemWriter())
                                .build();
        }

        @Bean
        public Job job() {
                return this.jobBuilderFactory.get("job")
                                .start(step1())
                                .build();
        }
```

```
        @Bean
        public DataSource dataSource() {
                return new EmbeddedDatabaseBuilder().build();
        }
}
...
```

There should be nothing too complex in Listing 13-20. It's a basic Spring configuration class using the same builders we've used throughout the book to create our step and job. The ItemReader is the ListItemReader provided by Spring Batch that will return the values foo, bar, and baz as the read() method is called. The ItemWriter is a lambda that delegates the writing of each item to System.out.println, writing each item out to standard out. The configuration finishes by configuring the step (a chunk based step with a chunk size of 10 using the reader and writer previously defined) and the job consisting of the step just defined.

The last piece of the puzzle is to test this job. To do this, we use the JobLauncherTestUtils to launch our job. In this case we don't have any job parameters, so we can use the JobLauncherTestUtils.launchJob() method to execute it. We don't need to specify the job to use because since we only have one job in our context, it will be autowired into the utility for us. This method call returns JobExecution which we can validate that the BatchStatus was COMPLETED. We can also inspect the StepExecution and validate that it also has a BatchStatus of COMPLETED as well as the number of items read and written. Listing 13-21 shows the code for the test method itself.

Listing 13-21. JobTests#test

```
...
@Test
public void test() throws Exception {
        JobExecution jobExecution =
                this.jobLauncherTestUtils.launchJob();

        assertEquals(BatchStatus.COMPLETED,
                jobExecution.getStatus());

        StepExecution stepExecution =
                jobExecution.getStepExecutions().iterator().next();

        assertEquals(BatchStatus.COMPLETED, stepExecution.getStatus());
        assertEquals(3, stepExecution.getReadCount());
        assertEquals(3, stepExecution.getWriteCount());
}
...
```

Summary

From unit-testing a single method in any component in your system all the way to executing batch jobs programmatically, you've covered the vast majority of testing scenarios you may encounter as a batch programmer. This chapter began with an overview of the JUnit test framework and the Mockito mock object frameworks for unit testing. You then explored integration testing using the classes and annotations provided by Spring, including executing tests in transactions. Finally, you looked at Spring Batch–specific testing by executing components that are defined in the step scope, individual steps in jobs, and finally the entire job.

Index

A

ActiveMQ, 19, 300–301
ActiveMQ/JMS, dependencies, 300
@AfterChunk, 86
@AfterJob, 63
afterJob method, 61
@AfterRead, 220
@AfterStep, 86–87, 222
Agile process
    CI environment, 32
    TDD, 32
    tenets, 30
    user stories
        acceptance criteria, 31
        narrative, 30
        title, 30
        universal remote control, 31
        vs. use cases, 31
    version control system, 32
Apache Geode, 245, 279, 286–291
Apache Maven, 329
application.properties, 46
AsyncItemProcessor, 381–384
AsyncItemWriter, 381–384

B

Backing service, 409
BatchConfiguration.java, 54
BatchConfigurer, 112, 116, 117, 200, 202, 275
Batch Job description
    data model, 40, 41
    statement-generation process, 39
        balance updation, 40
        bank statement jobflow, 39
        customer data, 39
        customer monthly statements, 40
        transaction data, 39
BATCH_JOB_EXECUTION_CONTEXT
    table, 68, 108, 109

BATCH_JOB_EXECUTION_PARAMS
    table, 49, 60, 108, 110
BATCH_JOB_EXECUTION
    table, 95, 108, 109
BATCH_JOB_INSTANCE table, 108, 109
Batch jobs scaling
    multithreaded steps
        MultithreadedJobApplication, 375, 376
        process, 374
    parallel steps
        bigtransactions.csv, 380–381
        configuration, 377–380
        parallelStepsJob logs, 381
        process flow, 377
    partitioning (*see* Partitioning,
        job scaling)
    remote chunking (*see* Remote chunking)
Batch processing
    challenges, 3
    history, 2, 3
    in java, 4, 5
    scaling, 4–5
    with Spring Batch, 9–10
Batch process profiling
    JVM arguments, 364
    Spring batch applications (*see* Spring
        batch applications)
    VisualVM (*see* VisualVM)
BatchStatus, 90, 93, 134, 454
BATCH_STEP_EXECUTION_CONTEXT
    table, 108, 111
BATCH_STEP_EXECUTION table, 95, 108,
    110, 388, 399, 405
BeanWrapperFieldExtractor, 249, 255
BeanWrapperFieldSetMapper, 161, 167,
    168, 172, 251, 320
@BeforeChunk, 86
beforeCommit method, 249
@BeforeJob, 63
beforeJob method, 61
@BeforeStep, 86

■ C

CallableTaskletAdapter, 72, 73
Chunk-based processing, 70
Chunk-based step, 79, 80
ChunkHandler, 400
ChunkListener, 86
Chunk oriented processing
    commit interval, 79–81
Chunk-size configuration, 80–84
Circuit breaker
    Hystrix, 418
@CircuitBreaker annotation, 418, 419
Circuit Breaker, cloud native
    attributes, 418
    @EnableRetry annotation, 419
    EnrichmentController, 420
    EnrichmentProcessor, 419
    fallback method, 420
    fault tolerant, 418
    Spring Retry, 418
ClassifierCompositeItemProcessor, 240–242
ClassifierCompositeItemWriter, 322
    classifierFormatJob, 324, 325
    configuration and dependencies, 323, 324
CloudFoundry, 11, 395, 396, 417, 427
Cloud native batch
    circuit breaker (*see* Circuit breaker, cloud
        native)
    twelve factor application (*see* Twelve factor
        application)
CloudNativeBatchApplication, 411, 424
CommandLineArgsProvider, 397
CommandLineJobRunner, 44, 45, 49
Comma-separated value (CSV), 135
Common Business Oriented Language
    (COBOL), 2, 4, 5, 158
CompletionPolicy interface, 81–84, 86
CompositeItemProcessor, 237–242
CompositeItemWriter
    customerWithEmail.csv, 319, 320
    output/step/job configuration, 320–321
    sequence diagram, 319
@ContextConfiguration, 450
Continuous integration (CI), 32
Create/read/update/delete (CRUD)
    operations, 209
Cursor *vs.* paging, 192
customerBatchWriter, 307
CustomerClassifier, 323
customer.csv file, 251
customerEmailFileReader, 307
CustomerItemPreparedStatementSetter.java, 266
CustomerItemValidatorIntegrationTests, 444
Customer.java, 250
customer.json, 189, 190

Customer relationship management (CRM)
    application, 310
CustomerRepository, 292
Customer_update.csv File, 331
Customer updates
    application.properties, 342
    CustomerUpdateClassifier, 341
    customer_update.csv File, 331
    customerUpdateItemReader, 334
    customerUpdateItemWriter, 342
    customerValidatingItemProcessor, 339, 340
    data model, 330, 331
    domain objects, 335, 336
    FieldSetMapper, 337, 338
    FlatFileItemReader, 333
    importCustomerUpdates, 333
    importJobConfiguration, 332, 333
    LineTokenizer, 334, 335
    MySql Dependency, 342
    record type, 331, 332
CustomerValidatingItemProcessor, 340
Custom record parsing
    CustomerFileLineTokenizer, 168, 169
    LineTokenizer Interface, 167

■ D

DailyJobTimestamper.java, 59
Data access object (DAO), 136
Database-based ItemWriters
    JdbcBatchItemWriter
        BeanPropertyItemSqlParameterSource
            Provider, 269, 270
        configuration options, 265
        customer table design, 264
        formatJob configuration, 267, 268
        formatJob execution, 268
        job results, 268, 269
        updation, 269
DefaultBatchConfigurer#getTransaction
    Manager() method, 200
Delimited files
    character configuration, 164
    CustomerFieldSetMapper, 167
    customerFile, 163
    DelimitedLineTokenizer, 163
    FieldSet Interface, 165, 166
    FieldSetMapper Interface, 165
    single street address, 164
Delimited file writing, formatJob
    results, 255
DelimitedLineTokenizer, 163
DeployerStepExecutionHandler, 397
Decision, 82
DOM parser, 183
DownloadingJobExecutionListener, 414–415

■ E

Email, 306–307, 332
@EnableBatchIntegration, 390, 393, 402
@EnableBatchProcessing, 27, 112, 126, 128, 330, 333, 411, 419, 448
@EnableTask, 430
EnrichmentController, 417
EnrichmentProcessor, 415
Error handling, 29
    empty input, 222, 223
    job failure
        ExitStatus, 150
        ParseException, 149
        TransactionReader, 148–149
    logging invalid records, 220–222
    NullPointerException, 148
    skipping records, 218–220
Eureka
    ApplicationContext, 424
    bootstrap.yml, 424
    @EnableDiscoveryClient
        annotation, 423, 424
    EnrichmentProcessor, 425
    REST API, 423, 426
    RestTemplate, 425
    service discovery, 423
ExecutionContext, 161, 176, 207, 217, 233, 384, 409, 415
    manipulation, 65–67
    persistence, 67–69
    relationship, 64
    vs. web application, 64
ExitStatus, 74, 78, 86, 88, 90, 93, 96, 128, 137, 139, 144, 145, 150
@ExtendWith, 450
Externalizing configuration
    service binding via Eureka (*see* Eureka)
    Spring Cloud Configuration (*see* Spring Cloud Config)
Extreme Programming (XP), 30

■ F

FieldExtractor, 249, 251
FieldSet, 139, 158, 161, 163–165, 168, 171, 173
FieldSetFactory, 161
FieldSetMapper, 158, 161, 164, 165, 167, 169, 172, 333, 337
File-Based ItemWriters
    FlatFileItemWriter
        configuration options, 248, 249
        customer.csv, 249
        delimited files, 254–256
        file management, 256–258

        pieces, 247
        text files, 249–252, 254
    StaxEventItemWriter
        attributes, 259, 260
        formatJob configuration, 260–262
        formatJob execution, XML, 263
        OXM library Maven dependency, 262
        XML, 260
Files
    delimited, 163–167, 247, 249, 254–256
    fixed width (*see* Fixed-width file)
    JSON, 188–191
    XML (*see* XML)
Filtering records, 242, 243, 338
FixedLengthTokenizer, 161
Fixed-width file
    BatchConfiguration, 160, 161
    copyFileJob, 162
    copyFileStep, 162
    customer file format, 158
    Customer.java, 159
    customer.txt, 159
    simple writer, 161
Fixed-width input file *vs.* formatted output file, 249
FlatFileItemReader, 137, 139, 142, 174, 181, 333, 403, 446
    configuration options, 157
    pieces, 156
FlatFileItemReaderBuilder, 160, 251, 333
FlatFileItemWriter, *see* File-Based ItemWriters, FlatFileItemWriter
FlatFileItemWriterBuilder, 251, 252, 255
flatFileOutputWriter Configuration, 255
Flat files, 156–158
    custom record parsing, 167–169
    delimited files, 163–167
    fixed-width files, 158–163
    multiline records, 174–176, 178, 179
    multiple record formats, 169–174
    multiple sources, 179–183
Flow, 304, 327
FlowBuilder, 380
FlowBuilder's split method, 377
Flow step configuration, 100–102
formatJob execution, 254
formattedCustomers.txt, 254
FormattedTextFileJob.java, 252–253

■ G

GemfireItemWriter, 288
getCustomer() method, 213
getJobExecutionContext() method, 65
getStepExecution method, 448
GUI-based programming, 3

■ H

HelloWorld.java, 46, 47
Hello, World Spring batch job
    DataSource, 27
    H2, 25
    HelloWorldApplication class, 26, 27
    Spring Intializr, 24, 26
    Tasklet, 27
HelloWorld Tasklet, 71
Hibernate
    annotations, 199, 200
    cursor processing, 199
    paged database, 202
    POM, 199
    properties, 200
    query options, 201, 202
HibernateBatchConfigurer, 200–201
HibernateBatchConfigurer.java, 271–273
HibernateCursorItemReader, 199, 201, 202
HibernateItemWriter
    configuration, 270
    Customer.java, 271
    HibernateImportJob.java, 273–275
    POM additions, 270
HibernatePagingItemReader, 202, 209
Hystrix, 418

■ I

If/Else logic, 88–92
Input and output options, 29
Integration tests
    AccountItemProcessor, 447
    CustomerItemValidator, 447
    database interaction, 443
    environment configuration
        CustomerItemValidatorIntegration
            Tests, 445
        HSQLDB, 443
        @JdbcTest, 444
    FlatFileItemReader, 446, 447
    job (*see* Testing jobs or steps)
    @SpringBatchTest, 448
    Spring bean interaction, 443
    step (*see* Testing a step)
    test file, 447
IntelliJ IDEA, 24
ItemPreparedStatementSetter Interface, 265
ItemProcessor, 18, 39, 40
    composite
        configuration, 239–242
        processing, 238
        validation, 238, 239
        ZipCodeClassifier, 240

    configuration, 235
    filtering items, 242–244
    input validation
        Customer class, 227, 228
        Customer object, 228, 229
        job, 230–234
    interface, 225, 226
    scripting languages, 236
    UpperCaseNameService, 234
    validating, 227
ItemProcessorAdapter, 234
ItemReader, 13, 40, 50, 69, 137, 139, 142, 155–223
ItemReaderAdapter, 211, 213, 226, 234
ItemReaderinterface
    custom input, 213, 215–218
    database input
        hibernate (*see* Hibernate)
        JDBC (*see* Java database connectivity
            (JDBC))
        JPA, 202–204
        Spring Data (*see* Spring Data)
        stored procedure (*see* Stored procedure)
    errors (*see* Error handling)
    file input
        flat files (*see* Flat files)
        XML, 183–188
    JSON, 188–191
ItemReader#read() method, 139
ItemStream, 67, 139, 215, 232, 233, 325–326, 438
ItemWriter, 13, 18, 40, 50, 69, 80, 139, 144, 162, 179,
            211, 225, 333, 346–347, 383, 414
    database-based (*see* Database-based
            ItemWriters)
    introduction, 246–247
    Spring Data (*see* Spring Data ItemWriters)
    step interaction, 246
    types of, 245
    See also specific ItemWriters
ItemWriterAdapter
    configuration, 295, 296
    CustomerService.java, 295
    dependencies, 294
    output, 296

■ J

JacksonJsonObjectReader, 190
Java database connectivity (JDBC)
    cursor *vs.* paging, 192
    cursor processing, 192, 193
    customer data model, 192
    customerItemReader, 194
    CustomerRowMapper, 193
    JdbcPagingItemReader configuration, 197, 198
    paged processing, 196

java–jar copyJob command, 194
Java Messaging Service (JMS), 299
Java Persistence API (JPA), 202, 275
JavaScript, 4, 53, 188, 206, 236, 238, 239
JAXB Dependencies, 186, 187
JdbcBatchItemWriter, 143, 307
jdbcBatchWriter's configuration, 266
JdbcCursorItemReader, 143, 195
JdbcOperations, 113, 449
JdbcPagingItemReader, 196–197, 209
JdbcTemplate, 191, 192, 195, 263, 264, 269, 450, 451
@JdbcTest, 450
JMockit framework, 439
Jms
    JmsItemReader, 302–303
    JmsItemWriter, 299–304
JmsItemWriter
    configuration, 301
    input and output, 301, 302
    jmsFormatJob, 300
    JmsItemReader, 302–303
    JmsTemplate, 301
    MessageConverter, 301
    sample output, 303–304
Job
    batch process, 43, 44
    configuration, 46, 47
    definition, 43
    ending
        completed state, 93–95
        failed state, 95, 96
        states description, 93
        stopped state, 96–98
    launching via REST, 125–134
    lifecycle, 44
    listeners (see Job listeners)
    metadata, 118–121
    parameters (see Job parameters
        (JobParameter))
    repository, 107–111
    with Spring Boot, 123–125
    stopping, 134–150
    runners, 44, 45
JobBuilder, 47, 54, 62, 100, 102, 150
JobBuilder.build() method, 47
JobBuilderFactory, 27, 47, 144
JobExecution, 17, 45, 61, 64, 65, 90, 111, 389, 445
JobExecutionListener, 413
JobExecutionListenerSupport, 415
JobExplorer, 27, 115–116, 118–121, 396
JobInstance, 16, 17, 45, 93, 121, 130, 131
JobInterruptedException, 147
Job/JobInstance/JobExecution, relationship of, 45
JobLauncher, 16, 27, 45, 116–117, 125
JobLauncherCommandLineRunner, 28, 45, 123, 124

JobLauncherTestUtils.launchJob() method, 454
Job listeners
    BatchConfiguration.java, 63, 64
    callbacks, 61
    creation, 61, 62
    output, 62, 63
JobLoggerListener.java, 61, 63
JobOperator, 107
Job parameters (JobParameter)
    accessing, 50–51
    CommandLineJobRunner, 49
    execution, 48, 49
    failure output, 56
    final output, 56, 57
    identification, 50
    increment, 57–60
    JobParametersIncrementer, 57, 58,
        129, 130, 134
    late binding, 52
    step scoped bean configuration, 53
    type specification, 49, 50
    validation, 53–57
JobParametersBuilder.getNextJobParameters(job)
        method, 130
jobParametersExtractor bean, 105
JobParametersIncrementer, 134
JobRegistry, 27, 44
JobRegistryBackgroundJobRunner, 45
Job repository
    definition, 107
    in memory, 111
    using relational database, 107–111
JobRepository, 15, 16, 27, 45, 98, 100, 107–114, 396,
        397, 410, 446
JobRepositoryFactoryBean
    in memory, 111
    table prefix, 114, 116
    transaction customization, 113
Job step configuration, 102–104
JpaItemWriter
    formatJob configuration, 277
    formatJob execution, 278
    JpaBatchConfigurer.java, 275, 277
    process, 275
JpaPagingItemReader
    PagingQueryProvider, 196
    SqlPagingQueryProviderFactoryBean, 196–197
JpaQueryProvider, 204
JsonItemReader, 189–191
JsonObjectReader, 189, 190
JUnit
    @AfterEach, 437, 438
    assert methods, 437
    @BeforeEach, 437, 438
    DAO, 438

JUnit (*cont.*)
    defined, 436
    lifecycle, 437
    @Test, 437, 438
    test cases, 436
    test method, 437

■ K

Kubernetes, 11, 395, 396, 417, 427

■ L

Launching a job, 125–134
LineAggregator
    PassthroughLineAggregator, 80
LineMapper
    DefaultLineMapper, 158, 160, 161, 169, 251
LineTokenizer, 158, 161, 163, 167, 168, 333, 334
Listeners
    ItemReadListener, 220
    ItemWriteListener, 317
    JobExecutionListener, 61, 63, 86, 413
    StepExecutionListener, 86, 145
logCustomerAddress method, 298

■ M

Marshaller, 259, 262
MetaDataInstanceFactory, 448
MethodInvokingTaskletAdapter, 73–76
Mockito, 435, 439
    CustomerItemValidator, 440
    CustomerStatementReaderTest, 440–441
    testInvalidCustomer, 442
    testValidCustomer(), 441
    validate() method, 440
MockitoAnnotations.initMocks method, 441
Mock objects
    MyMock.class, 439
    MyObject(), 439
    proxy object, 439
MongoDB
    dependencies, 206
    features, 206
    MongoItemReader, 206–208
    MongoItemWriter, 279–281
    output, 208
    Spring Boot Starter, 207
MongoItemReader, 208
Monthly statement
    AccountItemProcessor, 353, 354
    AccountResultSetExtractor, 354, 355
    individualStatementItemWriter, 358
    MultiResourceItemWriter, 358

    sample, 358, 359
    statement data, 350–353
    StatementHeaderCallback, 357
    statementItemWriter, 358
    StatementLineAggregator, 355–357
Multi-file input
    MultiResourceItemReader, 179, 181, 413, 414
Multiline records
    copyFileStep, 178
    CustomerFileReader, 174–178
    customer object updation, 175
    multiline job, 179
    toString() method, 178, 179
Multiple record formats
    comma delimited, 170
    copyJob, 173, 174
    customerFileReader, 171, 172
    flow of processing, 172
    issues, 169
    object code, 170
    TransactionFieldSetMapper, 173
    updated customerInputFile, 170
Multiple sources
    CustomerFileReader, 181, 183
    customer files processing, 180, 181
    multiline job, 183
MultiResourceItemReader, 181, 413
MultiResourceItemWriter
    configuration options, 311
    CustomerOutputFileSuffixCreator, 314, 315
    execution command, 313
    file name creation, 314
    header/footer flat file
        CustomerRecordCountFooter
            Callback, 317–318
        delegateCustomerItemWriter, 318
    header/footer XML fragments
        configuration, 316
        CustomerXmlHeaderCallback, 315–316
    process, 311
    ResourceSuffixCreator, 315
    step and job configuration, 312, 313
Multithreaded steps, 17–18, 374–376

■ N

Neo4j
    Neo4jItemWriter, 282, 284
Netflix, 418
next() method, 191

■ O

Object relational mapping (ORM)
        technologies, 198, 202
@OnReadError, 220, 221

org.springfamework.batch.item.file.LineMapper
    interface, 156
org.springframework.batch.core.ChunkListener
    interfaces, 86
org.springframework.batch.core.launch.
    JobLauncher interface, 45
org.springframework.batch.core.launch.
    support, 45
org.springframework.batch.core.
    StepExecutionListener, 86
org.springframework.batch.item.file.
    DefaultLineMapper, 160
org.springframework.batch.item.ItemReader<T>
    interface, 155
org.springframework.batch.item.ItemWriter
    interface, 246
org.springframework.core.task.SyncTaskExecutor, 45
org.springframework.core.task.TaskExecutor, 45

P

PagingAndSortingRepository, 209
Parallelization
    ItemProcessor/ItemWriter, 18
    multithreaded steps, 17, 18
    parallel steps, 18
    partitioning, 19
    remote chunking, 19
Parallel steps, 18, 376–381
Partitioning
    Partitioner, 384, 386, 391, 397
    PartitionHandler
        DeployerPartitionHandler, 394–399
        MessageChannelPartitionHandler,
            384, 385, 389–394
        TaskExecutorPartitionHandler, 385–388, 392
    PartitionStep, 387
Partitioning, job scaling
    DeployerPartitionHandler
        BatchConfiguration, 395, 396
        jps output, 398
        JVMs, 398
        partitioned job output, 399
        Spring Cloud Task, 395
        TaskLauncher, 394
        Unix ps command, 398
    MessageChannelPartitionHandler
        AMQP template, 391
        inbound flow, 391, 392
        IntegrationFlow, 391
        Java DSL, 391
        JobExecution, 389
        JVMs, 389
        MasterConfiguration, 390
        outbound flow, 390
        RabbitMQ, 389

        remote partitioned step, 389
        WorkerConfiguration, 392
    partitioner interface, 384
    TaskExecutorPartitionHandler
        BATCH_STEP_EXECUTION, 388
        configuration, 387
        ExecutionContext, 387
        fileTransactionReader, 386
        MultiResourcePartitioner, 386, 387
        partitionedMaster, 388
        SimpleAsyncTaskExecutor, 387
        SyncTaskExecutor, 387
PassThroughCommandLineArgsProvider, 397
PatternMatchingCompositeLineMapper, 172
peak method, 176
@PeerCacheApplication, 288
Pivotal Gemfire, 279, 286–291
PlatformTransactionManager, 27, 112, 115, 117, 249
pom.xml file, 199
@PostMapping, 128
Processing models, 69
PropertyExtractingDelegatingItemWriter
    customer service, 297
    output, 298–299

Q

Quartz, 130–134
QuartzJob class, 133
Quartz scheduler
    BatchScheduledJob, 132
    configuration, 133
    JobBuilder, 133
    JobDetails object, 131
    output, 133, 134
    scheduled job, 131, 132
    SchedulerFactory, 131
    SimpleScheduleBuilder, 133
    Spring Batch process, 131

R

RabbitMQ, 389, 394
RandomChunkSizePolicy, 85, 86
readDouble method, 173
read() method, 176, 194, 213
@Recover annotation, 419
Relational database, 107–111
Remote chunking
    batch processing, 399
    BOINC system, 399
    ChunkHandler, 400
    FlatFileItemReader, 403
    inbound flow, 403
    master configuration, 400–402
    output, 405

Remote chunking (*cont.*)
    RabbitMQ, 402
    RemoteChunkingWorkrerBuilder, 404
    structure, 400
    worker's configuration, 403, 404
RemoteChunkingMasterStepBuilder
    Factory, 402
RemotePartitioningWorkerStep
    BuilderFactory, 393
Repository
    CrudRepository, 291, 292
    PagingAndSortingRepository, 209, 291
RepositoryItemReader, 210
RepositoryItemWriter, 291–292
resilience4j, 418
Resource, 160, 181, 190, 236, 247, 252, 260,
    311, 384, 396
Restart control
    allowStartIfComplete() method, 152
    configuring restarts, 151, 152
    preventing rerun
        Nonrestartable Job, 151
        preventRestart() call, 150
        transactionJob, 150
Restarting a job, 16, 123, 129, 150, 153, 215
REST, job launching
    curl command, 128
    @EnableBatchProcessing, 128
    ExitStatus, 128
    HTTP POST, 128
    JobLauncher interface, 125
    JobLaunchingController, 126, 127
    JobParameters, 130
    JobParametersIncrementer, 129, 130
    Quartz (*see* Quartz scheduler)
    RunIdIncrementer, 130
    SimpleJobLauncher, 125, 126
    Spring Intializr, 126
    TaskExecutor, 125
Retry, 29, 40, 86
RowMapper, 156, 158, 193–194, 196, 205, 308, 353
RunIdIncrementer, 58

■ **S**

SAX parser, 183
Scalability, 29
Scheduling, 130–134
Scope
    job scope, 445–447
    step scope, 445–447
ScriptItemProcessor, 226, 236–237
Scrum, 30
Service discovery
    Eureka, 423

setResource method, 181
SimpleEnvironmentVariablesProvider, 397
SimpleMailMessageItemWriter
    customerImport job, 304
    Customer.java, 305–306
    customer table, 306
    customerWithEmail.csv, 305
    execution, 309
    ItemReader and ItemWriter, 307, 308
    Java mail dependencies, 306, 307
    job configuration, 308, 309
Simple Spring Batch application
    application.yml, 416
    JdbcBatchItemWriter, 414
    JobConfiguration, 411–413
    REST API, 417
    Spring Intializr, 411
    StepBuilderFactory, 414
Skip, 86, 110, 172, 218, 219
Spinnaker, 409
Split, 168, 377, 380, 386
Spring Batch
    ecosystem, 10
    ETL processing, 6
    features, 11
    framework, 8, 9
    Job, 9, 10
    local and remote parallelization, 10
    message processing, 8
    parallel processing, 6
    spring boot, 10
    spring cloud data flow, 6, 7, 11
    Spring Cloud Task, 11
    standardizing I/O, 10
    working, 11
    See also Batch processing
Spring batch applications
    CPU profiling
        AccountItemProcessor, 368, 370, 371
        bottlenecks, 367
        PricingTiersItemProcessor, 368, 369
        statement job, 367, 368, 370
    memory profiling
        AccountItemProcessor, 371
        memory usage, 373
        OutOfMemoryException, 372, 373
        PricingTierItemProcessor, 372
        statement job, 372, 373
Spring batch architecture
    batch job, 14
        chunk based step, 14
        decoupling features, 15
        interface, 15
        tasklet step, 14
    documentation, batch jobs sample, 20, 21

job execution
    JobLauncher, 16
    JobRepository, 15, 16
  parallelization (*see* Parallelization)
  StepExecution, 16
Spring Batch infrastructure
  BatchConfigurer interface, 112
  customization
    JobExplorer, 115, 116
    JobLauncher, 116, 117
    JobRepository, 112–114
  job metadata, 118–121
  Spring Boot, 117
  transaction manager, 114, 115
Spring Batch job, 14
Spring Batch project
  Github, 21
  IntelliJ IDEA, 24
  parameters, 21
  Spring Initializr, 21, 22
  STS, 22, 23
@SpringBatchTest, 450
Spring Beans, 27, 445
Spring Boot, 10, 27, 28, 117, 394
@SpringBootApplication annotation, 27
Spring Boot, job launching
  ApplicationRunner, 123
  CommandLineRunner, 123
  multiple jobs, 125
  Spring Batch job, 123
Spring Boot project
  current balance transaction
    applyTransactionsReader, 348
    applyTransactions Step, 347, 348
    applyTransactionsWriter, 349
    JdbcBatchItemWriter, 349
    JdbcCursorItemReader, 348
  customer id validation
    CustomerItemValidator, 338, 339
    customerValidatingItemProcessor, 339
  customer updates (*see* Customer updates)
  monthly statement (*see* Monthly statement)
  Spring Initializr, 328–330
  transactions
    domain object, 343, 344
    getTransactionAmount method, 344, 345
    importTransactions, 345
    JaxbDateSerializer, 344, 345
    Jaxb2Marshaller, 346
    transaction file, 343
    transactionItemReader, 346, 347
spring-boot-starter-data-mongodb, 280
Spring Boot über jar, 408
Spring Cloud, 420, 427
Spring Cloud CLI, 421, 422, 426

Spring Cloud Config
  client, 421
  encryption, 422
  git repository, 422
  REST API, 423
  Spring Cloud CLI, 421
  spring.cloud.config.failFast, 421
Spring Cloud Config Server, 407, 420
Spring Cloud Data Flow, 427
  applications orchestration, 429
  architecture, 428
  Audit Records, 431
  GUI, 431
  job execution, 433
  Maven repository, 430
  REST APIs, 427
  shell, 430
  Streams, 431
  step execution, 434
  task and executions, 431, 432
  wget commands, 428
Spring Cloud Deployer, 384–385
Spring Cloud Netflix, 423
Spring Cloud project, 407
Spring Cloud Task, 11, 61, 395, 429–434
Spring Data, 206, 220, 291
  MongoDB, 206–208
  repository, 209, 210
Spring Data ItemWriters
  Apache Geode, 291
  customer mapping, 291, 292
  importJob, 292–294
  MongoDB
    customer.java, 279, 280
    mongoFormatJob, 280–281
    Robot 3T, 282
  Neo4J
    customer mapping, 283
    dependencies, 283, 284
    importJob, 284, 285
    output, 286
  Pivotal Gemfire
    configuration, 288
    GemfireImportJob, 288–291
    Neo4jImportJob, 287
    pom.xml, 287, 288
  repository, 291
Spring Initializr, 21, 328–330
Spring Retry library, 418
Spring test utilities, 436
Spring Tool Suite (STS), 22, 23
SqlPagingQueryProviderFactoryBean, 196, 197
Statement job, 327, 328
  batch job, 35
  batch process, 34

Statement job (*cont.*)

plain text, 34

printed bank statement, 32, 33

user stories, 35

account summary, 38

import transactions, 36

print statement, 37

record formats, 35, 36

update transactions, 37

StaxEventItemReader, 185, 188, 189, 259, 346

StaxEventItemWriter, 259–263

StAX parser, 183

StepBuilder, 27

StepBuilderFactory, 27, 47, 142–144

StepContext.getJobExecutionContext()
method, 65

StepExecution, 17, 90, 445, 446, 448, 454

StepExecution.setTerminateOnly()
method, 145, 146

Step flow

conditional logic, 88–90, 92, 93

ending a job, 93–98

externalizing flows, 98–105

RandomDecider, 91

wild cards, 90

Steps

configuration, 70

flow (*see* Step flow)

listeners

configuration, 87, 88

logging step start, 86, 87

tasklet, 70

@StepScope, 413

StepScopeTestExecutionListener, 446, 448

Stopping a job

natural completion, 134

programmatic end

error handling (*see* Error handling)

StepExecution, 146–148

stop transition (*see* Stop
transition, job)

Stop transition, job

applyTransactionsStep, 142, 143

CSV files, 135

custom ItemReader, 137

data model, 135

domain objects, 136, 137

generateAccountSummaryStep, 143

importTransactionFileStep, 140, 142

ItemStream, 139

JdbcBatchItemWriter, 142

Spring Intializr, 136

StepExecutionListener#AfterStep, 139

steps, 135

TransactionApplierProcessor, 140

TransactionDao, 139, 140

transactionJob, 144, 145

TransactionReader, 138, 139

Stored procedure

definition of, 204

ItemReader, 205, 206

JpaQueryProvider, 205

StoredProcedureItemReader, 205

Streaming API for XML (StAX)
implementation, 259

StreamUtils, 415

Stubs, 439

SystemCommandTasklet, 76–79

■ T

TaskExecutor, 45, 78

Tasklet, 69, 71

TaskletAdapter, 72–79

Tasklet *vs.* chunk processing, 69

Tasklet.execute method, 69

Tasklet step, 70, 71

Test-driven development (TDD), 32

TestExecutionListener, 445

Testing jobs or steps

code, 453, 454

infrastructure, 452

JobLauncherTestUtils, 448, 450–454

test method, 454

Testing job or step scoped components

JobScopeTestExecutionListener, 446

StepScopeTestExecutionListener, 446

Testing a step

importCustomerUpdates, 451, 452

infrastructure, 449

test values, 450, 451

Tomcat, 410

toString() method, 178

@Transactional, 450

Transition, 14, 47, 70, 88, 90, 135,
145, 147, 407

Travis CI, 32

Twelve factor application

administration processes, 411

backing service, 409

build, release, run, 409

codebase, 408

concurrency, 410

configuration, 409

dependencies, 408

dev/prod parity, 410

disposability, 410

log files, 410

port binding, 409

processes, 409

■ U

Unit test, 435
Unmarshaller, 186, 188, 189, 259
User stories, 30

■ V, W

Validating input, 227
ValidatingItemProcessor
    validator, 227, 229, 233
VisualVM
    Java process, 363
    Monitor tab, 364
    overview, 363, 364

    Sampler tab, 365, 366
    Threads tab, 364, 365

■ X, Y, Z

XML
    copyFileStep, 188
    customer sample file, 184
    fragment, 184
    JAXB, 186, 187
    Jaxb2Marshaller, 188
    parsers, 183, 184
    Spring Batch, 185
    StaxEventItemReader, 185, 186
XML header, 316

Printed in the United States
By Bookmasters